*INTERNATIONAL SERIES IN*
**NATURAL PHILOSOPHY**
GENERAL EDITOR: D. TER HAAR

VOLUME 94

# GASEOUS ELECTRONICS AND GAS LASERS

## Some other Pergamon Titles of Interest

1. CLARKE & GRAINGER:
Polarized Light and Optical Measurement

2. AUVRAY & FOURRIER:
Problems in Electronics

3. MATHIEU:
Optics

4. WILLETT:
Gas Lasers: Population Inversion Mechanisms

*An Important Review Journal*
Progress in Quantum Electronics
Edited by T. S. Moss and S. Stenholm
Free specimen copy available on request

***A full list of other titles in the International Series in Natural Philosophy follows the Index***

# GASEOUS ELECTRONICS AND GAS LASERS

BY

BLAKE E. CHERRINGTON, Ph.D.

Department of Electrical Engineering
University of Florida, Gainesville

PERGAMON PRESS

OXFORD · NEW YORK · TORONTO · SYDNEY · PARIS · FRANKFURT

U.K. Pergamon Press Ltd., Headington Hill Hall, Oxford OX3 0BW, England

U.S.A. Pergamon Press Inc., Maxwell House, Fairview Park, Elmsford, New York 10523, U.S.A.

CANADA Pergamon of Canada, Suite 104, 150 Consumers Road, Willowdale, Ontario M2J 1P9, Canada

AUSTRALIA Pergamon Press (Aust.) Pty. Ltd., P.O. Box 544, Potts Point, N.S.W. 2011, Australia

FRANCE Pergamon Press SARL, 24 rue des Ecoles, 75240 Paris, Cedex 05, France

FEDERAL REPUBLIC OF GERMANY Pergamon Press GmbH, 6242 Kronberg-Taunus, Pferdstrasse 1, Federal Republic of Germany



**British Library Cataloguing in Publication Data**

Cherrington, B E

Gaseous electronics and gas lasers.

(International series in natural philosophy; vol. 94)

Includes bibliographical references.

1. Electric discharges through gases. 2. Gas lasers. I. Title.

QC711.C553 537.5'32 77-4137

ISBN 0-08-020622-0



*In order to make this volume available as economically and as rapidly as possible the author's typescript has been reproduced in its original form. This method unfortunately has its typographical limitations but it is hoped that they in no way distract the reader.*



*Printed and bound in Great Britain by William Clowes (Beccles) Limited, Beccles and London*

To

Marilyn, Graham and Andrea

# CONTENTS

# PREFACE

This book has evolved from a set of class notes prepared for a beginning graduate level course that I have taught at the University of Illinois. The original purpose of the notes and one of the purposes of this book is to allow readers with a good basic background in physical science to learn the fundamental principles and methods of analysis of weakly ionized gas discharges and to be able to apply them to current research efforts in gas discharges and gas lasers. This book should also be suitable for self-study and reference by professionals interested in working with weakly ionized discharges.

The book stresses the fundamental processes occurring in gas discharges and develops the analytical methods used to calculate important process rates. This background is then applied to detailed analyses of a variety of gas discharges with atomic, ionic and gas lasers being used as primary illustrations as well as important topics of study in their own right. However, the general processes and methods of analysis considered are applicable to any aspect of weakly ionized gas discharges.

This book proceeds from the simplest models of a gas discharge through the most complex as the need for greater sophistication becomes apparent. Chapters one through six follow this sequence and complete the analytical framework for the study of gas discharges. Chapters seven through nine blend this framework with a thorough consideration of fundamental rate processes in order to establish the temporal and spatial behavior of charged, neutral and excited species in pulsed and D.C. discharges. Chapters ten through twelve then proceed to an in depth analysis of selected atomic, ionic and molecular gas lasers that are of significant current interest.

In an effort to be consistent, the equations and data have been expressed almost entirely in MKS units. I caution the reader to keep this in mind, since much of the literature presents data in CGS units.

# ACKNOWLEDGEMENTS

During the past fifteen years and particularly during the preparation of this book, I have been fortunate enough to have advice and assistance from a large number of people. It would be impossible to mention them all here. However, there are a few that have been particularly helpful and who have certainly influenced my ideas and what I have written.

Professor Ladislas Goldstein, a truly remarkable gentleman, guided my doctoral education in the Gaseous Electronics Laboratory. Professor Joseph T. Verdeyen, a close friend and colleague for some seventeen years, has been my partner in vigorous (and sometimes noisy) discussions on all aspects of gaseous electronics. Professor Julius Cahn introduced me to the mysteries of the Boltzmann equation and Professor H. W. Wyld, Jr. introduced me to many aspects of kinetic theory. I owe a particular debt to all these gentlemen.

My many friends from government and industry, including graduates of the Gaseous Electronics Laboratory, have also been extremely helpful during the writing of this book. To mention but a few, Dr. L. A. Weaver, Dr. Lou Denes and Dr. Lyle Taylor of the Westinghouse Research Laboratories. Dr. Colin Willett, author of a companion volume on gas lasers by Pergamon Press, was also kind enough to take the time to read over the manuscript and offer numerous suggestions. Their assistance is greatly appreciated.

In spite of all of my efforts and the efforts of my colleagues, none of this would have come to fruition if it had not been for the labors of three remarkable ladies. Many years ago Mrs. Marilena Stone turned my scrawl into the first set of typed classnotes. Since then Mrs. Katharine Flessner has labored through a number of revisions until the final manuscript was ready and began the really hard work of typing the camera ready copy. Then as our will was weakening, Mrs. Janice Rogers lent her considerable skill to the task of completing this book. No words can express my thanks to Marilena and Janice, and especially to Kay, for the cheerful way in which they handled the hard part of this whole enterprise.

# Chapter 1

# INTRODUCTION

When we speak of an ionized gas, we imply that through some mechanism we have separated one or more electrons from some of the gas atoms or molecules, resulting in a gaseous medium containing electrons, ions and neutral atoms or molecules. If such a medium is produced by applying an electric field, then the resulting ionized medium conducts current and is called a gas discharge.

The typical gas discharges that we shall consider have an additional important property for purposes of our discussion - they are <u>weakly</u> ionized gas discharges. This means that typically one atom or molecule in $10^6$ (or more) has been ionized. The medium consists primarily of neutral species. This is an important distinction to make since fully ionized gases consisting entirely of charged particles (electrons and ions) have fundamental mechanisms and behavior that are quite different from that of weakly ionized gases. Such media constitute a separate area of study that will not be touched upon here.

One of the great problems in analyzing gas discharges involves the vast number of species that can occur, even in very simple gas discharges, and the interactions between the various species. Indeed, gas discharges are very complex chemical systems. Therefore, let us try some initial categorization of the species we can expect to find.

## 1.1 GAS DISCHARGE SPECIES

### 1.1.1 Neutrals

Starting with the initially un-ionized gas, we can have simple atoms, X, such as we obtain from the rare gases and metal vapors, or relatively simple molecules XY or

XYZ, such as $N_2$, $O_2$, or $CO_2$, or quite complex molecules such as methyl fluoride, $CH_3F$. The pressure range can be enormous. Typical pressures lie in the few torr range (1 torr is equivalent to a pressure of 1 mm Hg), but can range from pressures as low as a few microns ($10^{-3}$ torr) up to atmospheric pressure (760 torr) and even as high as tens or hundreds of atmospheres. The densities of neutrals then can vary over a range of about $10^8$, but for a few torr pressure the neutral densities will be in the $10^{22}$-$10^{23}$ $m^{-3}$ range. Specifically, the gas density at 1 torr and 0°C, $N = 3.54 \times 10^{22}$ $m^{-3}$.

### 1.1.2 Charged Particles

Since the medium is ionized, we will have electrons and several species of ions. For typical gas discharges the electron densities we expect will fall in the range of $10^{16}$ to $10^{20}$ $m^{-3}$. We will have positive ions, naturally, but we may have negative ions as well, due to the chemical processes occurring. Moreover, the ionic species, both positive and negative, may be quite different from the original neutral species, especially when the gases are molecular. As an example, discharges in He will result in $He_2^+$ ions being present. Furthermore, discharges in $N_2$ and $O_2$ (air) give a vast number of species including $N^+$, $N_2^+$, $O^+$, $O_2^+$, $NO^-$, $O_2^-$, $NO_2^-$, $O_3^-$, and so on. Each of the ion species will affect the electrical properties of the discharge, however, in this regard it is the electrons that usually dominate.

As one additional point, we are usually interested in what is known as "quasi-neutral" plasmas, where strong electrostatic fields are not present. This implies that the net charge density is approximately zero, or the sum of the densities of electrons, $n_e$, plus negative ions, $n_-$, approximately equals the sum of the densities of the (singly ionized) positive ions, $n_+$.

$$n_e + n_- - n_+ << n_e. \tag{1.1}$$

### 1.1.3 Excited Species and Photons

The neutral and charged species usually dominate the analysis of the electrical properties of gas discharges, but obviously the presence of excited neutral species ($X^*$ and $YZ^*$) and possibly excited ionized species ($X^{+*}$) is of prime importance for lasers, light sources or other devices where quantum effects occur. The resulting emission of photons is important as is the possibility of absorption of photons causing photo excitation, photo ionization or stimulated emission.

## 1.2 INTERACTIONS BETWEEN SPECIES

The real basis for our study of gas discharges involves the interaction between these species. So let us now briefly summarize some of the important interactions that occur.

| | |
|---|---|
| (1) $e + X \rightarrow X^{*} + e$ <br> $e + X^{*} \rightarrow X^{**} + e$ | electron collisional excitation of an atom or molecule, |
| (2) $e + X^{*} \rightarrow X + e$ | <u>superelastic</u> collision of an excited atom and an electron resulting in a de-excited atom and a fast electron. |

Process (2) is the inverse of process (1) and is often called a collision of the second kind. Every collisional or interaction process obviously has its inverse, and these processes need to be evaluated to determine their relative importance.

| | |
|---|---|
| (3) $e + X \rightarrow X^{+} + e + e$ <br> $e + X^{*} \rightarrow X^{+} + e + e$ | electron collisional ionization, |
| (4) $X^{+} + e + e \rightarrow X + e$ | collisional recombination, |
| (5) $e + YZ \rightarrow e + Y^{+} + Z^{-}$ <br> $e + YZ \rightarrow e + Y + Z$ <br> $e + YZ \rightarrow Y + Z^{-}$ | this represents many possible forms of dissociations and ionizations of molecules, |
| (6) $e + X \rightarrow X^{-} + h\nu$ | radiative attachment to form a negative ion, |
| (7) $X^{-} + h\nu \rightarrow e + X$ | photo-detachment, |
| (8) $e + YZ^{+} \rightarrow Y + Z^{*}$ | dissociative recombination. |

Of the processes shown above, photons can play many of the same roles as electrons in photo excitation, photo ionization and photo detachment. In fact, so can ions and neutrals, only much higher ion or neutral energies are required and these are rarely found in the gas discharges of interest to us.

There are some important ionic processes, however, such as:

| | |
|---|---|
| (9) $X + Y^{+} \rightarrow X^{+} + Y$ | charge transfer, |
| (10) $X^{-}+Y^{+}\rightarrow XY$ | ion-ion recombination. |

Similarly neutrals play important roles in processes such as:

| | |
|---|---|
| (11) $X^{*} + Y \rightarrow X + Y^{*}$ <br> $X^{*} + Y \rightarrow X + Y + KE$ <br> (KE represents kinetic energy) | excitation exchange and collisional de-excitation (of particular importance in lasers), |
| (12) $X^{*} + Z \rightarrow X + Z^{+} + e$ | Penning ionization. |

All of the processes mentioned imply that the collision partners do not have the same internal energy state after the collision as before - or they suffered an <u>inelastic</u> collision. The elastic collisions are also important in determining the electrical properties of the discharge, for instance:

(13) $e + X \rightarrow (e - KE) + (X + KE)$,

(14) $e_1 + e_2 \rightarrow (e_1 - KE) + (e_2+KE)$,

(15) $X^{+}+Y\rightarrow(X^{+}-KE) + (Y+KE)$,

are all elastic collisions that involve an exchange of energy and momentum between the partners (analogous to billiard ball collisions).

The fifteen collisional processes shown give some idea, although incomplete, of the complexity of the systems we wish to study. A more thorough listing of 103 such processes is given on pages 3-4 of Hasted's book on atomic collisions (1964).

## 1.3 BASIC CHARACTERIZATION OF ELECTRONS

Where do we start to analyze a gas discharge? We start with the electrons. This is the species which is usually of overwhelming importance in determining the electrical and optical properties of a discharge. In fact, the behavior of electrons dominates most of the other processes of interest in the discharge.

Figure 1.1 indicates some of the important collisional processes involving electrons. We will, therefore, initially concentrate on examining the role of electrons in gas discharges.

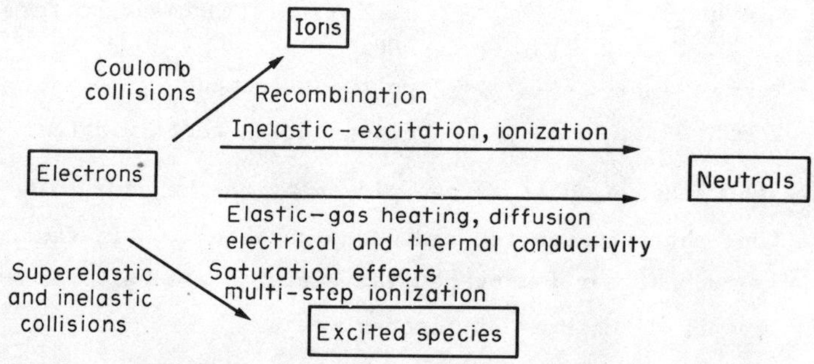


*Fig. 1.1. Important electron collisional interactions*

One of the most significant characterizations of a gas discharge is given by the electron number density, $n_e$, and the electron temperature, $T_e$, which is related to the energy of the electrons. Figure 1.2 shows the general ranges occupied by different classifications of ionized gases.

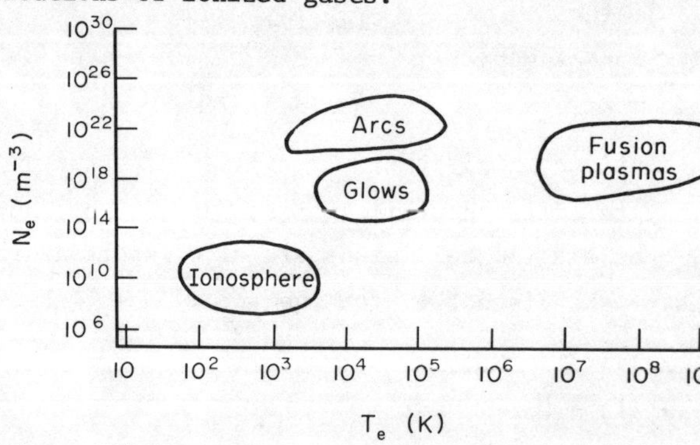


*Fig. 1.2. Electron density and temperature ranges for commonly known ionized media*

The electron density parameter is straightforward and in most gas discharges it is linearly related to the electric current through the medium. Moreover, as alluded to previously, the quasi-neutrality condition requires that there be approximately equal densities of positive and negative charge.

The electron temperature parameter is a consequence of the "statistical" nature of the electron gas. Because of the large numbers of particles considered, the electrons are characterized as a gas whose statistical nature is governed by classical Maxwell-Boltzmann statistics. In thermodynamic equilibrium, the distribution of electron velocities (or energies) is given by the Maxwellian distribution which allows an electron temperature, $T_e$, to be defined such that the mean electron energy is directly related to $T_e$ by:

$$m \langle v^2 \rangle /2 = 3k\, T_e/2, \tag{1.2}$$

where $\langle v^2 \rangle$ is the mean square electron velocity (v being the electron velocity which is largely the thermal or random velocity of the electrons), k is Boltzmann's constant, m is the electron mass, and $T_e$ is the electron temperature in degrees Kelvin.

Some further reasonably general characteristics of ionized gases are:

1. The particle velocities of any population are usually much more random than directed. In other words, the gas species considered are usually in approximate local thermodynamic equilibrium, with the thermal energy being much greater than the directed energy. This is not to say that the entire medium is in thermodynamic equilibrium. The species within a discharge need not be in even approximate equilibrium with each other and are rarely in equilibrium with the radiation. For example, glow discharges may have electrons with kinetic temperatures of tens of thousands of degrees Kelvin while the ions, neutrals and excited species may be cold with kinetic temperatures of 300 K.

2. The "Debye length" (see the collected works of Debye, 1954) is usually small with respect to the dimensions of the discharge and there are many particles within a Debye sphere. This concept relates to the effect of surfaces or electrodes on the behavior of a discharge since the Debye length is basically a measure of the distances over which significant deviations from quasi-neutrality are possible. If this condition is indeed satisfied, then the nature of the ionized gas should be independent of the nature of the boundaries since the condition of the surface will not materially affect the behavior of the ionized gas except very close to the walls. Since the Debye length is an important parameter, a simple derivation will be used to determine the minimum size that a discharge tube must have if the previous condition is to be satisfied.

### 1.3.1 Debye Shielding

Let us assume that we start with a uniform charge neutral medium into which we insert a positive point test charge + Q. (see Fig. 1.3). In free space we know

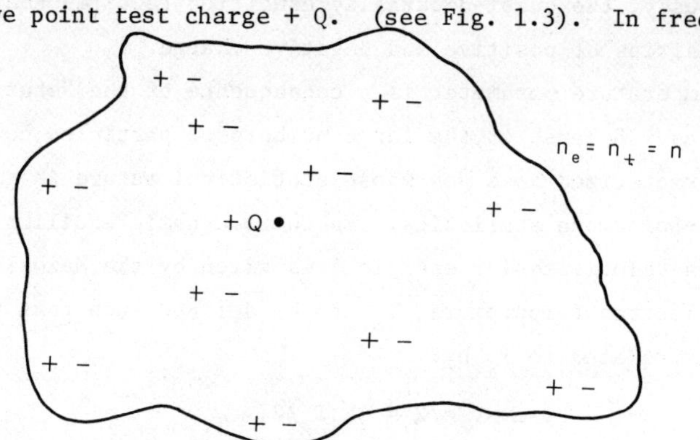


*Fig. 1.3. Ionized medium with a positive charge perturbation*

that the Coulomb potential due to the test charge is

$$V(r) = \frac{Q}{4\pi\varepsilon_o r} \,. \tag{1.3}$$

This defines for free space the region of influence of the positive charge, or how deeply into the medium an external influence can be felt. In the ionized medium, the field will not be strictly Coulombic since the electrons will tend to move towards the test charge and the ions will move away. This will tend to neutralize or "shield" the test charge from the rest of the medium.

To solve this problem, we will assume that the electrons are in thermodynamic equilibrium at a temperature $T_e$, and the ions are in thermodynamic equilibrium at a temperature $T_+$. If the potential near the charge Q is V(r), then

$$n_e(r) = n \exp\,(eV/kT_e)$$

and

$$n_+(r) = n \exp\,(-eV/kT_+), \tag{1.4}$$

where e is the electronic charge.

The screened potential is then determined by Poisson's equation

$$\nabla^2 V = n[\exp\,(eV/kT_e) - \exp(-eV/kT_+)]\,\frac{e}{\varepsilon_o} \,. \tag{1.5}$$

If the test charge is small so that particles within an interparticle distance do not suffer changes in potential energy comparable with their thermal energy, then

$$eV/kT_e \ll 1 \quad , \quad eV/kT_+ \ll 1.$$

Poisson's equation can then be simplified to give

$$\frac{1}{r^2}\frac{d}{dr}\, r^2 \frac{dV}{dr} = \frac{ne^2}{k\,\varepsilon_o} V\left[\frac{1}{T_e} + \frac{1}{T_+}\right] = \frac{V}{\lambda_D^2} \,, \tag{1.6}$$

where $\lambda_D = \left[\dfrac{k\,\varepsilon_o}{ne^2} \Big/ \left[\dfrac{1}{T_e} + \dfrac{1}{T_+}\right]\right]^{1/2}$ is the Debye length. (1.7)

The solution to this equation, taking into account that the solution at small radius approaches the Coulomb solution, is

$$V(r) = \frac{Q}{4\pi\varepsilon_o r} e^{-r/\lambda_D} . \quad (1.8)$$

The exponential form shows the effect of charge shielding on the penetration of a field within an ionized medium. If the ions are considered to be stationary, $n_+ = n$, then

$$\lambda_D = \left[\frac{k T_e \varepsilon_o}{ne^2}\right]^{1/2} = 69\sqrt{T_e/n} \text{ meters,} \quad (1.9)$$

where $T_e$ is in K and n is in $m^{-3}$.

If the dimensions of the ionized medium are comparable to or smaller than $\lambda_D$, then wall effects will play an important role on the behavior of the medium. If the dimensions are larger than $\lambda_D$, then the moving charges will shield the medium from wall effects. Also, in order for the shielding to be effective, there must be a large number of particles within the Debye sphere, i.e. $n_D$ must be large where

$$n_D = \frac{4}{3} n\pi \lambda_D^3 = 1.38 \times 10^6 \frac{T_e^{3/2}}{n^{1/2}} \quad (1.10)$$

and where $T_e$ is in K and n is in $m^{-3}$.

A much more complete discussion of Debye shielding is given by Tanenbaum (1967).

1.3.2 Plasma Frequency

The concept of charge screening indicates that the electrons and ions will adjust their positions to nullify the effect of a charge perturbation. The question we now ask is - how quickly can this occur? This will give an idea of the dynamic behavior of a plasma. We can examine this question in two ways. In both ways we use the most simple conceivable model of an ionized gas. We will assume that all of the charged particles are cold i.e. have no thermal motion. We will consider only the directed motion of the electrons and assume that we can consider the motion of one electron (or ion) to be characteristic of the motions of all of the electrons.

In the first approach to the plasma frequency, we will momentarily perturb the density in a plasma and then calculate the restoring force and the natural frequency of oscillation.

Take a uniform neutral plasma ($n_e = n_+ = n$) and shift all the electrons in a certain width W by an amount $\delta$. Assume that the ions are stationary. This will result in a charge imbalance as shown in Fig. 1.4.

The surface charge density in each thickness $\delta$ is $\rho_s = ne\,\delta$ and the electrostatic field intensity in the region W is

$$E = \frac{\rho_s}{\varepsilon_o} = \frac{ne\delta}{\varepsilon_o} . \quad (1.11)$$

The resulting restoring force on the electrons is

$$F = -eE = \frac{-ne^2\delta}{\varepsilon_o} \quad (1.12)$$

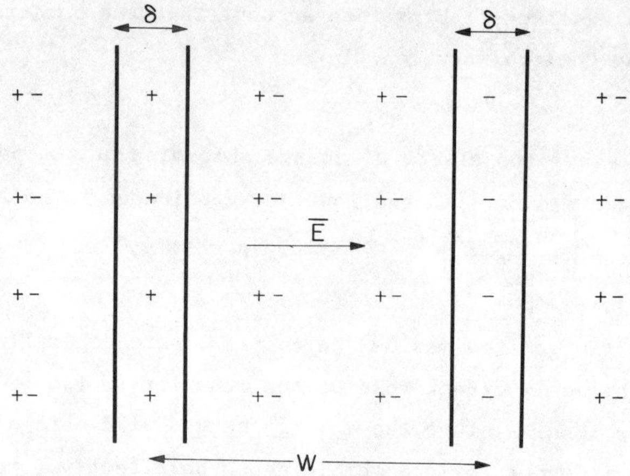


*Fig. 1.4. A perturbed uniform neutral plasma*

and the equation of motion of the electrons is

$$m \frac{d^2\delta}{dt^2} = \frac{-ne^2}{\varepsilon_o} \delta \, . \qquad (1.13)$$

This indicates an oscillatory solution. If $\delta$ is taken to be

$$\delta = \delta_o \, e^{j\omega_p t} \, , \qquad (1.14)$$

where $\omega_p$ is the frequency of oscillation, which we define to be the plasma frequency, then we find

$$\omega_p^2 = \frac{ne^2}{m\,\varepsilon_o} = 3.18 \times 10^3 \, n \qquad (1.15)$$

where n is in $m^{-3}$. This characteristic frequency of oscillation defines the time that it takes the electrons to move under the influence of a perturbation; and, therefore, defines how quickly an external perturbation can be screened out.

Another instructive method of considering this problem comes about by examining the motion of one electron under the influence of an alternating electromagnetic field. The electric field is assumed to vary as

$$\bar{E} = \bar{E}_o \, e^{j\omega t} \, . \qquad (1.16)$$

The equation of force is

$$m \frac{d\bar{v}}{dt} = m \frac{d^2 \bar{x}}{dt^2} = -c\bar{E} - c\bar{v} \times \bar{B} \, , \qquad (1.17)$$

where B is the magnetic flux density. In the absence of an external magnetic field, the RF magnetic force contribution is down by v/c and so is negligible in all but relativistic cases. Therefore,

$$m \frac{d^2 \bar{x}}{dt^2} = -e\bar{E}_o \, e^{j\omega t} \, . \qquad (1.18)$$

The solution must be of the form

$$\bar{x} = \bar{x}_o \, e^{j\omega t} \quad , \tag{1.19}$$

therefore,

$$\bar{x} = \frac{e\bar{E}}{m\omega^2} \quad . \tag{1.20}$$

The resulting induced dipole moment is

$$\bar{p} = -e\bar{x} = \frac{-e^2\bar{E}}{m\omega^2} \quad . \tag{1.21}$$

Assuming a uniform electron density n and that the electrons as a whole act essentially as the sum of the actions of n individual electrons, the total polarization

$$\bar{P} = \frac{-ne^2}{m\omega^2}\,\bar{E} \quad . \tag{1.22}$$

This can now be related to the dielectric constant of the electron gas

$$\bar{D} = \varepsilon\bar{E} = \varepsilon_o\,\varepsilon_R\,\bar{E} = \varepsilon_o\bar{E} + \bar{P} \,, \tag{1.23}$$

$$\bar{D} = \varepsilon_o\left[1 - \frac{ne^2}{m\varepsilon_o\omega^2}\right]\bar{E}.$$

Therefore, the relative dielectric constant

$$\varepsilon_R = 1 - \frac{ne^2}{m\varepsilon_o\omega^2} = 1 - \frac{\omega_p^2}{\omega^2} \tag{1.24}$$

Once again, this formalism containing the plasma frequency allows us to interpret its physical significance. Considering the electron gas to be a dielectric, if $\omega > \omega_p$ the relative dielectric constant is positive and less than 1. An electromagnetic wave of frequency $\omega$ can, therefore, propagate through the electron gas. The perturbation varies more quickly than the plasma can shield it out so the perturbation can penetrate the electron gas. If $\omega < \omega_p$, the relative dielectric constant is negative. Interpreting this in terms of a dielectric, this means that an electromagnetic wave cannot propagate through the electron gas. In a physical sense, the plasma is capable of responding quickly enough to shield the electron gas from the electromagnetic perturbation. Take the particular case when $\omega = \omega_p$. This means that $\varepsilon_R \to 0$. What does this mean? Well, inside the medium

$$\bar{E}_{internal} = \frac{\bar{D}}{\varepsilon_o\,\varepsilon_R} \quad .$$

Outside the medium $\bar{E}_{external} = \bar{D}/\varepsilon_o$. Therefore,

$$\bar{E}_{internal} = \frac{\bar{E}_{external}}{\varepsilon_R} \quad .$$

When $\varepsilon_R \to 0$, this means that a field can exist inside the electron gas with no external excitation. Obviously, this means that self-sustained oscillations can occur at this frequency in the plasma and that is consistent with the previous approach to calculating the plasma frequency.

## 1.4 REFERENCES

The Collected Papers of Peter J. W. Debye (1954), Interscience Publishers, New York, pp. 226-228.

Hasted, J. B. (1964), Physics of Atomic Collisions, Butterworth's, London.

Tanenbaum, B. Samuel (1967), Plasma Physics, McGraw-Hill, New York, pp. 203-209.

# Chapter 2

# ELEMENTARY THEORY OF A GAS DISCHARGE

In the derivation of the plasma frequency, it was assumed that the electrons had no random velocity and no interactions either with themselves or with other species within the medium. The first assumption is clearly wrong; however, if we consider only the average directed velocity, $\bar{u}$, of the electrons under the influence of an external field, the random motions cancel out so we can persist in ignoring the random motion until the full kinetic theory approach is developed. It is necessary at this point to begin to include particle interactions on an elementary basis if we are to progress beyond a simple equation for the dielectric constant of an electron gas (1.24).

## 2.1 THE LANGEVIN EQUATION

The simplest model that includes interactions is the Lorentz gas. In this model it is assumed that the electrons are free to move in a stationary uniform background of ions and neutrals which provide a viscous damping force. The electrons are assumed to interact with each other only through the space charge fields. Fortunately, this is a very good model of the medium of particular interest to us; one which is weakly ionized and where electron-neutral collisions dominate. We continue to use Newton's laws to describe the external forces acting on the electron, but we add a term which represents the momentum loss of the electrons due to collisions with background ions and neutrals. The inclusion of this term gives us the Langevin equation as follows:

$$\frac{d}{dt} m\bar{u} = -e\,[\bar{E} + \bar{u} \times \bar{B}] - m\bar{u}\,\nu_m. \qquad (2.1)$$

$\nu_m$ is defined as the collision frequency for momentum transfer. This means that on the average $\nu_m$ times per second the electron loses all its directed motion (momentum). A model of a billiard ball collision in which an electron would lose all its momentum is shown in Fig. 2.1.

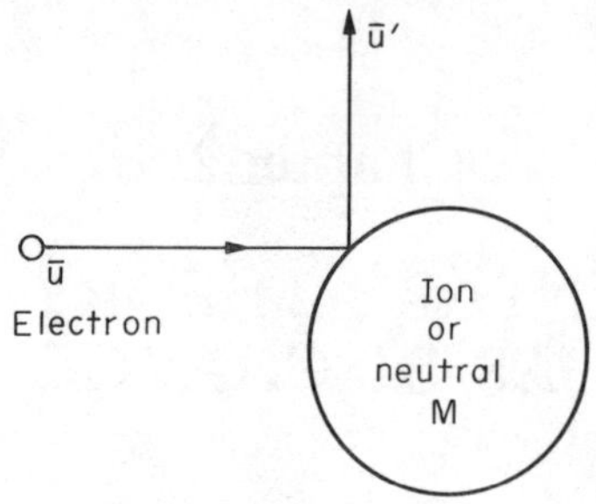


*Fig. 2.1. A model of a collision in which an electron loses all its momentum.*

All of the motion in the x direction is lost in this collision but note that the magnitude of the electron velocity (energy) remains the same except for a small fraction that is lost in the recoil of the ion or neutral. As will be seen later, in such an elastic collision a fraction $\delta = 2m/M$ of the original electron energy $(m\langle v^2\rangle/2)$ is lost if the ions or neutrals are motionless.

The solution of equation (2.1) provides the directed motion of an "average" electron which is assumed to be representative of all of the electrons in the discharge. The results for one electron will then be multiplied by the density of electrons and the result will be assumed to represent the gas discharge as a whole. As will be seen later, this approach is rigorously correct only if $\nu_m$ is independent of the electron velocity.

## 2.2 MOBILITY, CONDUCTIVITY AND DIELECTRIC CONSTANT

Initially, the magnetic field can be ignored and the electric field can be considered to be oscillatory

$$\bar{E} = \bar{E}_o e^{j\omega t} \quad .$$

Ignoring transients, the solution for $\bar{u}$ must have this same time dependence so, from (2.1)

$$j\omega m\bar{u} = -e\bar{E} - m\bar{u}\nu_m \quad . \tag{2.2}$$

Therefore,

$$\bar{u} = \frac{-e\bar{E}}{m(\nu_m + j\omega)} \quad . \tag{2.3}$$

$\bar{u}$ is the average directed velocity, $\langle\bar{v}\rangle$, under the influence of the electric field, and it is known as the drift velocity. The electron drift velocity is linearly proportional to the applied electric field and the proportionality constant is called the <u>electron mobility</u> $\mu_e$.

Defining

$$\bar{u} = -\mu_e \bar{E} \quad , \tag{2.4}$$

$$\mu_e = \frac{e}{m(\nu_m + j\omega)} \quad . \tag{2.5}$$

In the DC case ($\omega \to 0$)

$$\mu_e = \frac{e}{m\nu_m} \quad . \tag{2.6}$$

Obviously, the collision frequency varies directly with the number of neutrals with which the electrons collide so $\nu_m$ varies directly with gas pressure (for a constant gas temperature). In the case of helium and hydrogen, for which $\nu_m$ is relatively independent of electron velocity, from Brown (1959),

$$\nu_m \simeq 4.8 \times 10^9 \ P(H_2) \tag{2.7}$$

and

$$\nu_m \simeq 2.3 \times 10^9 \ P(He) \tag{2.8}$$

where P is the gas pressure in torr at 300K. Using these values to determine the mobility gives the following relationship between the drift velocity and electric field when electron-neutral collisions dominate:

For hydrogen

$$u(m/sec) = 37 \ \frac{E}{P}(V/m\text{-}torr) \tag{2.9}$$

and for helium

$$u(m/sec) = 76 \ \frac{E}{P}(V/m\text{-}torr) . \tag{2.10}$$

Similar equations can be derived for the motion of the ions in the medium simply by changing the sign of the charge, changing the mass and using the ion-neutral collision frequency. (More exact expressions will be developed in a later section).

It was stated previously that the drift velocity was usually much less than the thermal velocity in an ionized gas. Let us check that assumption for the case of a helium discharge at a few torr pressure. A typical E/P would be 400 V/m-torr and a typical electron temperature would be 60,000K.

The thermal velocity

$$c = \sqrt{\langle v^2 \rangle} = \sqrt{\frac{3kT_e}{m}} = 1.65 \times 10^6 \ m/sec \quad .$$

The drift velocity

$$u = \mu_e E \simeq 3 \times 10^4 \ m/sec \quad .$$

Therefore, it can be seen that the previous assumption is indeed valid.

Once the mobility is known, the conductivity can be readily obtained. The conduction current density due to the electric field is:

$$\bar{J} = -n_e e\bar{u} = n_e e\mu_e \bar{E} \quad . \tag{2.11}$$

Therefore,

$$\bar{J} = \frac{n_e e^2 \bar{E}}{m(\nu_m + j\omega)} \tag{2.12}$$

and the conductivity of the medium due to the electrons is:

$$\sigma = \frac{n_e e^2}{m(\nu_m + j\omega)} \quad . \tag{2.13}$$

Since the drift velocity of the electrons is much greater than that of the ions, this term accurately reflects the conductivity of the ionized gas.

If the previous derivation of the conductivity is retraced, it is possible to obtain the expression for the dielectric constant of the medium.

From equation (2.3)

$$\bar{u} = \frac{d\bar{x}}{dt} = j\omega\bar{x} = \frac{-e\,\bar{E}}{m(\nu_m + j\omega)} \quad . \tag{2.14}$$

The dipole moment

$$\bar{p} = -e\bar{x} = \frac{e^2\,\bar{E}}{j\omega m(\nu_m + j\omega)} \tag{2.15}$$

and the polarization

$$\bar{P} = \frac{ne^2\,\bar{E}}{j\omega m(\nu_m + j\omega)} \quad . \tag{2.16}$$

Therefore,

$$\bar{D} = \varepsilon_o\varepsilon_R\bar{E} = \varepsilon_o\bar{E} + \bar{P} = \varepsilon_o\bar{E}\left[1 + \frac{ne^2}{j\omega\varepsilon_o m(\nu_m + j\omega)}\right] \tag{2.17}$$

from which

$$\varepsilon_R = 1 - \frac{\omega_p^2/\omega^2}{(1 - j\nu_m/\omega)} \quad . \tag{2.18}$$

In fact the rather general relationship between dielectric constant and conductivity is

$$\varepsilon_R = 1 - j\frac{\sigma}{\omega\varepsilon_o} \quad . \tag{2.19}$$

Having defined the dielectric constant and conductivity, we are now in a position to consider the significance of the complex nature of both terms. Generally speaking, a complex dielectric constant simply implies a lossy dielectric, i.e. one in which an electromagnetic wave is attenuated as it propagates. The degree of attenuation, however, depends upon the frequency of the wave, or more accurately, the ratio $\nu/\omega$.

From equation (2.18), it is seen that as $\nu/\omega \to 0$, equation (1.24) is reconstructed. The wave propagates with no loss provided $\omega > \omega_p$. As $\nu/\omega$ increases the degree of attenuation increases; the dielectric is no longer perfect. In the DC limit, from equation (2.13), the conductivity becomes

$$\sigma = \frac{n_e e^2}{m\nu_m} \quad . \tag{2.20}$$

We have a pure conductor. The way in which the lossiness of the medium varies can be obtained by obtaining the expression for the ohmic loss in the medium

$$P_{\text{dissipated}} = \bar{J} \cdot \bar{E} = (R_e\sigma)\, E^2 \quad . \tag{2.21}$$

$R_e\sigma$ is the real part of the conductivity

$$R_e\sigma = \frac{ne^2\nu_m}{m(\nu_m^2+\omega^2)} = \frac{ne^2}{m\omega}\;\frac{(\nu_m/\omega)}{(1+\nu_m^2/\omega^2)} \quad . \tag{2.22}$$

This shows the explicit dependence of the dissipation on $\nu/\omega$.

The equations obtained for the conductivity and dielectric constant, even though very simple, are very useful for obtaining good approximations of the actual behavior of a gas discharge, particularly in the area of investigation involving the propagation of electromagnetic waves through ionized media. Probably the greatest use, however, is in the study of electromagnetic wave propagation through the ionosphere, a region where magnetic fields are important.

The conductivity of a magneto-ionic medium can be obtained from the complete Langevin equation,

$$(\nu_m + j\omega)m\bar{u} = -e\bar{E} - e\bar{u} \times \bar{B}_o \tag{2.23}$$

This is most easily solved if it is assumed that the magnetic field is in the $\hat{Z}$ direction and $\bar{E}$ can have components in any direction.

$$\begin{aligned} (\nu_m + j\omega)mu_x &= -eE_x - eu_yB_o, \\ (\nu_m + j\omega)mu_y &= -eE_y + eu_xB_o, \\ (\nu_m + j\omega)mu_z &= -eE_z \quad . \end{aligned} \tag{2.24}$$

Now $J_x = -n_e e u_x$ , etc., so that the entire solution can be written as:

$$\begin{bmatrix} J_x \\ J_y \\ J_z \end{bmatrix} = \begin{bmatrix} \sigma_\perp & \sigma_x & 0 \\ -\sigma_x & \sigma_\perp & 0 \\ 0 & 0 & \sigma_{||} \end{bmatrix} \begin{bmatrix} E_x \\ E_y \\ E_z \end{bmatrix} \tag{2.25}$$

or

$$\bar{J} = \bar{\bar{\sigma}} \cdot \bar{E}$$

where $\bar{\bar{\sigma}}$ is the tensor conductivity.

The resultant components are:

$$\sigma_\perp = \frac{ne^2}{m} \frac{\nu_m + j\omega}{(\nu_m + j\omega)^2 + \omega_c^2} ,$$

$$\sigma_x = \frac{-ne^2}{m} \frac{\omega_c}{(\nu_m + j\omega)^2 + \omega_c^2} , \tag{2.26}$$

$$\sigma_{||} = \frac{ne^2}{m} \frac{1}{\nu_m + j\omega} ,$$

where $\omega_c = eB_o/m$ is the electron cyclotron frequency.

Similarly, the dielectric constant can be derived from the generalized relationship

$$\overline{\overline{\varepsilon}}_R = \overline{\overline{I}} - j \frac{\overline{\overline{\sigma}}}{\omega \varepsilon_o} \tag{2.27}$$

where $\overline{\overline{I}}$ is the unit tensor.

$$\overline{\overline{\varepsilon}}_R = \begin{bmatrix} \varepsilon_\perp & \varepsilon_x & 0 \\ -\varepsilon_x & \varepsilon_\perp & 0 \\ 0 & 0 & \varepsilon_{||} \end{bmatrix} \tag{2.28}$$

where

$$\varepsilon_\perp = 1 - \frac{(\omega_p^2/\omega^2)\ (1-j\nu_m/\omega)}{[(1-j\nu_m/\omega)^2 - (\omega_c/\omega)^2]} ,$$

$$\varepsilon_x = -j \frac{(\omega_p^2/\omega^2)\ (\omega_c/\omega)}{[(1-j\nu_m/\omega)^2 - (\omega_c/\omega)^2]} , \tag{2.29}$$

$$\varepsilon_{||} = 1 - \frac{\omega_p^2/\omega^2}{(1-j\nu_m/\omega)} .$$

In the case of the magneto-ionic medium, the ion terms play a different role than the electrons and so the additional ion terms in the conductivity or dielectric constant (which are just added to the electron terms) can be obtained from the electron terms by substituting

$$\Omega_p^2 = \frac{m}{M} \omega_p^2 \quad \text{for} \quad \omega_p^2$$

and

$$-\Omega_c = \frac{-c\ B_o}{M} \quad \text{for} \quad \omega_c$$

where M is the ion mass. For $\varepsilon_\perp$ and $\varepsilon_{||}$, however, the factor of one is not repeated when the ion terms are added as can be seen if the dielectric constant is derived from the conductivity which includes the ion terms.

Now, the applications of the tensor dielectric constant in the study and analysis of electromagnetic wave propagation through magneto ionic media has been thoroughly developed. The books by Davies (1965), by Allis, Buchsbaum and Bers (1963), and by Heald and Wharton (1965), cover this area thoroughly so we will limit our further consideration to a few additional comments on the interpretation of the conductivity in a magnetic field.

If an RF field is used to excite a gas discharge or to heat the electrons, then the energy absorbed is proportional to the real part of the conductivity. From equation (2.26), then, it is useful to determine the real part of $\sigma_\perp$. After a little algebra, this can be expressed as:

$$\mathrm{Re}\sigma_\perp = \frac{ne^2\nu_m}{2m}\left[\frac{1}{\nu_m^2 + (\omega-\omega_c)^2} + \frac{1}{\nu_m^2 + (\omega+\omega_c)^2}\right]. \tag{2.30}$$

If the collision frequency is reasonably small with respect to the frequency of the electromagnetic field ($\nu_m^2 << \omega^2$) and if the electromagnetic frequency approaches the cyclotron frequency, then

$$\mathrm{Re}\sigma_\perp \simeq \frac{ne^2\nu_m}{2m}\left[\frac{1}{\nu_m^2 + (\omega-\omega_c)^2}\right] \tag{2.31}$$

As can be seen, if $\omega = \omega_c$ the conductivity peaks in a resonant manner and close to $\omega = \omega_c$, the conductivity exhibits a Lorentzian line shape with a full width at half maximum of $2\nu_m$.

The peak in the conductivity can be understood as the condition under which the motion of the electron in the magnetic field is in synchronism with the driving electromagnetic field, setting up a condition under which the electron continuously gains energy from the electric field. This is analogous to the conditions achieved in a DC field where the velocity of the electrons (or the current) is impeded by collisions ($\mathrm{Re}\ \sigma \sim 1/\nu_m$) hence collisions limit the amount of energy absorbed.

## 2.3 ENERGY BALANCE, ELECTRON TEMPERATURE AND ENERGY RELAXATION

Now that we have shown that the Langevin equation gives useful expressions for mobility, conductivity and dielectric constant, we can apply this simple approach to answering several more questions of interest in an ionized medium. These questions are:

1. If a weak electric field is applied to a plasma, by how much does the electron temperature change to represent the energy deposited in the plasma by the field?

and

2. If this perturbing field is removed, how quickly does the electron temperature (or average energy) relax back to its equilibrium value?

These answers can be obtained by equating the energy gain and loss terms in the plasma.

If we assume that the plasma is weakly ionized and that only elastic electron-neutral collisions are important, we can say that the energy gained per unit time from the field is

$$\bar{J} \cdot \bar{E} = \sigma E^2 = \frac{ne^2}{m\,\nu_m} E^2 \quad \text{for a DC field.} \tag{2.32}$$

Similarly, the energy lost per unit time due to electron neutral collisions is

$$n\,\delta\,\nu_m\,3k(T_e-T)/2 \tag{2.33}$$

where $3k\,(T_e-T)/2$ is the excess energy of the electrons above the background gas energy ($3kT/2$) and $\delta$ ($= 2m/M$) is the fraction of the excess energy lost per collision. If $T_e = T$, there are as many collisions giving energy to the electrons as there are collisions taking energy away and so the total energy loss would be zero.

We may now write a rate equation for the electron energy

$$\frac{d}{dt}\left(\frac{3}{2}\,n\,kT_e\right) = \sigma E^2 - n\,\delta\,\nu_m\,3k(T_e-T)/2. \tag{2.34}$$

Let us consider the steady state electron temperature with an applied electric field. (This is good for a DC field or a high frequency AC electric field.) For a DC field

$$\frac{ne^2}{m\,\nu_m} E^2 = n\,\delta\,\nu_m \frac{3}{2} k(T_e-T). \tag{2.35}$$

In this case

$$T_e = T + \frac{2\,e^2\,E^2}{3k\,m\,\delta\,\nu_m^2} \tag{2.36}$$

is the temperature of the electrons when a DC electric field is applied.

Note that $T_e$ is a function of the variable $(E/\nu_m)^2$. Since $\nu_m$ varies directly with the neutral gas pressure, P, or the neutral gas density, N, we can see that the electron temperature (or average electron energy) is a function of the parameter E/P or E/N. This is quite a general result, as will be seen later.

Now let us determine how quickly the electron temperature will relax back to the gas temperature if the field is removed. In this case

$$\frac{d}{dt}(3n\,kT_e/2) = -\,n\,\delta\,\nu_m\,3k(T_e-T)/2 \quad . \tag{2.37}$$

The solution, therefore, is

$$T_e = T + (T_e-T)_o\,e^{-\delta\,\nu_m\,t}. \tag{2.38}$$

The electron energy relaxation time constant (due to elastic collisions) is therefore

$$\tau_{e\ell} = \frac{1}{\delta\, \nu_m} \quad . \tag{2.39}$$

This is the time of energy equilibration between the electrons and the neutral gas.

These equations, as well as the equations for the electron temperature in DC fields, are good for small increases in the electron temperature above gas temperature. For large changes in temperature, some corrections should be made since $\nu_m$ is, in fact, $\nu_m(T_e)$. We could also generalize these equations for AC fields by substituting

$$\frac{E_o^2}{\omega^2 + \nu_m^2} \quad \text{for} \quad \frac{2E^2}{\nu_m^2} \, ,$$

where $E_o$ is the peak AC field, ie $\bar{E} = \bar{E}_o\, e^{j\omega t}$.

## 2.4 REFERENCES

Allis, William P., Buchsbaum, Solomon J., and Bers, Abraham (1963), Waves in Anisotropic Plasmas, MIT Press.

Brown, Sanborn C. (1959), Basic Data of Plasma Physics, Wiley & Sons.

Davies, Kenneth (1965), Ionospheric Radio Propagation, U.S. National Bureau of Standards Monograph 80.

Ginzburg, V.L., and Gurevich, A.V. (1960), "Nonlinear Phenomena in a Plasma Located in an Alternating Electromagnetic Field," Sov. Phys. Uspekhi 3, 115-146.

Heald, M.A., and Wharton, C.B. (1965), Plasma Diagnostics with Microwaves, Wiley & Sons.

# Chapter 3
# COLLISIONS

In our discussion of the Langevin equation and its applications, we expressed the interaction of electrons (or other particles) with the other constituents of the plasma in terms of a "collision frequency for momentum transfer" which represents the viscous loss in the medium. For particular applications of the Langevin equation, we need to be able to determine $\nu_m$ for the gas constituents in question. Since $\nu_m$ will, in general, depend upon the species with which the electrons (or other particles) collide, the density of the collision partners and the relative velocities of the colliding particles, we need to look into the more fundamental aspects of collisions in plasmas. When we look into the more exact theory (Boltzmann equation) of gas discharges we also need to know some of the details of the collisional interactions. We will therefore try to tie some of these basic concepts together at this point.

Our approach will be to relate the collision frequency to the collision cross section $\sigma$ and then to the differential scattering cross section, $d\sigma/d\Omega$, which contains the essential features of the collisional interaction between two particles. Schematically,

$$\nu \leftrightarrow \sigma \leftrightarrow \frac{d\sigma}{d\Omega} \leftrightarrow \text{classical or quantum scattering theory.}$$

## 3.1 CROSS SECTION, MEAN FREE PATH AND COLLISION FREQUENCY

Let us start with a fundamental quantity, the collision cross section $\sigma$. In a very simple manner, if we consider an atom as essentially a "billiard ball" of radius a, then an electron travelling towards this atom will undergo a collision if its trajectory falls within the "shadow" of the billiard ball. The collision cross section is then the area obscured by the atom or

$$\sigma = \pi a^2 \text{ for this simple model.}$$

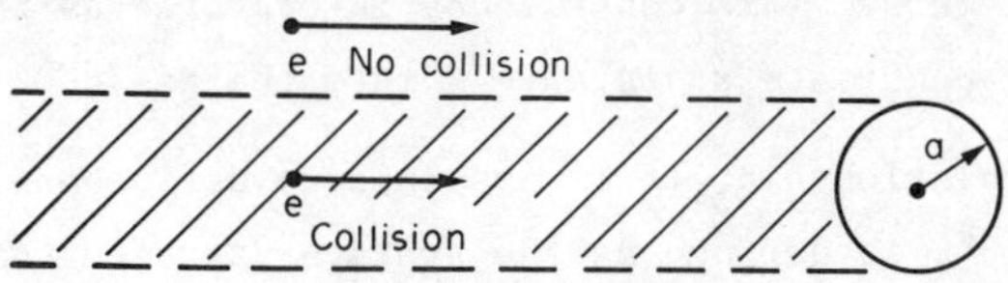


*Fig. 3.1. Concept of collision cross section*

This can be easily extended to the behavior of a monoenergetic electron beam entering a gas. Assume that a uniform electron beam of cross sectional area A and a flux $\Gamma = nv$ is incident upon a gas of density N atoms/$m^3$ and each atom has a

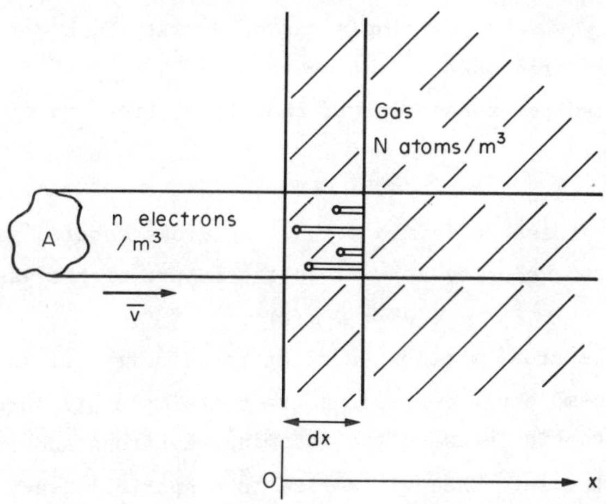


*Fig. 3.2. Collisions of an electron beam in a gas*

cross section $\sigma$ $m^2$. We then ask ourselves, what fraction of the flux is scattered out of the beam as the beam penetrates a distance dx into the gas? The answer seems fairly obvious. The fraction of the flux scattered out of the beam is equal to the fraction of the area of the beam which is obscured by the scatterers within the distance dx, or

$$\frac{d\Gamma}{\Gamma} = \frac{\text{number of scatterers in dx} \times \text{scattering cross section}}{\text{area of the beam}} \quad .$$

Therefore, the amount by which the flux is changed as the beam penetrates dx is:

$$\frac{d\Gamma}{\Gamma} = - \frac{(N\ A\ dx)\ \sigma}{A} = - N\ \sigma\ dx \quad . \tag{3.1}$$

The flux (or intensity) of the electron beam will diminish as the electron beam propagates through the medium, and, by solving the above equation, the flux is found to vary as

$$\Gamma\ (x) = \Gamma_o\ e^{-N\ \sigma\ x} \tag{3.2}$$

where, of course, this represents the statistical average for a large number of electrons and scatterers. This can also be interpreted as the probability that any given electron will penetrate a distance x into the gas.

From this basic relationship, it is possible to define one other common term. The mean free path, $\ell$, is defined as the average distance that an electron travels before colliding with a gas atom. From equation (3.2), this can be found to be

$$\ell = \langle x \rangle = \frac{\int_0^\infty x\, e^{-N\sigma x}\, dx}{\int_0^\infty e^{-N\sigma x}\, dx} = \frac{1}{N\sigma} \quad . \tag{3.3}$$

The final connection that we want to make is back to the collision frequency. If an electron of velocity v is traversing a gas of density N, how many collisions per second will such an electron make on the average?

The distance travelled per second is vt, therefore, from equation (3.3), the collision frequency is:

$$\nu = N \sigma v \quad . \tag{3.4}$$

Keep in mind in all of these definitions that the cross section is, in general, a function of the electron velocity as well as the nature of the gas, so that:

$$\nu(v) = N \sigma(v) v \quad . \tag{3.5}$$

The dependence of the cross section on electron velocity (or energy) is due to the fact that electron-molecule collisions are not simple billiard ball collisions, but represent an interaction between the incoming electrons and the electrons plus the nucleus of the molecule. This gives rise to a spatially varying interaction potential between the electron and the molecule which, as will be seen later, causes the cross section to vary with the relative velocities of the two collision partners.

Keep in mind as well that the velocity used in equation (3.5) to calculate the collision frequency is the total electron velocity, which is equal to the sum of the drift velocity and the random or thermal velocity. In most cases, then, the velocity used to calculate $\nu$ is simply the thermal velocity of the electron.

To obtain some rough ideas of the magnitudes of $\sigma$ we can refer again to our simple idea of a cross section being approximately the area of the collision partner. For atoms, then, we are talking of areas $\simeq 10^{-19}$ $m^2$ plus or minus a few orders of magnitude. As examples, we will give some plots of $\sigma$ as a function of electron energy for various gases (assuming elastic collisions only). For the rare gases see Fig. 3.3.

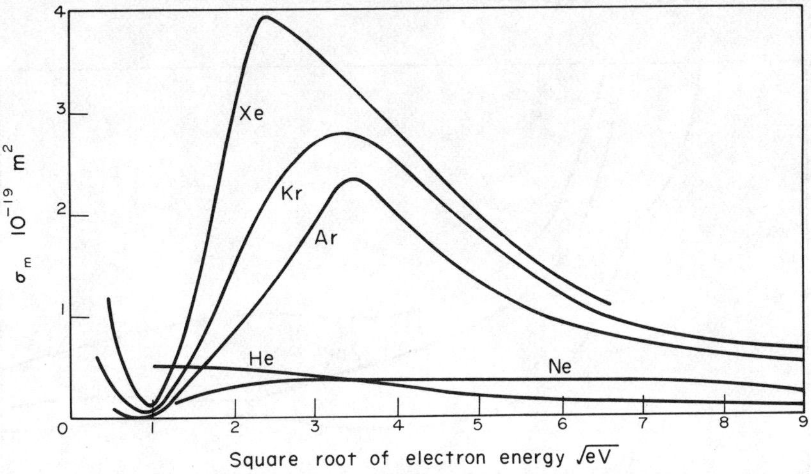


*Fig. 3.3. Elastic collision cross sections for the rare gases (from Brode, 1933)*

For some atmospheric molecular gases, see Fig. 3.4.

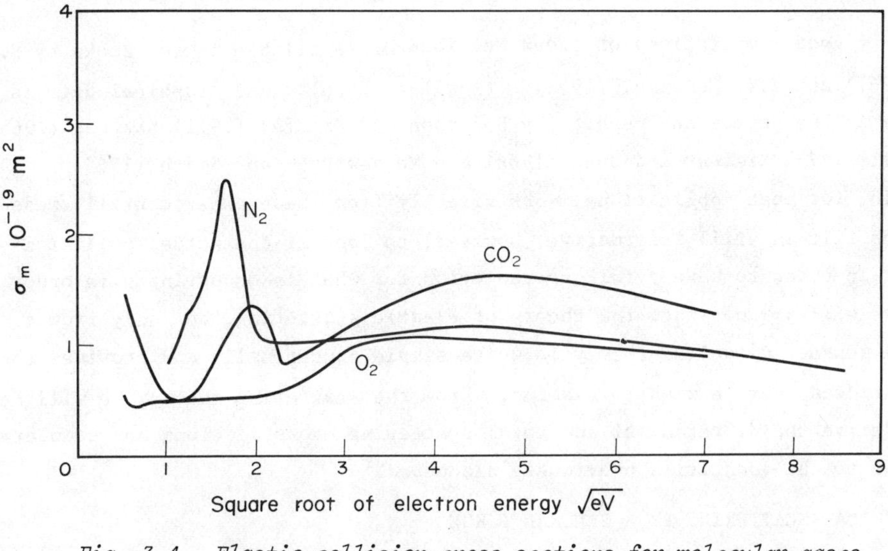


*Fig. 3.4. Elastic collision cross sections for molecular gases (from Brode, 1933)*

For some metal vapors, see Fig. 3.5

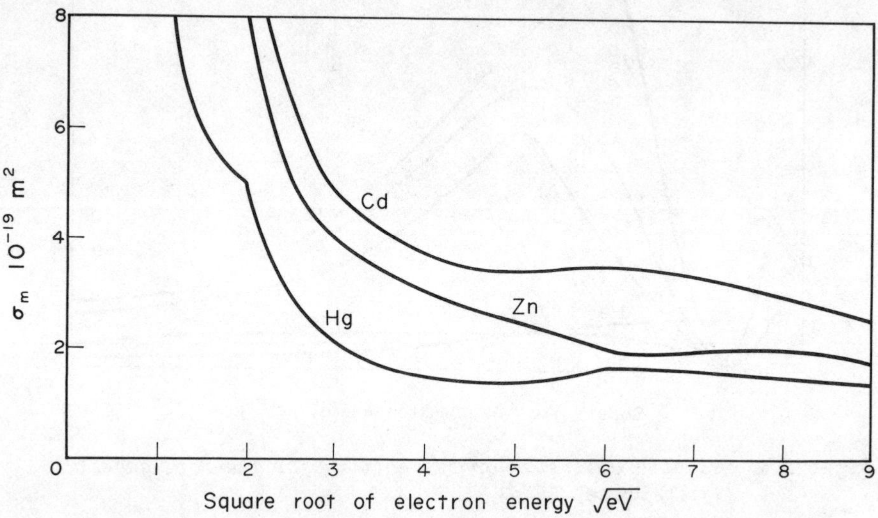


*Fig. 3.5. Elastic collision cross sections for metal vapors (from Brode, 1933)*

Some very good compilations of cross sections are available in the books by S.C. Brown (1967) and E.W. McDaniel (1964). Excellent tabular and graphical data is available in the papers and reports by Bederson and Kieffer (1971) Kieffer (1965, 1969, 1971, 1973) Kieffer and Dunn (1966) and Moisewitsch and Smith (1968).

We could, for most applications, work directly from these experimentally measured quantities. It is still informative, however, to look at the actual collision processes in order to have a full appreciation for what is occurring. In order to do this we will now consider the theory of elastic scattering, but only from a classical sense. Classical theory is quite simple conceptually and provides the essential ideas that we wish to examine. From the scattering theory, we will be able to derive the differential and total scattering cross sections and complete the connection to the quantities previously discussed.

## 3.2 CLASSICAL SCATTERING BY A CENTRAL FORCE

We are going to consider the scattering of a particle by a central force. That is, the potential due to the scattering partner is purely radial (spherically symmetric) $V = V(r)$. In general, there can be angular dependences of V but this tends to make the solutions more difficult and is probably better handled through the quantum theory where the angular momentum of the quantum states of atoms can be more explicitly treated. We shall deal with scattering in the system in which the scattering center is considered to be stationary (the Laboratory coordinate system).

This is a good assumption for electron-heavy particle collisions. If the motion of the scattering centers is significant, then corrections will have to be made to the theory.

Consider the following case. The scattering center is fixed at r = o and the electron comes in from the left ($r = \infty$) with a velocity $\bar{v}_o$ and at a distance b (impact parameter) away from a line going through the center of the scatterer. If we assume a repulsive potential, the electron will follow the orbit shown in Fig. 3.6 and will be deflected by an angle $\theta$, the scattering angle. Since we are assuming a truly elastic collision (no energy exchange) the speed of the receding particle will be the same as when it entered the region (conservation of energy) and it will leave with the same separation (b) from the line going through the scattering center (conservation of angular momentum). What we wish to do is to determine the scattering angle $\theta(b,v_o)$ in terms of the initial velocity ($v_o$) and impact parameter (b) for a given V(r) (type of scattering center).

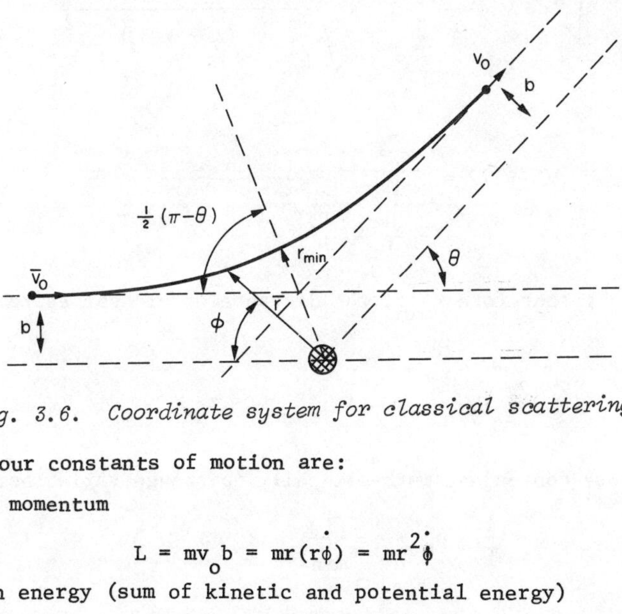


*Fig. 3.6. Coordinate system for classical scattering*

In this case, our constants of motion are:

1. The angular momentum

$$L = mv_o b = mr(r\dot{\phi}) = mr^2\dot{\phi} \tag{3.6}$$

2. The electron energy (sum of kinetic and potential energy)

$$\varepsilon = mv_o^2/2 = m(\dot{r}^2 + r^2\dot{\phi}^2)/2 + V(r) \quad . \tag{3.7}$$

From the angular momentum,

$$\dot{\phi} = \frac{L}{mr^2} \quad \text{and} \quad \dot{r} = \frac{dr}{dt} = \frac{dr}{d\phi}\dot{\phi} \quad . \tag{3.8}$$

The energy equation can then be re-written as

$$\varepsilon = \frac{L^2}{2mr^4}\left[\left(\frac{dr}{d\phi}\right)^2 + r^2\right] + V(r) \tag{3.9}$$

and

$$\left(\frac{dr}{d\phi}\right)^2 = \frac{2m}{L^2} r^4 \, [\varepsilon - V(r)] - r^2 \quad . \tag{3.10}$$

Therefore

$$\frac{d\phi}{dr} = \frac{\pm 1}{\sqrt{\frac{2m}{L^2} r^4 \, (\varepsilon - V(r)) - r^2}} \quad . \tag{3.11}$$

This is the equation for determining the trajectory. The minus sign corresponds to the first half of the orbit and the plus sign corresponds to the second half of the orbit. Since the orbit is symmetric from $r = \infty$ into $r_{min}$ and from $r = r_{min}$ out to $r = \infty$, we only have to solve for $r = \infty$ to $r = r_{min}$.

Now, at $r = \infty$, $\phi = 0$ and at $r = r_{min}$, $\phi = (\pi-\theta)/2$, therefore

$$\int_0^{(\pi-\theta)/2} d\phi = -\int_\infty^{r_{min}} \frac{dr}{\sqrt{\frac{2m}{L^2} r^4 \, (\varepsilon - V(r)) - r^2}} \tag{3.12}$$

or

$$(\pi-\theta)/2 = \int_{r_{min}}^{\infty} \frac{dr}{\sqrt{\frac{2m}{L^2} r^4 \, (\varepsilon - V(r)) - r^2}} \quad . \tag{3.13}$$

At $r = r_{min}$, $\frac{dr}{d\phi} = 0$, therefore $r_{min}$, the distance of closest approach, is determined by

$$\frac{2m}{L^2} r_{min}^{\ 4} \, [\varepsilon - V(r_{min})] - r_{min}^{\ 2} = 0 \quad . \tag{3.14}$$

It is sometimes more convenient mathematically to change variables. If we take

$$u = \frac{1}{r}, \quad u_{max} = \frac{1}{r_{min}} \quad \text{and} \quad du = \frac{-1}{r^2} dr$$

then

$$(\pi-\theta)/2 = \int_0^{u_{max}} \frac{du}{\sqrt{\frac{2m}{L^2} (\varepsilon - V(\frac{1}{u})) - u^2}} \, ,$$

$$\frac{2m}{L^2} (\varepsilon - V(\frac{1}{u_{max}})) - u_{max}^{\ 2} = 0 \quad . \tag{3.15}$$

If we express

$$\frac{2m}{L^2} = \frac{2m}{m^2 v_o^2 b^2} = \frac{1}{\varepsilon b^2} ,$$

then

$$(\pi-\theta)/2 = \int_o^{u_{max}} \frac{du}{\sqrt{\frac{1}{b^2}(1 - \frac{V}{\varepsilon}) - u^2}} ,$$

$$(1 - V/\varepsilon) - b^2 u_{max}^{\;2} = 0 . \qquad (3.16)$$

In general then, if V(r) is known we can solve for $\theta$ as a function of $\varepsilon(v_o)$ and b. In relatively few cases can an exact analytical solution be found. As shown by Goldstein (1959), if we take V(r) to be of the form

$$V(r) = ar^n$$

exact solutions can be found for n = 2, -1, -2 in terms of circular functions and n = 6, 4, 1, -3, -4, -6 in terms of elliptic functions. Some fractional exponents can also lead to elliptic solutions.

Let us now apply this theory to two particularly interesting cases for gas discharges.

### 3.2.1 Electron-molecule Hard-sphere Collisions

As a first model of electron molecule or electron atom elastic collisions, we take the atom to be an impenetrable sphere of radius r = a as shown in Fig. 3.7.

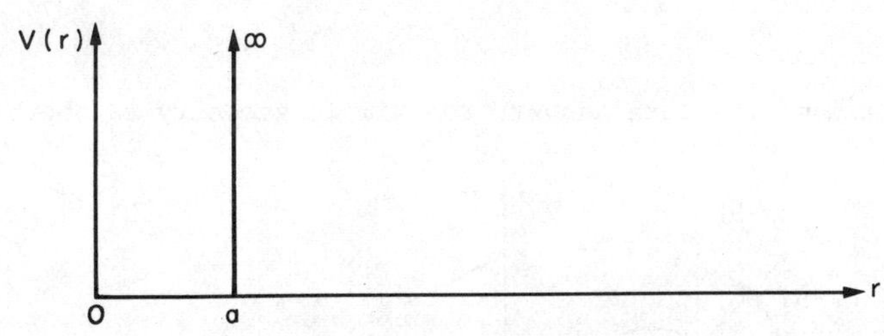


*Fig. 3.7. Hard sphere interaction potential*

This is a reasonable model of an actual potential for a molecule which is approximately as shown in Fig. 3.8.

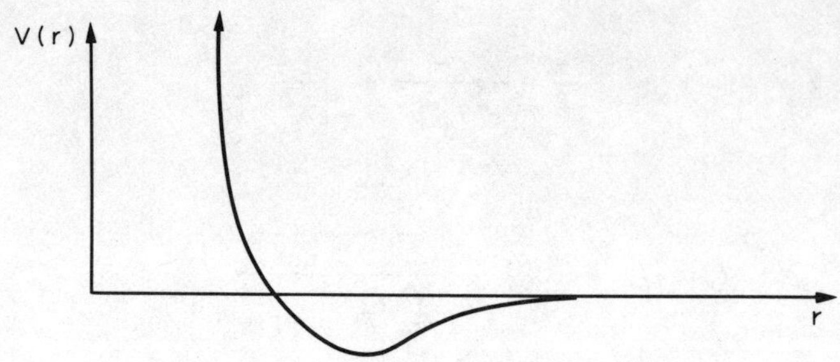


*Fig. 3.8. Realistic electron-molecule interaction potential*

Now $r_{min} = a$ and $u_{max} = \frac{1}{a}$, therefore

$$(\pi-\theta)/2 = \int_0^{1/a} \frac{du}{\sqrt{\frac{2m}{L^2}\varepsilon - u^2}} = \int_0^{1/a} \frac{du}{\sqrt{\frac{1}{b^2} - u^2}} \quad . \tag{3.17}$$

If $b > a$, there are no collisions. If $b < a$

$$(\pi-\theta)/2 = \sin^{-1} \frac{u}{1/b} \Big|_0^{1/a} = \sin^{-1} \frac{b}{a} \quad , \tag{3.18}$$

therefore

$$\frac{b}{a} = \sin\left(\frac{\pi}{2} - \frac{\theta}{2}\right) = \cos\frac{\theta}{2} \tag{3.19}$$

and

$$\theta = \begin{cases} 2\cos^{-1} b/a & b < a \\ 0 & b > a \quad . \end{cases} \tag{3.20}$$

We could have obtained the same answer from simple geometry as shown in Fig. 3.9.

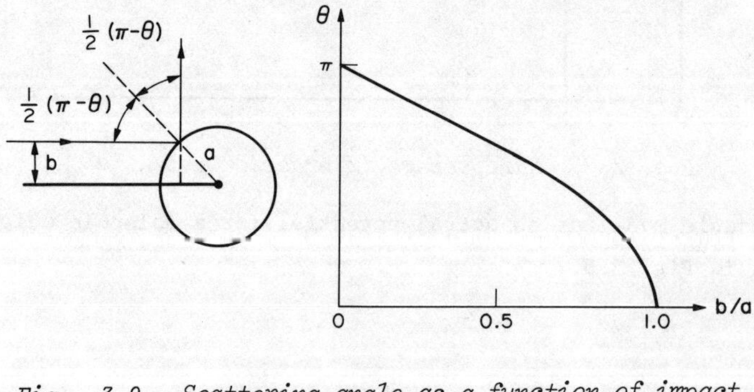


*Fig. 3.9. Scattering angle as a function of impact parameter for a hard sphere*

In this case $\theta = \theta(b)$ since the scattering is independent of the electron velocity.

### 3.2.2 Coulomb Collision Scattering

Another interesting case is the scattering of electrons by ions or other charged particles. This case becomes increasingly important as the density of the charged particles in the plasma increases. For this case, the interaction potential

$$V(r) = \frac{e_1 e_2}{4\pi \varepsilon_o r} \tag{3.21}$$

where $e_1$ is the electron charge and $e_2$ is the ion charge if we are specifically interested in electron-ion scattering. For this case

$$(\pi - \theta)/2 = \int_o^{u_{max}} \frac{du}{\sqrt{\frac{1}{b^2} - \frac{2 e_1 e_2}{4\pi \varepsilon_o mv^2 b^2} u - u^2}} \tag{3.22}$$

where

$$u_{max} = \frac{-e_1 e_2}{4\pi \varepsilon_o mv^2 b^2} + \sqrt{\left(\frac{e_1 e_2}{4\pi \varepsilon_o mv^2 b^2}\right)^2 + \frac{1}{b^2}} \; . \tag{3.23}$$

Solving, we find

$$\theta = 2 \sin^{-1} \frac{\left(\frac{e_1 e_2}{4\pi \varepsilon_o mv^2 b^2}\right)}{\sqrt{\left(\frac{e_1 e_2}{4\pi \varepsilon_o mv^2 b^2}\right)^2 + \frac{1}{b^2}}} \; . \tag{3.24}$$

This defines a simple geometrical relation shown in Fig. 3.10 and also allows θ to

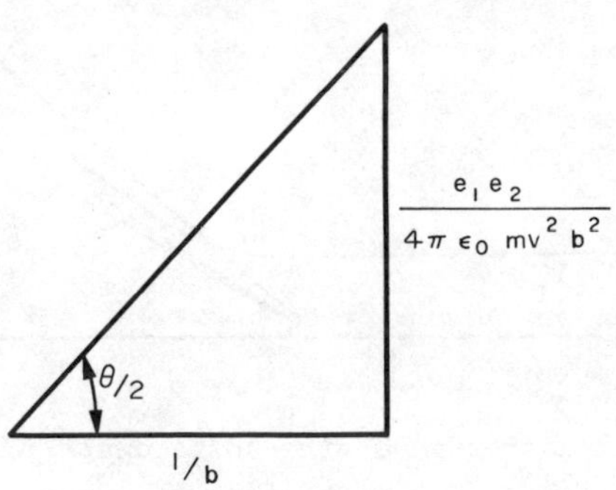


*Fig. 3.10. Scattering triangle for Coulomb collisions*

be expressed a little more simply as

$$\theta = 2 \tan^{-1} \frac{e_1 e_2}{4\pi \varepsilon_o mv^2 b} \quad \text{(see Fig. 3.11).} \tag{3.25}$$

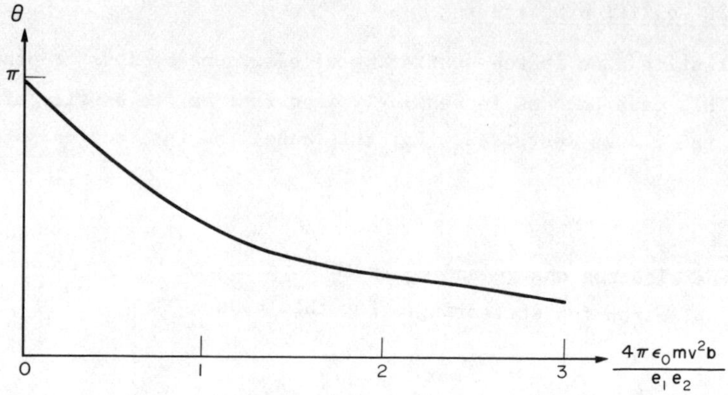


*Fig. 3.11. Scattering angle as a function of impact parameter and energy for Coulomb collisions*

As we see for this case in particular, and most cases in general, $\theta = \theta(b,v)$. As could be expected, the higher the energy of the electron, the less it is scattered by the Coulomb field.

On the basis of these two simple, but relevant, cases, let us determine the scattering cross section. To do this, we will first introduce the concept of the differential scattering cross section.

### 3.2.3 Differential Scattering Cross Section

By definition,

$$d\sigma = \frac{\text{total number of particles scattered out through the solid angle } d\Omega \text{ at } \theta}{\text{Incident flux per m}^2} .$$

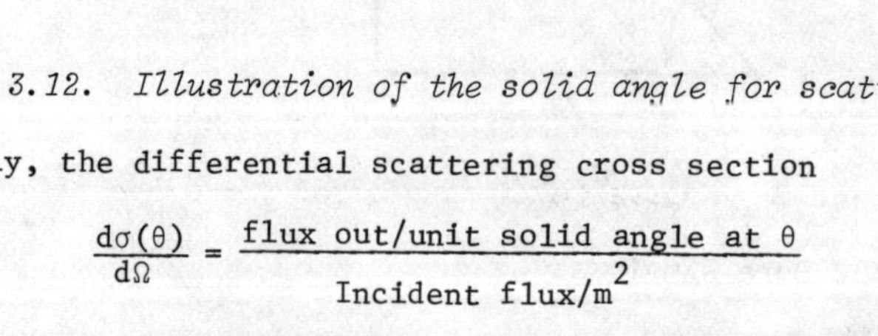


*Fig. 3.12. Illustration of the solid angle for scattering*

More appropriately, the differential scattering cross section

$$\frac{d\sigma(\theta)}{d\Omega} = \frac{\text{flux out/unit solid angle at } \theta}{\text{Incident flux/m}^2}$$

If we assume that the scattering is azimuthally symmetric (central potential) we can obtain a rather simple formulation of this problem. Considering the region between the cones shown in Fig. 3.13. The particles scattered into $d\theta$ at $\theta$ were in db at b before the scattering. Therefore, the flux out through $d\theta$ at $\theta$ = flux in through db at b. The flux in through db at b = incident flux density × $2\pi b\,db$

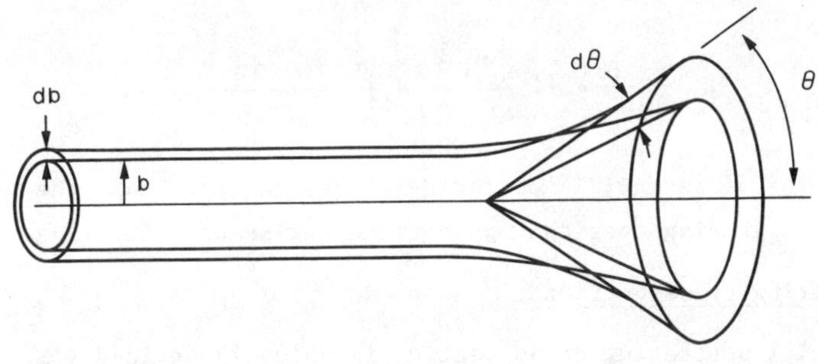


*Fig. 3.13. Azimuthally symmetric scattering*

and the incident flux density equals the number of particles/$m^3$ in the incident beam times the velocity of the beam. Therefore, the flux in through db at $b = nv\,2\pi b\,db$ equals the flux out through $d\theta$ at $\theta$. The solid angle defined by $d\theta$ at $\theta$ is

$$d\Omega = 2\pi \sin\theta \; d\theta \quad ,$$

therefore, the differential scattering cross section

$$\frac{d\sigma\,(\theta)}{d\Omega} = \frac{2\pi\, b\, db}{2\pi \sin\theta\, d\theta} = \frac{b}{\sin\theta}\left|\frac{db}{d\theta}\right| \quad . \tag{3.26}$$

The magnitude $|db/d\theta|$ has been introduced since an increase in the impact parameter (db) gives a decrease in the scattering angle ($d\theta$), however, $d\sigma/d\Omega$ is always defined to be a positive quantity.

3.2.3.1 Hard spheres. For the example we previously analyzed

$$\theta = 2\cos^{-1}\frac{b}{a} \qquad b < a$$

$$= 0 \qquad b > a$$

therefore,

$$b = a\cos\frac{\theta}{2} \quad \text{and} \quad \left|\frac{db}{d\theta}\right| = \frac{a}{2}\sin\frac{\theta}{2} \quad .$$

The differential scattering cross section then becomes

$$\frac{d\sigma(\theta)}{d\Omega} = \frac{b}{\sin\theta}\left|\frac{db}{d\theta}\right| = \frac{a\cos\theta/2}{\sin\theta}\cdot\frac{a}{2}\sin\theta/2 \quad .$$

Since $\sin\theta = 2\sin\theta/2\cos\,\theta/2$ ,

$$\frac{d\sigma\,(\theta)}{d\Omega} = \frac{a^2}{4} \quad . \tag{3.27}$$

3.2.3.2 Coulomb collisions. In this case

$$\theta = 2 \tan^{-1} \frac{e_1 e_2}{4\pi\varepsilon_o m v^2 b}$$

and

$$\frac{d\sigma(\theta)}{d\Omega} = \left(\frac{e_1 e_2}{8\pi \varepsilon_o m v^2}\right)^2 \frac{1}{\sin^4(\theta/2)} \quad . \tag{3.28}$$

Once we have the differential scattering cross, we can obtain the scattering cross sections by integrating over the appropriate angles of scattering.

### 3.2.4 Scattering Cross Section

The differential scattering cross section is uniquely defined for a particular interaction potential, however, there are a number of different scattering cross sections that can be defined depending on the importance of the scattering angle.

3.2.4.1 Total scattering cross sections. In referring to the propagation of a monoenergetic collimated electron beam through a gas, it does not matter whether the electron is scattered by 2°, 70° or 150°. The electron is still scattered out of the beam. In this case we can obtain the total scattering cross section $\sigma_c$ by simply integrating the differential scattering cross section over all possible scattering angles.

The total scattering cross section (cross section for all collisions) is, then,

$$\sigma_c = \int \frac{d\sigma}{d\Omega} d\Omega = \int_o^\pi \frac{d\sigma(\theta)}{d\Omega} 2\pi \sin\theta \, d\theta \tag{3.29}$$

for collisions with azimuthal symmetry.

3.2.4.2 Cross section for momentum transfer. If we wish to obtain the cross section for momentum transfer, then each collision must be weighed according to the momentum loss per collision. For a collision between a particle of mass m and a scattering center of mass M, we can calculate the fractional energy or momentum loss per collision from energy and momentum conservation (Fig. 3.14).

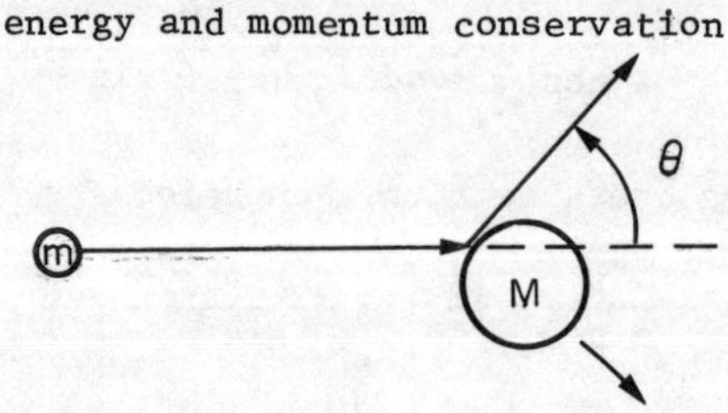


*Fig. 3.14. Collision geometry*

The fractional energy loss for M >> m is

$$\delta_\varepsilon \simeq \frac{2m}{M}(1 - \cos\theta) \quad . \tag{3.30}$$

Similarly for the fractional momentum loss

$$\delta_m \simeq 1 - \cos\Theta \quad . \tag{3.31}$$

Therefore, the <u>cross section for momentum transfer</u>

$$\sigma_m = \int_0^\pi \frac{d\sigma(\theta)}{d\Omega}(1 - \cos\theta)\, 2\pi \sin\theta\, d\theta \tag{3.32}$$

and the <u>cross section for energy transfer</u>

$$\sigma_\varepsilon = \int_0^\pi \frac{d\sigma(\theta)}{d\Omega}\frac{2m}{M}(1 - \cos\theta)\, 2\pi \sin\theta\, d\theta \quad . \tag{3.33}$$

Now we can apply these to our previous examples.

3.2.4.3 <u>Hard spheres</u>.

$$\sigma_c = \int_0^\pi \frac{a^2}{4}\, 2\pi \sin\theta\, d\theta = \pi a^2 \tag{3.34}$$

and

$$\sigma_m = \int_0^\pi \frac{a^2}{4}\, 2\pi \sin\theta\,(1 - \cos\theta)\, d\theta = \pi a^2 \quad . \tag{3.35}$$

For the hard sphere case $\sigma_m = \sigma_c$ which is due to the fact that the scattering is isotropic so that the differences in momentum loss cancel exactly. For this reason we often do not differentiate between $\sigma_c$ and $\sigma_m$ in calculating the collision frequency for momentum transfer $\nu_m$.

3.2.4.4 <u>Coulomb collisions</u>. Now consider Coulomb collisions (electron-ion collisions).

$$\sigma_c = \int_0^\pi \left(\frac{e_1 e_2}{8\pi \varepsilon_o m v^2}\right)^2 \frac{2\pi \sin\theta}{\sin^4(\theta/2)}\, d\theta \to \infty \quad . \tag{3.36}$$

The cross section diverges due to the fact that scattering at small angles can occur for very large impact parameters. This suggests that Coulomb collisions are effective everywhere.

We can, however, modify the classical theory and obtain fairly realistic results by invoking the idea of Debye shielding and simply saying that the Coulomb potential should not be effective if the impact parameter, b, is greater than $\lambda_D$. This sets a lower limit on $\theta$.

Using the cross section for momentum transfer

$$\sigma_m = \int_{\theta_m}^{\pi} \left(\frac{e_1\, e_2}{8\pi\ \varepsilon_o\ mv^2}\right)^2 \frac{2\pi\ \sin\theta\ (1 - \cos\theta)}{\sin^4\ (\theta/2)}\, d\theta \tag{3.37}$$

and

$$\sigma_m = 8\pi \left(\frac{e_1\, e_2}{8\pi\ \varepsilon_o\ mv^2}\right)^2 \ \ell n \left(\frac{2}{1 - \cos\theta_m}\right) \tag{3.38}$$

where $\theta_m$ is a minimum scattering angle determined by the Debye length.

Now, from the triangle defining the Coulomb scattering angle (Fig. 3.10)

$$1 - \cos\theta = \frac{2}{1 + \left[{}^{b}/\left(\dfrac{e_1\, e_2}{4\pi\ \varepsilon_o\ mv^2}\right)\right]^2} \ . \tag{3.39}$$

Therefore, the minimum scattering angle $\theta_m$ can be related to the maximum impact parameter.

If we assume that $\theta_m << \pi/2$, then we can approximate

$$1 - \cos\theta_m \simeq 2\left[\left(\frac{e_1\, e_2}{4\pi\ \varepsilon_o\ mv^2}\right)/b_{max}\right]^2 \tag{3.40}$$

and, if we take $b_{max} = \lambda_D$,

$$\sigma_m = 16\pi \left(\frac{e_1\, e_2}{8\pi\ \varepsilon_o\ mv^2}\right)^2 \ell n\left[\frac{4\pi\ \varepsilon_o\ mv^2\ \lambda_D}{e_1\, e_2}\right] \ . \tag{3.41}$$

Since the logarithmic term varies slowly, we shall use the thermodynamic relationship

$$mv^2/2 = 3kT_e/2$$

where $T_e$ represents the electron temperature for an electron gas in thermodynamic equilibrium.

Therefore

$$\sigma_m = 16\pi \left(\frac{e^2}{8\pi\ \varepsilon_o\ mv^2}\right)^2 \ell n\left[\frac{4\pi\ \varepsilon_o}{e^2}\ 3kT_e \left(\frac{k\ \varepsilon_o}{n_e\, e^2}\ \frac{T_e\, T_+}{T_e + T_+}\right)^{1/2}\right] \ . \tag{3.42}$$

We can also write

$$\sigma_m = \frac{e^4\ \ell n\ \Lambda}{4\pi\ \varepsilon_o^2\ (mv^2)^2} \tag{3.43}$$

where

$$\Lambda = \frac{12\pi\,(\varepsilon_o\, kT_e/e^2)^{3/2}}{n_e^{1/2}} \tag{3.44}$$

when the ions are considered to be stationary.

We could avoid the problem of the diverging cross section by explicitly solving the scattering by a shielded charge. Unfortunately, this problem cannot be solved analytically. Computer solutions have been obtained, however, by Everhart, Stone and Carbone (1955) and Lane and Everhart (1960).

### 3.2.5 Attractive Potentials

For the hard-sphere and Coulomb collisions considered, it has been somewhat implied that repulsive potentials were being considered. In some cases, the potential is attractive and the incoming particle can be "captured" in a sense. Of course, for a truly elastic collision, a capture is not possible and the incoming particle will eventually recede with very close to its initial energy. The orbits, however, are very complex and do lead to an idea of what could happen under conditions of an inelastic collision.

To understand the various possibilities, the scattering equations can be cast into an essentially one-dimensional form. Recalling:

$$\varepsilon = mv_o^2/2 = m[\dot{r}^2 + r^2\dot{\phi}^2]/2 + V(r) \quad , \tag{3.7}$$

$$L = mv_o b = mr^2\dot{\phi} \quad . \tag{3.6}$$

We can rewrite the energy equation as:

$$\varepsilon = m\dot{r}^2/2 + \left[\frac{mv_o^2 b^2}{2r^2} + V(r)\right] \quad . \tag{3.45}$$

The quantity in the square bracket is now an "effective" potential $V_{eff}(r)$ and the term

$$\frac{mv_o^2 b^2}{2r^2} = V_c(r) \tag{3.46}$$

is known as the centrifugal potential. $V_{eff}(r)$ is plotted in Fig. 3.15 for a repulsive potential. When $\varepsilon = V_{eff}(r)$, we have the classical distance of closest approach.

$$\frac{mv_o^2 b^2}{2r_{min}^2} + V(r_{min}) = mv_o^2/2 = \varepsilon$$

or

$$1 - \frac{V(r_{min})}{mv_o^2/2} - \frac{b^2}{r_{min}^2} = 0 \quad . \tag{3.47}$$

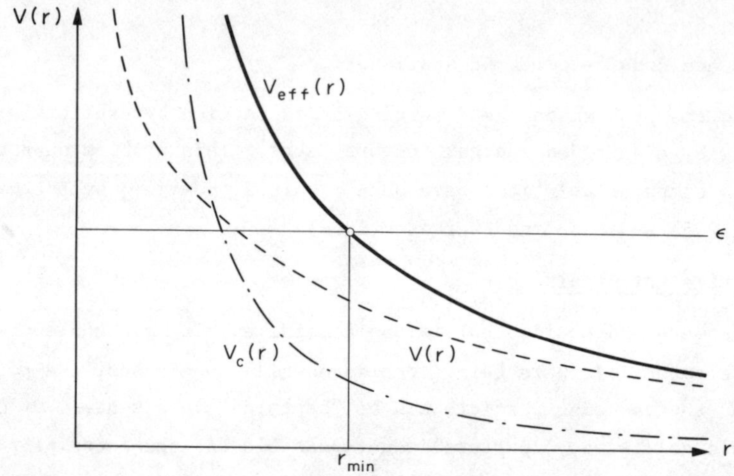


*Fig. 3.15. Effective potential for a repulsive interaction*

The largest real solution of this equation is the distance of closest approach. Keeping in mind that $V_c(r) \sim 1/r^2$, real solutions for $r_{min}$ will always occur for repulsive solutions as well as for attractive solutions of the form:

$$V(r) \sim - \frac{1}{r^n}$$

for $n < 2$. In all these cases, the $1/r^2$ term will dominate near the origin. $n = 2$ is a borderline case, but if $n \geq 3$, then there are values of $\varepsilon$ for which no real solution for $r_{min}$ exists. The attractive potential dominates the centrifugal potential and the collision partner will approach the scattering center until it is reflected by some core repulsive force. An example is shown in Fig. 3.16. For $mv^2/2 = \varepsilon_a$, reflection will occur at $r_a$. For $mv^2/2 = \varepsilon_c$, no reflection will occur. The critical energy is $\varepsilon_b$. Above that value, no reflection will take place and the incoming particle will spiral into the center. For energies below $\varepsilon_b$, reflection will occur, and very close to $\varepsilon_b$, the incoming particle will essentially enter a long lived orbit at a radius close to $r_b$.

An interesting example of an attractive potential with regard to ion-molecule collisions is the inverse fourth power polarization interaction for which

$$V(r) = - \frac{Ke^2}{r^4} \tag{3.48}$$

where K is related to the polarizability of the molecule.

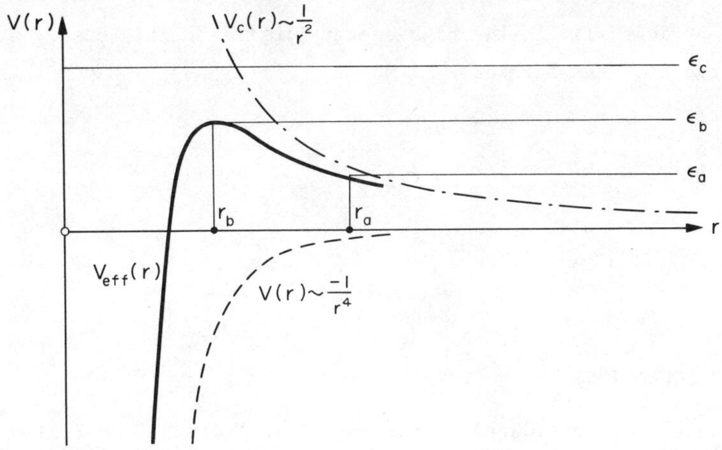


*Fig. 3.16. Effective potential for an attractive interaction*

The equation for $r_{min}$ is

$$1 + \frac{2\,Ke^2}{mv_o^2\, r_{min}^4} - \frac{b^2}{r_{min}^2} = 0 \qquad (3.49)$$

from which

$$r_{min}^2 = \frac{b^2 + \sqrt{b^4 - \dfrac{8\,Ke^2}{mv_o^2}}}{2} \; . \qquad (3.50)$$

The critical value of the impact parameter $b_c$, <u>below</u> which no real root exists is

$$b_c = \left[\frac{8\,Ke^2}{mv_o^2}\right]^{1/4} \; . \qquad (3.51)$$

At this value of impact parameter

$$r_{min} = b_c/\sqrt{2} \; . \qquad (3.52)$$

An approximate model for inelastic collisions can be constructed from these considerations if it is assumed that an inelastic process will occur if $r_{min} < r_c$, some critical radius.

If $r_c < b_c/\sqrt{2}$, then, of course, any particle coming in with an impact parameter $b \leq b_c$ will ultimately lead to a reaction. The cross section for this reaction will, therefore, be:

$$\sigma = \pi\, b_c^2 = \frac{\pi e}{v_o}\left(\frac{8K}{m}\right)^{1/2} \; . \qquad (3.53)$$

Note the inverse velocity dependence for this type of reaction.

This equation is not valid in the high energy limit. In this case, returning to the equation for $r_{min}$ and letting $r_{min} = r_c$

$$1 - \frac{V(r_c)}{mv_o^2/2} - \frac{b_c^2}{r_c^2} = 0 \tag{3.54}$$

and

$$\sigma = \pi b_c^2 = \pi r_c^2 \left[ 1 - \frac{V(r_c)}{mv_o^2/2} \right] . \tag{3.55}$$

3.3 INELASTIC COLLISIONS

Most of what have been considered so far have been "elastic" collisions; those in which the internal energy of the collision partners is essentially unchanged. These collisions are very important in determining the transport coefficients in a gas, such as the electrical and thermal conductivities, diffusion coefficients, mobilities, etc. In many ways, the more important class of collisions are the "inelastic collisions" that are crucial in the determination of such important discharge parameters as the electron temperature or energy, the electron number density and the populations of the various excited species within the gas. In these collisions, there are large exchanges of energies between the collision partners resulting in a change of the internal energy of the collision partners. Many of the types of processes are summarized in the introduction and further listings of the important processes are given by Hasted (1964) and by Delcroix (1968). An excellent collection of information on these collisional processes is contained in the four volume second edition of Electronic and Ionic Impact Phenomena by Massey, Burhop and Gilbody (1969).

3.3.1 Ionization

Electrons of sufficient energy colliding with neutral or excited atoms or molecules can ionize the target atom or molecule. For molecules this can result in the production of a molecular ion

$$e + XY \rightarrow XY^+ + e + e \quad ,$$

or in some cases dissociative ionization can take place

$$e + XY \rightarrow X^+ + Y + e + e \quad .$$

For the atom, of course, the positive atomic ion is produced

$$e + X \rightarrow X^+ + e + e$$

although ion conversion processes can eventually result in the production of molecular ions.

Since ionization products can usually be measured fairly easily, a great deal of cross section information is available. Kieffer and Dunn (1966), Kieffer (1965) and Kieffer (1973) have provided tabulations and graphs of critically evaluated cross sections for ionization. The cross section for ionization of helium is a good example and is characteristic of ionization cross section behavior (Fig. 3.17)

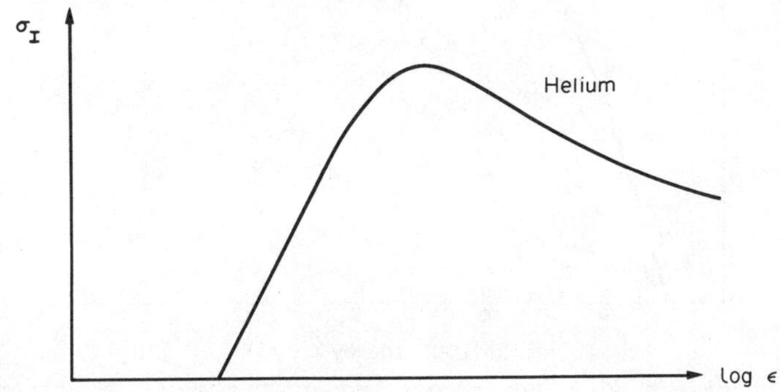


*Fig. 3.17. Ionization cross section for helium*

A number of general features are apparent. The ionization cross section varies approximately linearly with excess energy just above the ionization threshold. The cross section is rather broad in energy, with the peak value (for He) occurring at around four to five times the threshold energy. In the high energy range, the cross section drops off approximately as

$$\sigma\ (\varepsilon) \sim \ell n\ \varepsilon/\varepsilon \quad .$$

The review paper by Rudge (1968) discusses the theory of electron impact ionization and also provides a useful discussion of empirical formulas that can be used to esimate ionization cross sections in the absence of experimental or theoretical data.

One area in which it is very difficult to obtain experimental data is for electron impact ionization of excited species. This turns out to be an important process in the evaluation of excited state densities in many gas discharges and lasers, so theoretical extimates are very useful. Vriens (1964) has provided theoretical estimates for the ionization of metastable excited states of He, Ne, Ar and Hg. In Fig. 3.18, Vriens' theoretical curve for the ionization of He ($2^3S$) excited atoms is compared with the experimentally measured values obtained by Long and Geballe (1970). The theoretical estimates are seen to be quite good.

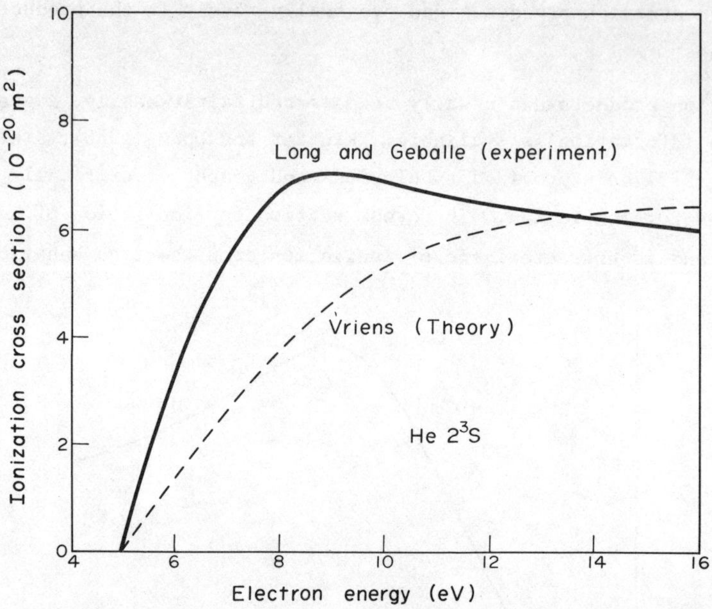


*Fig. 3.18. Theoretical and experimental ionization cross sections for He($2^3S$) excited atoms (from Vriens, 1964 and Long and Geballe, 1970)*

### 3.3.2 Electronic Excitation

The general shape of the excitation cross section is quite similar in nature to the cross section for ionization provided that the state to which the electron excites the ground state atom is optically connected (has an allowed optical transition) to the ground state. If the excited state has the same multiplicity as the ground state (i.e.singlet-singlet transition) but is not optically allowed, then the shape is still similar, but the cross section drops off as

$$\sigma\ (\varepsilon) \sim 1/\varepsilon \quad .$$

If the ground state is a singlet, the excitation cross section to the triplet states (as in helium) has a markedly different energy dependence. A comparison for some helium states is shown in Fig. 3.19. For the triplet excitation, the cross section rises extremely steeply and the peak is quite close to the threshold. The cross section then drops off approximately as:

$$\sigma\ (\varepsilon) = 1/\varepsilon^3 \quad .$$

This type of excitation process is almost "resonant"in nature and this rapid variation of the cross section in He can be attributed to the fact that excitation of the triplet state involves an electron exchange. The incoming electron is captured so that its spin is parallel to the remaining electron and the anti-parallel valence electron is ejected. Another important example of this effect is

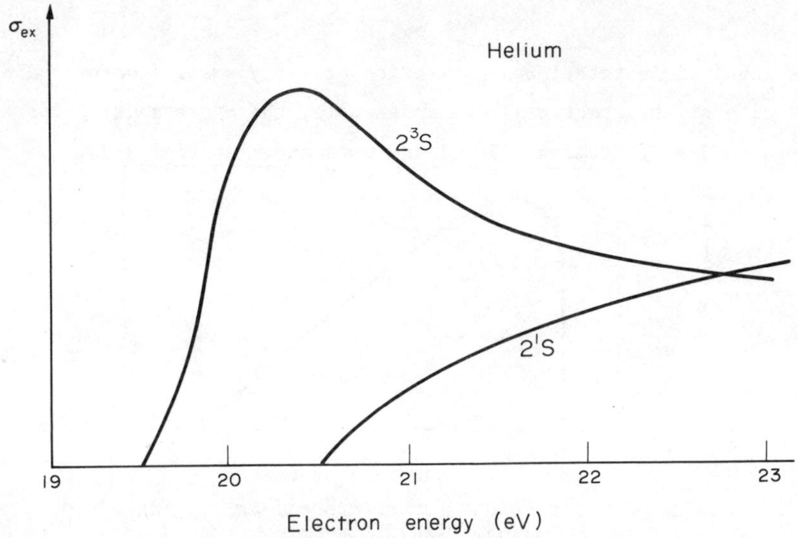


*Fig. 3.19. Cross sections for excitation to the $He(2^3S)$ and $He(2^1S)$ excited states*

Hg, where the excitation cross sections to the $^3P_0$, $^3P_1$ and $^3P_2$ states vary rapidly with energy while the excitation cross section for the $^1P_1$ state has a much broader structure with varying energy.

A very good review of many aspects of electron impact excitation is given by Moiseiwitsch and Smith (1968).

### 3.3.3 Vibrational and Rotational Excitation of Molecules

In many gaseous discharges, particularly $CO_2$ and CO lasers, it is the excitation of the vibrational rather than the electronic levels that is of prime importance. The discussion of molecular structure will be deferred until later, however, in a somewhat pictorial manner, rotational excitation of a diatomic molecule involves spinning it around like a dumbbell.

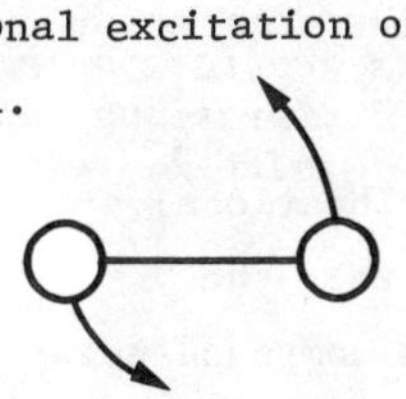

*Fig. 3.20. Rotational excitation of a molecule*

Vibrational excitation involves setting the atoms in motion along the internuclear axis.

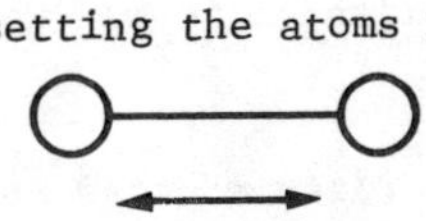

*Fig. 3.21. Vibrational excitation of a molecule*

Two excellent reviews are by Phelps (1968) and Golden, Lane, Temkin and Gerjuoy (1971).

The energies involved in rotational excitation are very small (corresponding to the microwave portion of the spectrum). As an example, the cross section for excitation from the J = 4 to J = 5 rotational level in CO is shown in Fig. 3.22.

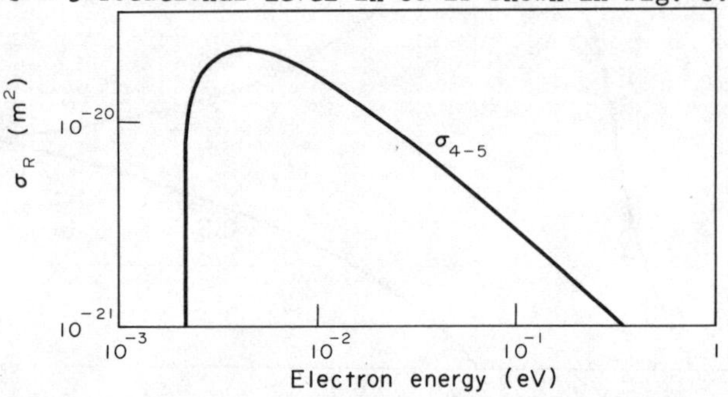


*Fig. 3.22. Cross section for rotational excitation of CO (from Phelps, 1968).*

An example for the excitation of the vibrational levels of $N_2$ as obtained by Schulz (1964) is shown in Fig. 3.23. An interesting note is that it takes a 2 eV electron

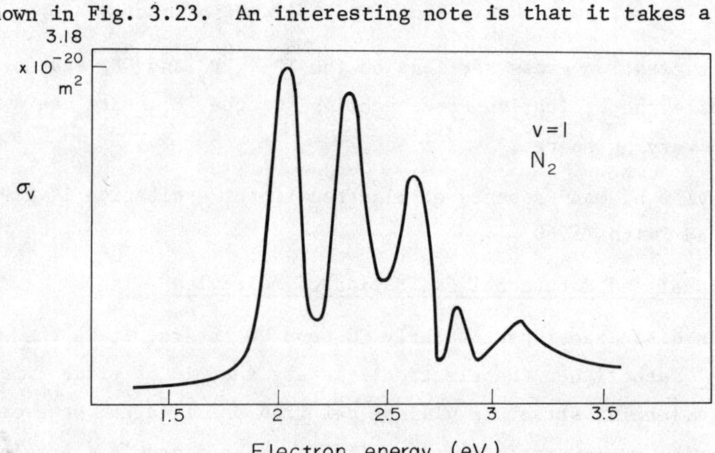


*Fig. 3.23. Cross section for excitation of the v = 1 vibrational state of $N_2$ (from Schulz, 1964)*

to excite nitrogen to the v = 1 vibrational state. This state has an energy of about 0.2 eV. This indicates that the v = 1 state is not excited directly, but rather through the formation of a compound negative ion state which subsequently results in the production of vibrationally excited nitrogen.

An indication of the relative amplitudes of the various cross sections for $CO_2$ is given in Fig. 3.24.

Keep in mind that even though the elastic cross section is larger than the inelastic cross sections, the energy loss is much smaller than for the inelastic processes so

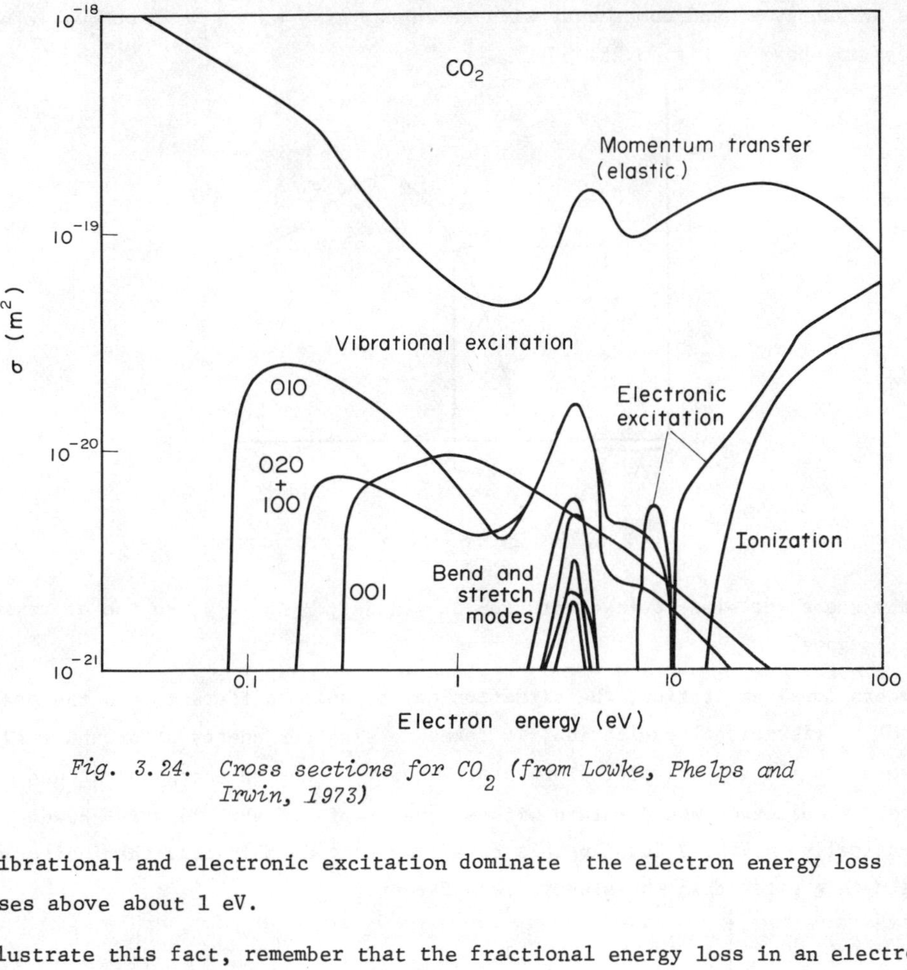


*Fig. 3.24. Cross sections for $CO_2$ (from Lowke, Phelps and Irwin, 1973)*

that vibrational and electronic excitation dominate the electron energy loss processes above about 1 eV.

To illustrate this fact, remember that the fractional energy loss in an electron-neutral elastic collision is:

$$\delta = \frac{2m}{M}$$

which for $CO_2$, for example, is

$$\delta \simeq 2.5 \times 10^{-5} \quad .$$

So even through the elastic cross sections are large, the cross section for energy loss is small.

For the inelastic fractional energy losses, there are two interesting cases. For electronic excitation and ionization, it is usually assumed that the electron loses <u>all</u> its energy in excitation, i.e. $\delta \simeq 1$. To be a little more careful, consider the cross section for excitation of an atom to an energy state with a fixed discrete potential energy $\varepsilon_1$. The cross section for such a process has a threshold at $\varepsilon_1$. How much energy is lost by an electron with energy $\varepsilon_2 > \varepsilon_1$? This electron loses an energy $\varepsilon_1$ and comes away with an energy $\Delta\varepsilon = \varepsilon_2 - \varepsilon_1$. A schematic of this process is shown in Fig. 3.25.

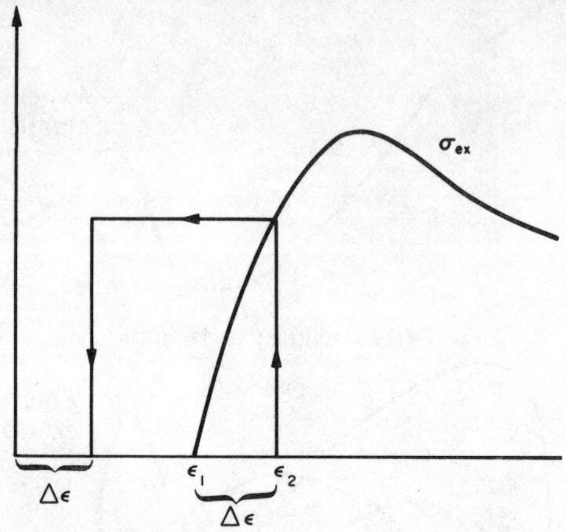


*Fig. 3.25. Energy flow in excitation*

In many cases for electronic excitation or ionization $\Delta\varepsilon << \varepsilon_1$ so the approximation $\delta \simeq 1$ is good.

For vibrational excitation, the situation may be quite different. In the previous case of $N_2$ vibrational excitation, it takes an electron energy of around 2 eV to excite $N_2$ to the v = 1 state. This state, however, has an energy of around 0.2 eV, so the 2 eV electron would return with an energy of 1.8 eV. This is shown schematically in Fig. 3.26. In this case $\delta$ is considerably less than unity but still much greater than the elastic loss factor.

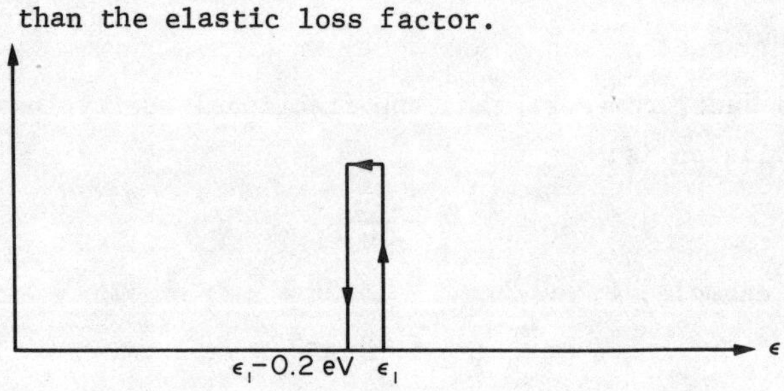


*Fig. 3.26. Energy flow in vibrational excitation of* $N_2$

These energy loss considerations are particularly important in distribution function calculations.

### 3.3.4 Recombination

In electron-ion recombination, a positive ion captures an electron to form a neutral atom. Since this is a Coulomb collision, the cross section should decrease rapidly with increasing electron energy. Walls and Dunn (1974) have demonstrated this for $O_2^+$ and $NO^+$ as shown in Fig. 3.27.

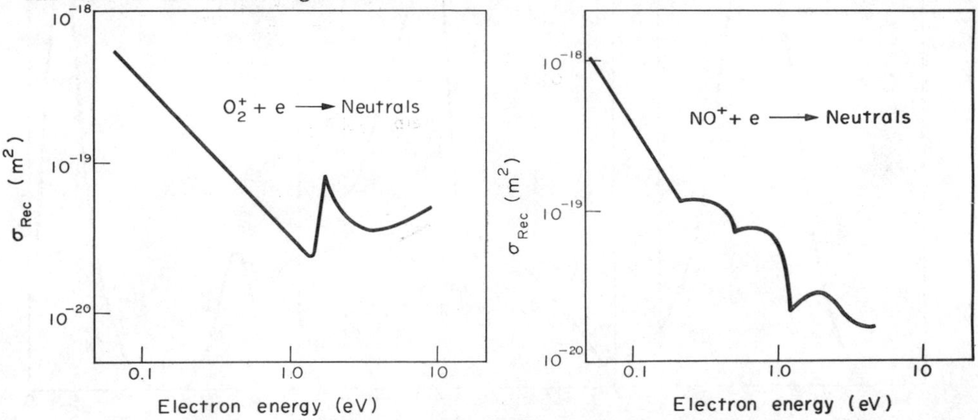


*Fig. 3.27. Recombination cross sections for $O_2^+$ and $NO^+$ (from Walls and Dunn, 1974, reproduced with permission, copyrighted by American Geophysical Union)*

### 3.3.5 Attachment

In electronegative gases, an electron can be captured by a neutral to form a negative ion. The exact processes can be quite complicated, proceeding quite often through the formation of an unstable molecular negative ion which subsequently dissociates. As a result, the cross section can be strongly dependent on electron energy. As an example Schulz (1962) has obtained the cross sections shown in Fig. 3.28 for the formation of $O^-$ from $O_2$, CO and $CO_2$ by electron impact.

Similarly Melton (1972) has found the cross sections shown in Fig. 3.29 for the production of negative ions in water vapor.

The cross sections provided so far are used primarily for illustrative processes. There are many more processes of importance such as vibrational-vibrational energy exchange collisions or vibrational-rotational energy exchange collisions in molecules. Quite often, however, these processes are specified in terms of collision or reaction rates (one of which is the collision frequency for momentum transfer). In order to consider reaction rates, it is necessary to specify the distribution in the energies of the particles participating in the reaction. All additional collision processes will therefore be referred to as the necessity arises after the concept of the distribution function is considered.

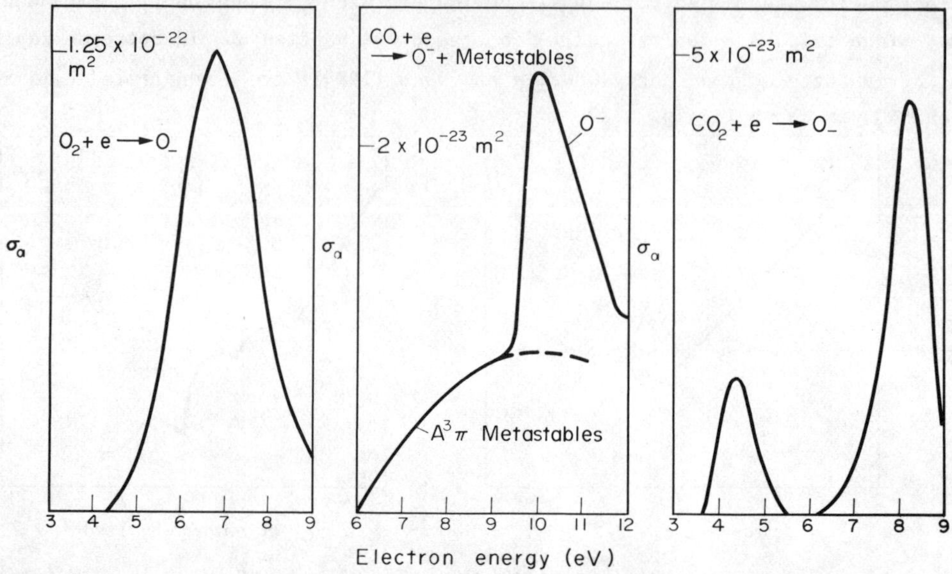


*Fig. 3.28. Cross sections for formation of* $O^-$ *from* $O_2$, *CO and* $CO_2$ *(from Schulz, 1962)*

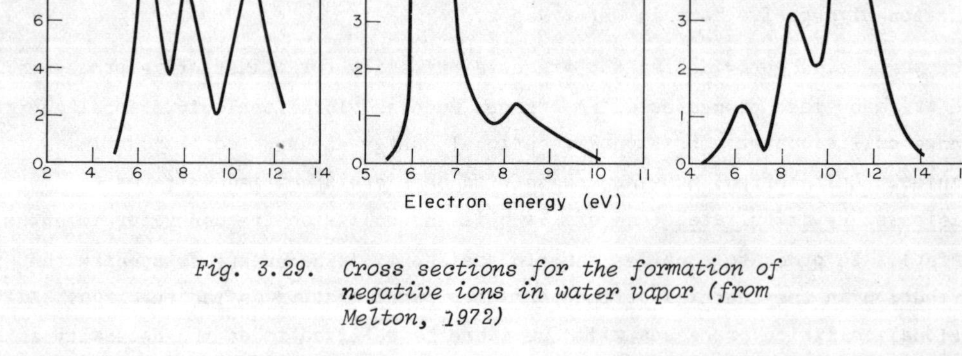


*Fig. 3.29. Cross sections for the formation of negative ions in water vapor (from Melton, 1972)*

## 3.4 REFERENCES


Bederson, Benjamin and Kieffer, L.J. (1971), "Total Electron-Atom Collision Cross Sections at Low Energies - A Critical Review," Rev. Mod. Phys. 43, 601-640.

Brode, Robert B. (1933), "The Quantitative Study of the Collisions of Electrons with Atoms," Rev. Mod. Phys. 5, 257-279.

Brown, S.C. (1967), Basic Data of Plasma Physics, 1966 MIT Press.

Delcroix, J.L. (1968), Plasma Physics, Vol. 2, Wiley & Sons.

Everhart, Edgar, Stone, Gerald, and Carbone, R.J. (1955), "Classical Calculation of Differential Cross Section for Scattering from a Coulomb Potential with Exponential Screening," Phys. Rev. 99, 1287-1290.

Golden, D.E., Lane, Neal F., Temkin, A., and Gerjuoy, E. (1971), "Low Energy Electron-Molecule Scattering Experiments and the Theory of Rotational Excitation," Rev. Mod. Phys. 43, 642-679.

Goldstein, H. (1959), Classical Mechanics, Addison-Wesley.

Hasted, J.B. (1964), Physics of Atomic Collisions, Butterworths, London.

Kieffer, L.J. (1965), "A Compilation of Critically Evaluated Electron Impact Ionization Cross Section Data for Atoms and Diatomic Molecules," JILA Report #30, University of Colorado, Boulder, Colorado.

Kieffer, L.J. (1969), "Compilation of Low-Energy Electron-Collision Cross-Section Data - Part I Ionization, Dissociation, Vibrational Excitation," Atomic Data 1 19-89.

Kieffer, L.J. (1969), "Compilation of Low-Energy Electron-Collision Cross-Section Data - Part II Electronic Excitation Cross Sections," Atomic Data 1 121-287.

Kieffer, L.J. (1971), "Compilation of Low-Energy Electron-Collision Cross-Section Data - Part III Total Scattering, Differential Elastic Scattering," Atomic Data 2 293-391.

Kieffer, L.J. (1973), "A Compilation of Electron Collision Cross-Section Data for Modeling Gas Discharge Lasers," JILA Information Center Report 13, University of Colorado, Boulder, Colorado.

Kieffer, L.J. and Dunn, Gordon H. (1966), "Electron Impact Ionization Cross-Section Data for Atoms, Atomic Ions, and Diatomic Molecules: I. Experimental Data," Rev. Mod. Phys. 38, 1-35.

Lane, G.H. and Everhart, E. (1960), "Calculations of Total Cross Sections for Scattering from Coulomb Potentials with Exponential Screening," Phys. Rev. 117, 920-924.

Long, Daniel R. and Geballe, Ronald (1970), "Electron Impact Ionization of He($2s\,^3S$)," Phys. Rev. A1, 260-265.

Lowke, J.J., Phelps, A.V., and Irwin, B.W. (1973), "Predicted Electron Transport Coefficients and Operating Characteristics of $CO_2$-$N_2$-He Laser Mixtures," J. Appl. Phys. 44, 4664-4671.

Massey, H.S.W., Burhop, E.H.S., and Gilbody, H.B. (1969), Electronic and Ionic Impact Phenomena, Second Edition in Four Volumes, Oxford.

McDaniel, E.W. (1964), Collision Phenomena in Ionized Gases, Wiley & Sons.

Melton, Charles H. (1972), "Cross Sections and Interpretation of Dissociative Attachment Reactions Producing $OH^-$, $O^-$ and $H^-$ in $H_2O$," J. Chem. Phys. 57, 4218-4225.

Moiseiwitsch, B.L. and Smith, S.J. (1968), "Electron Impact Excitation of Atoms," Rev. Mod. Phys. 40, 1-116; also National Standard Reference Data Series - National Bureau of Standards 25.

Phelps, A.V. (1968), "Rotational and Vibrational Excitation of Molecules by Low-energy Electrons," Rev. Mod. Phys. 40, 399-410.

Rudge, M.R.H. (1968), "Theory of the Ionization of Atoms by Electron Impact," Rev. Mod. Phys. 40, 564-590.

Schulz, G.J. (1962), "Cross Sections and Electron Affinity for $O^-$ ions from $O_2$, CO and $CO_2$ by Electron Impact," Phys. Rev. 128, 178-186.

Schulz, G.J. (1964), "Vibrational Excitation of $N_2$, CO, and $H_2$ by Electron Impact," Phys. Rev. 135A, 988-994.

Vriens, L. (1964), "Calculation of Absolute Ionization Cross Sections of He, $He^*$, $He^+$, Ne, $Ne^*$, $Ne^+$, Ar, $Ar^*$, Hg and $Hg^*$," Phys. Letters 8, 260-261.

Walls, F.L. and Dunn, G.H. (1974), "Measurement of Total Cross Sections for Electron Recombination with $NO^+$ and $O_2^+$ Using Ion Storage Techniques," J. Geophys. Res. 79, 1911-1915.

# Chapter 4

# DISTRIBUTION FUNCTIONS AND THE BOLTZMANN EQUATION

In all of the discussion up to now, we have dealt with essentially monoenergetic electrons. In the Langevin equation we have ignored the thermal (random) velocity components and have assumed that all electrons have the same velocity. In our examination of collisions we have dealt with the scattering of monoenergetic electrons but we saw that the collision cross section can be a very strong function of electron velocity (particularly for inelastic collisions). To obtain a realistic idea of the processes occurring within a plasma we should therefore determine the distribution of velocities of the electrons (or other particles) within the plasma.

To be precise, we should solve the equations for each and every particle separately. In that case we would have on the order of $10^{15}$ to $10^{26}$ separate equations to describe the plasma behavior in 1 $m^3$ of space. Beyond the fact that this is completely out of the question, it is also unnecessary. The properties of a plasma that we can observe are due to the contributions of very large numbers of particles and so a statistical description is perfectly adequate. This statistical description is provided by the distribution function.

In order to give a complete specification of the properties of a system of particles, it is necessary to know the position and velocity of each particle; x, y, z, $v_x$, $v_y$, $v_z$. These six axes define "phase space". The entire system at any instant of time is specified by the positions of the points in phase space where each point represents a particle. As a necessity for the validity of the statistical approach, each elemental volume in phase space must contain a large number of particles. The number of particles in any elemental volume of the phase space is then specified by a <u>distribution function</u> f. Therefore, $dn = f(\bar{r},\bar{v},t)d^3\bar{r}d^3\bar{v}$ represents the number of particles in a volume of phase space $d^3\bar{r}d^3\bar{v} = (dxdydz)(dv_x dv_y dv_z)$ at a point specified by $\bar{r}(x,y,z)$ and $\bar{v}(v_x,v_y,v_z)$ (note that $\bar{v}$ and $\bar{r}$ are independent variables in phase space). Since this is a time varying quantity, the distribution function f is such a function of time.

As can be seen from the above equations, the distribution function, f is such that

$$\int f(\bar{r},\bar{v},t)d^3\bar{v} = n(\bar{r},t), \qquad \text{the number density,} \tag{4.1}$$

and the normalized distribution function

$$\frac{f(\bar{r},\bar{v},t)}{n(\bar{r},t)}$$

represents the fraction of particles in $d^3\bar{r}$ at $\bar{r}$ that have velocities in the unit volume $d^3\bar{v}$ at $\bar{v}$.

## 4.1 AVERAGES AND COLLISIONAL RATES

The distribution function is particularly useful in computing averages of quantities representing the entire collection of particles. For example, the average velocity of all particles is:

$$<\bar{v}> = \frac{1}{n}\int \bar{v}f(\bar{r},\bar{v},t)d^3\bar{v} \quad . \tag{4.2}$$

The average speed,

$$<v> = \frac{1}{n}\int vf(\bar{r},\bar{v},t)d^3\bar{v} \tag{4.3}$$

and the average energy

$$<\varepsilon> = \frac{1}{n}\int \frac{mv^2}{2} f(\bar{r},\bar{v},t)d^3\bar{v} \quad . \tag{4.4}$$

Of particular interest is the use of the distribution function to calculate collisional rates, such as the collision frequency for momentum transfer.

For monoenergetic electrons of velocity $\bar{v}$, the collision frequency for momentum transfer

$$\nu_m (v) = N\, \sigma(v)\, v \quad . \tag{4.5}$$

When we have a distribution of electron velocities, we want to calculate the average collision frequency, simply called $\nu_m$, where

$$\nu_m = N <\sigma(v)v> = \frac{N}{n}\int \sigma(v)v\, f(\bar{r},\bar{v},t)d^3\bar{v} \quad . \tag{4.6}$$

The term $<\sigma(v)\, v>$ represents the number of collisions per second per electron per neutral atom and is commonly referred to as a collisional rate coefficient. For general collisional processes, whether elastic or inelastic, the total number of collisional reactions occurring per unit volume per second is given by:

$$R = N_1 N_2 \langle \sigma v \rangle \tag{4.7}$$

where $N_1$ and $N_2$ represent the densities of the respective interacting species (electrons and neutrals, for example) and $\langle \sigma v \rangle$ is the reaction rate per pair of particles in $m^3$/sec where $\sigma$ is the cross section for the reaction and v is the relative speed of the reacting particles.

Much of the study of gas discharges involves the evaluation of the collisional rate coefficients, so to proceed further, the specific form of the distribution function must be known.

## 4.2 EQUILIBRIUM DISTRIBUTIONS AND RATES

It is useful, first, to consider the distribution function of a system which is not subject to any external forces and is in thermodynamic equilibrium. Most texts on Statistical Mechanics and Kinetic Theory give very good treatments of the derivation of such equilibrium distributions based on the determination of the "most probable" distribution. A very readable account is given by Hemenway, Henry and Caulton (1962). The derivation will not be repeated but if you are familiar with this approach you will remember that the equilibrium distribution is given by

$$f(\varepsilon) = Ae^{-\beta\varepsilon} \qquad \text{(Maxwell-Boltzmann)}$$

where $\varepsilon$ is the electron energy $(mv^2/2)$ and A and $\beta$ are constants. From thermodynamic considerations (Kennard, 1938) we find that $\beta$ is related to the equilibrium temperature T by $\beta = 1/kT$. The distribution function for a gas in thermodynamic equilibrium therefore becomes, the Maxwellian distribution

$$f(\bar{v}) = n \left(\frac{m}{2\pi\, kT}\right)^{3/2} e^{-mv^2/2kT} \quad . \tag{4.8}$$

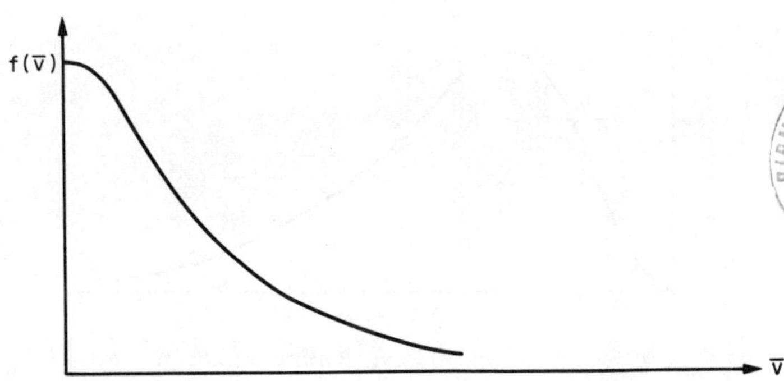




*Fig. 4.1. The Maxwellian distribution function for velocity*

Incidentally the temperature T can only be defined rigorously for such an equilibrium situation.

As we can see, and as we would expect, the equilibrium distribution function is isotropic in velocity space. In other words, there is no preferred direction of motion of the electrons under such circumstances. In this case, and for any other <u>isotropic</u> distribution function, it is worthwhile to define the distribution function f(v) that describes the distribution of the <u>speeds</u> of the particles. To do this, we convert $d^3\bar{v}$ to spherical coordinates in velocity space

$$d^3\bar{v} = dv_x\, dv_y\, dv_z = v^2 \sin\theta\, d\theta\, d\phi\, dv \quad . \tag{4.9}$$

Therefore,

$$n(\bar{r},t) = \int_0^\infty \int_0^{2\pi} \int_0^\pi f(\bar{r},\bar{v},t)\, v^2 \sin\theta\, d\theta\, d\phi\, dv \quad . \tag{4.10}$$

Since we are considering distribution functions that have no θ or ϕ dependence in velocity space

$$n(\bar{r},t) = \int_0^\infty 4\pi\, v^2\, f(\bar{r},\bar{v},t) dv = \int_0^\infty f(\bar{r},v,t) dv$$

so that

$$f(\bar{r},v,t) = 4\pi\, v^2\, f(\bar{r},\bar{v},t) \tag{4.11}$$

can be considered as a distribution function for <u>speed</u> and represents the number of particles in the unit volume $d^3\bar{r}$ at $\bar{r}$ that have speeds in the unit speed interval dv at v. In a slightly simpler notation, neglecting the $\bar{r}$ and t dependence,

$$f(v) = 4\pi\, v^2\, f(\bar{v}) \tag{4.12}$$

<u>if</u> and only if, $f(\bar{v})$ is isotropic in velocity space.

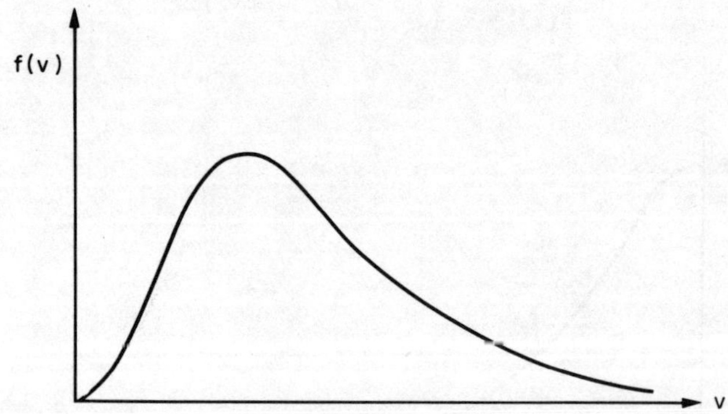


*Fig. 4.2. The Maxwellian distribution for speed*

This definition is particularly useful in determining the mean values of <u>scalar</u> quantities such as the speed (v), the energy ($mv^2/2$), and collisional excitation rates.

In this case, then

$$<P(v)> = \frac{1}{n} \int_0^\infty P(v)\ f(v)\ dv \qquad (4.13)$$

where P(v) is some scalar function of v (such as $\sigma(v)$ v). For the mean values of vector quantities, we retain the "vector" distribution function $f(\bar{v})$. In this case

$$<Q(\bar{v})> = \frac{1}{n} \int_{-\infty}^{\infty}\int_{-\infty}^{\infty}\int_{-\infty}^{\infty} Q(\bar{v})\ f(\bar{v})\ dv_x\ dv_y\ dv_z \ . \qquad (4.14)$$

With the equilibrium distribution function, all of the interesting properties of the equilibrium gas can be calculated with relative ease.

### 4.2.1 Collisional Rates for a Maxwellian Distribution

Of course, a distribution which is completely Maxwellian is not expected since this would preclude, for example, any current flow in the discharge. Nevertheless, we often find that electron distribution functions are very close to Maxwellian, so this form can be used to compute some important collisional rates. For instance, for a Maxwellian distribution

$$<\sigma(v)v> = \frac{1}{n} \int_0^\infty \sigma(v)\ v\ f(v)\ dv = 4\pi \left(\frac{m}{2\pi\ kT}\right)^{3/2} \int_0^\infty v^3\ \sigma(v)\ e^{-mv^2/2kT}\ \mathbf{dv}\ . \qquad (4.15)$$

Usually the cross sections are expressed in terms of the relative energies of the reacting particles rather than their relative speeds, so it is convenient to re-express the Maxwellian speed distribution as a Maxwellian energy distribution

$$f(\varepsilon) = \frac{2\sqrt{\varepsilon}}{\sqrt{\pi}\ (kT)^{3/2}}\ e^{-\varepsilon/kT} \qquad (4.16)$$

where $f(\varepsilon)$ has been normalized so that

$$\int_0^\infty f(\varepsilon)\ d\varepsilon = 1$$

($\varepsilon$ is the energy in joules).

As an example of a specific rate calculation, consider the electron impact excitation of the $6^3P_1$ radiating state in mercury. The actual cross section is shown in Fig. 4.3. The mean electron energy is usually fairly low in mercury

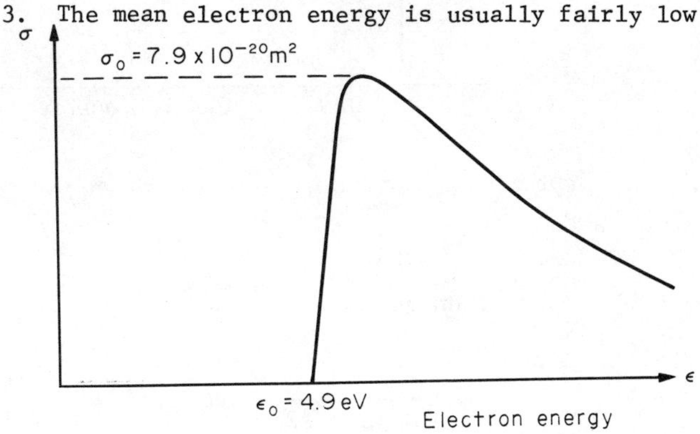


*Fig. 4.3. Excitation cross section for the mercury $6^3P_1$ state*

discharges (fluorescent lamps), on the order of about 1 eV, so $f(\varepsilon)$ has the relative shape shown in Fig. 4.4. Obviously, only those few electrons in the tail of the distribution do any exciting. For this purpose, the exponential character of the distribution function dominates the slower decay of the excitation cross section, and for simplicity we can approximate the cross section as shown in Fig. 4.5.

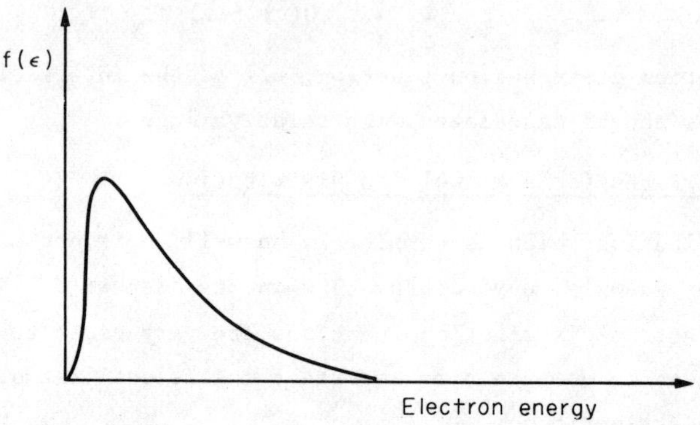


*Fig. 4.4. Relative Maxwellian electron energy distribution*

*Fig. 4.5. Approximate excitation cross section for the mercury $6^3P_1$ state*

The integral in equation (4.15) then becomes

$$\langle \sigma v \rangle = \sigma_o \int_{\varepsilon_o (\text{joules})}^{\infty} \frac{2}{\sqrt{\pi}(kT_e)^{3/2}} \sqrt{\varepsilon}\, e^{-\varepsilon/kT_e} \cdot \frac{\sqrt{2\varepsilon}}{\sqrt{m}}\, d\varepsilon \qquad (4.17)$$

which becomes

$$<\sigma v> = \left(\frac{8e}{m\pi}\right)^{1/2} \sigma_o \left(\frac{kT_e}{e}\right)^{1/2} \left[1 + \frac{\varepsilon_o}{kT_e}\right] e^{-\varepsilon_o/kT_e} \quad . \qquad (4.18)$$

In this expression, $\varepsilon_o$ is in joules and $kT_e/e$ is in volts. As in most integrations, this function is a relatively smooth function of mean electron energy ($kT_e$) even though $\sigma$ is a sharp function of $\varepsilon$. The resultant $<\sigma v>$ ($m^3$/sec) as a function of $kT_e/e$ (volts) is shown in Fig. 4.6. A further examination of equilibrium distributions also allows us to calculate the rates for inverse processes with only the knowledge of the cross section or rate for the direct process by invoking the principle of detailed balance.

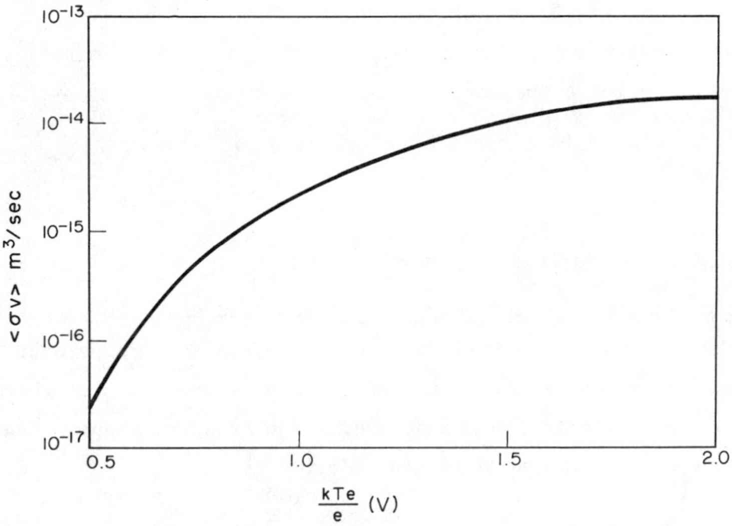


*Fig. 4.6. Excitation rate as a function of characteristic electron energy from equation (4.18)*

### 4.2.2 Detailed Balance and Inverse Processes

If we have a system in thermodynamic equilibrium, then the energy distributions of all the free particles are known and the populations of all excited species are known. This is not a particularly interesting condition in itself, but simply invoking this condition allows us to determine the relationship between any given process (say collisional excitation) in a gas discharge and its inverse (collisional de-excitation). For example, consider two energy levels of a particular atomic species in a medium (Fig. 4.7). The energy difference between the levels is $\Delta\varepsilon$ and the degeneracies are $g_1$ and $g_o$ as shown. From the Boltzmann distribution, in thermodynamic equilibrium, we have

$$n_1 = n_o \frac{g_1}{g_o} e^{-\Delta\varepsilon/kT_e} \quad . \qquad (4.19)$$

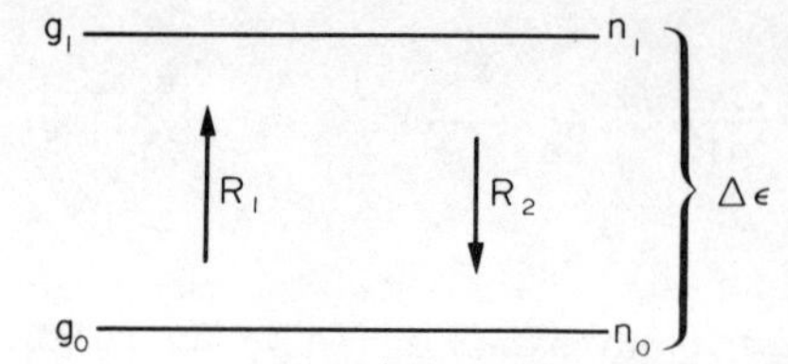


*Fig. 4.7. A simple two level system*

At thermodynamic equilibrium, we also know that the processes forming $n_1$ from $n_o$ ($R_1$) exactly equal the processes forming $n_o$ from $n_1$ ($R_2$). Therefore, in terms of the rate expressions, if we consider electrons to be responsible for the excitation and de-excitation processes

$$R_1 = n_e n_o \langle\sigma_{01} v_e\rangle = n_e n_1 \langle\sigma_{10} v_e\rangle = R_2 \quad . \tag{4.20}$$

This allows us to determine the de-excitation rate $\langle\sigma_{10}\ v_e\rangle$ from the excitation rate $\langle\sigma_{01}\ v_e\rangle$ without having to measure the actual processes,

i.e. $\langle\sigma_{10} v_e\rangle = \frac{n_o}{n_1} \langle\sigma_{01} v_e\rangle$ in thermodynamic equilibrium

or

$$\langle\sigma_{10} v_e\rangle = \frac{g_o}{g_1} e^{\Delta\varepsilon/kT_e} \langle\sigma_{01} v_e\rangle \quad . \tag{4.21}$$

Now this last statement is completely general provided that the electron distribution function can be described by a Maxwellian distribution at a temperature, $T_e$. To be even more general, we can use the detailed balance of processes at thermodynamic equilibrium to relate excitation and de-excitation cross sections. Consider an excitation cross section as shown in Fig. 4.8.

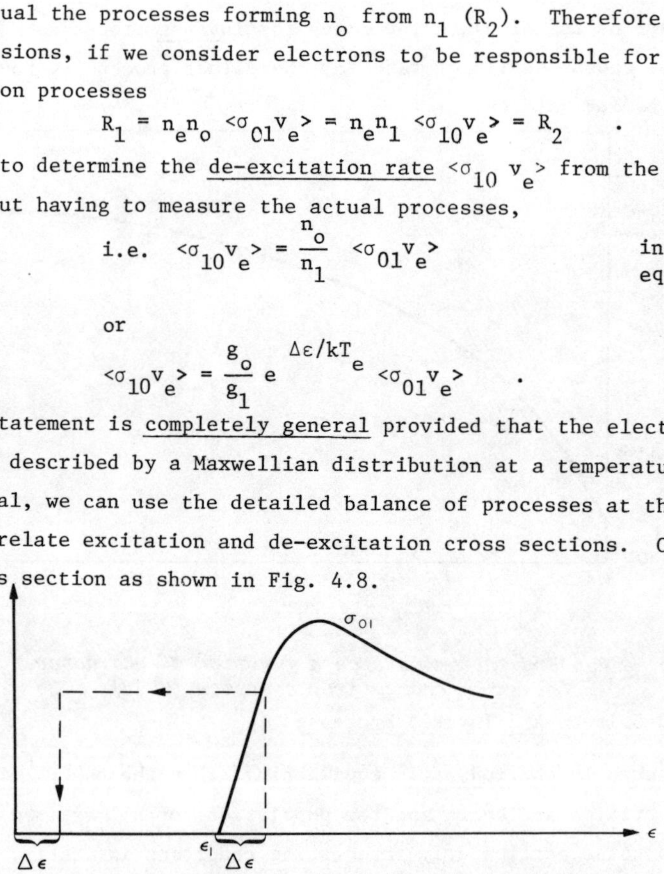


*Fig. 4.8. Example excitation cross section and energy loss*

For thermodynamic equilibrium, the number of excitations (01) exactly balances the number of de-excitations (10) for electrons in each energy interval. Therefore

$$n_e n_o\ \sigma_{01}(\varepsilon_1 + \Delta\varepsilon)\ f(\varepsilon_1 + \Delta\varepsilon)\ v(\varepsilon_1 + \Delta\varepsilon) = n_e n_1\ \sigma_{10}\ (\Delta\varepsilon)\ f(\Delta\varepsilon)\ v(\Delta\varepsilon) \tag{4.22}$$

where f is the Maxwellian distribution at temperature $T_e$ and v is the corresponding electron velocity. Therefore,

$$\sigma_{10}(\Delta\varepsilon) = \frac{n_o}{n_1}\frac{f(\varepsilon_1 + \Delta\varepsilon)}{f(\Delta\varepsilon)}\frac{v(\varepsilon_1 + \Delta\varepsilon)}{v(\Delta\varepsilon)}\sigma_{01}(\varepsilon_1 + \Delta\varepsilon)$$

$$= \frac{g_o}{g_1} e^{\varepsilon_1/kT} \cdot \frac{\sqrt{\varepsilon_1 + \Delta\varepsilon}}{\sqrt{\Delta\varepsilon}} \frac{e^{-(\varepsilon_1 + \Delta\varepsilon)/kT}}{e^{-\Delta\varepsilon/kT}} \cdot \frac{\sqrt{\varepsilon_1 + \Delta\varepsilon}}{\sqrt{\Delta\varepsilon}}\sigma_{01}(\varepsilon_1 + \Delta\varepsilon)$$

or

$$\sigma_{10}(\Delta\varepsilon) = \left(\frac{\varepsilon_1 + \Delta\varepsilon}{\Delta\varepsilon}\right)\frac{g_o}{g_1}\sigma_{01}(\varepsilon_1 + \Delta\varepsilon) \; . \tag{4.23}$$

The relationship between excitation and de-excitation cross sections is therefore known and this result can now be applied to a non-equilibrium situation to calculate excitation and de-excitation rates for arbitrary distribution functions.

## 4.3 THE BOLTZMANN EQUATION

By far the most interesting case (and important) is the determination of the properties of a discharge under the influence of external fields. What we need, therefore, is an "equation of motion" of the distribution function due to external fields, or some method of describing the change in the number of particles in $d^3\bar{r}\, d^3\bar{v}$ as a function of time. This is known as the Boltzmann Equation.

The time change in the distribution function is due to particle flows across the six bounding surfaces of the elemental volume in phase space (Fig. 4.9).

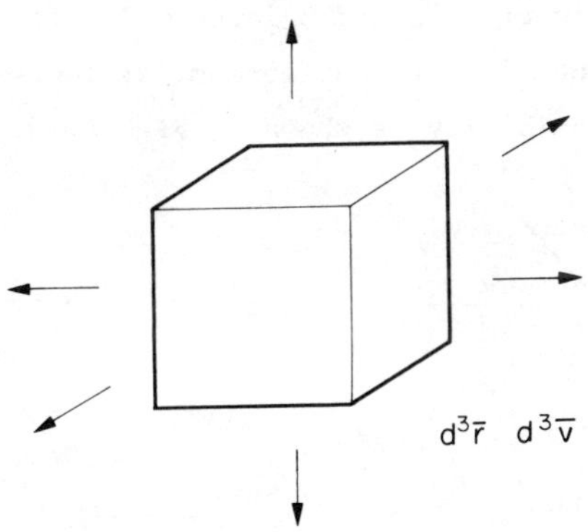


*Fig. 4.9. Particle flows in phase space*

Boltzmann's equation can, thefore, be regarded as an <u>equation of continuity in phase space</u>. Or, in other words, the total time rate of change of f is zero

$$\frac{df}{dt} = 0 = \frac{\partial f}{\partial t} + \frac{\partial f}{\partial \bar{r}}\frac{\partial \bar{r}}{\partial t} + \frac{\partial f}{\partial \bar{v}}\frac{\partial \bar{v}}{\partial t} \; . \tag{4.24}$$

Rewriting in three space and three velocity dimensions

$$\frac{\partial f}{\partial t} + \bar{v} \cdot \nabla_r f + \frac{\bar{F}}{m} \cdot \nabla_v f = 0 \tag{4.25}$$

where $\nabla_r$ is a gradient in x, y, z space and $\nabla_v$ is a gradient in velocity space. $\bar{F}/m$ is the acceleration or $\partial\bar{v}/\partial t$.

In this formulation we have ignored discontinuous changes in f which arise due to interparticle collisions. Short range collisions result in abrupt changes in the velocity and position coordinates of the colliding particles and so this effect cannot be described in terms of a particle flow in phase space but rather as the disappearance of a particle from one element of phase space and its reappearance in some distinctly different element of phase space. This effect is usually expressed, in a rather indefinite sense, as

$$\frac{\partial f}{\partial t} + \bar{v} \cdot \nabla_r f + \frac{\bar{F}}{m} \cdot \nabla_v f = S = \left(\frac{\delta f}{\delta t}\right)_{coll} \tag{4.26}$$

where $S = (\delta f/\delta t)_{coll}$ is the collision integral. Equation (4.26) is the Boltzmann equation of interest to us.

### 4.3.1 The Collision Integral

Since the collision integral plays a very important role in the determination of the form of the distribution function, we shall want to determine its explicit form. To do this we can refer back to the collision concepts discussed in the previous chapter. A good discussion is given by Kennard (1938).

In determining $(\partial f/\partial t)_{coll}$, there are essentially two contributions: the number of particles/second scattered out of $d^3\bar{v}$ at $d^3\bar{r}$ by collisions (a loss term) and the number of particles/second scattered into $d^3\bar{v}$ at $d^3\bar{r}$ by collisions.

Now, since the principle of detailed balance holds for collisions, these two processes are equally probable (at equilibrium, of course, they balance exactly). Schematically, this reversibility is shown in Fig. 4.10.

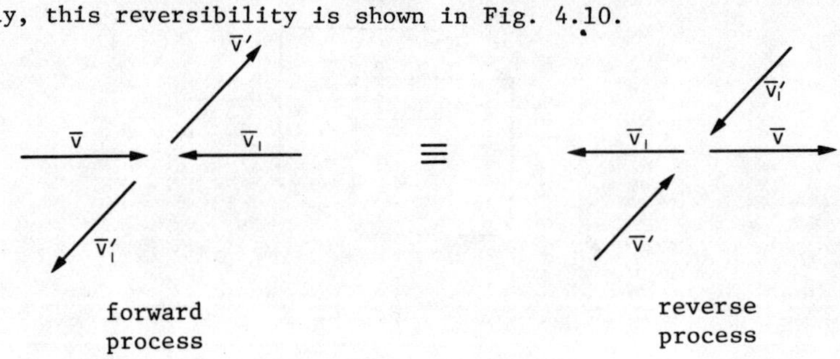


*Fig. 4.10. Microscopic reversibility of scattering*

In this figure, for the forward process, $\bar{v}$ is the velocity of the particle under consideration and $\bar{v}_1$ is the velocity of the scattering partner before the collision. $\bar{v}'$ and $\bar{v}_1'$ are the corresponding velocities after the collision.

We also consider a range of initial and final velocities $\bar{v}$ in $d^3\bar{v}$, $\bar{v}_1$ in $d^3\bar{v}_1$, $\bar{v}'$ in $d^3\bar{v}'$ and $\bar{v}_1'$ in $d^3\bar{v}_1'$. If we consider the scatterings that occur in the unit volume $d^3\bar{r}$ at $\bar{r}$, from our previous discussion of collisions,

$$\left[\begin{array}{c}\text{The flux scattered}\\ \text{out through } d\Omega \text{ at } \theta\end{array}\right] = \left[\begin{array}{c}\text{Number of scattering}\\ \text{centres}\end{array}\right] \times \left[\begin{array}{c}\text{Incoming}\\ \text{flux}\end{array}\right] \times \frac{d\sigma(\theta)}{d\Omega} \quad . \tag{4.27}$$

Therefore, the flux scattered out of $d^3\bar{v}$ at $\bar{v}$ through $d\Omega$ at $\theta$ is

$$\left[d^3\bar{r}\ d^3\bar{v}_1\ F(\bar{r},\bar{v}_1,t)\right] \times \left[d^3\bar{v}\ f(\bar{r},\bar{v},t)\,|\ \bar{v} - \bar{v}_1\ |\right] \times \frac{d\sigma(\theta)}{d\Omega} \quad . \tag{4.28}$$

Similarly, the flux scattered into $d^3\bar{v}$ is

$$\left[d^3\bar{r}\ d^3\bar{v}_1'\ F(\bar{r},\bar{v}_1',t)\right] \times \left[d^3\bar{v}'\ f(\bar{r},\bar{v}',t)\ |\ \bar{v}' - \bar{v}_1'\ |\right] \times \frac{d\sigma(\theta)}{d\Omega} \tag{4.29}$$

where F represents the distribution function of the scattering partner. Now for the forward and reverse processes the relative velocities are equal

$$|\ \bar{v} - \bar{v}_1\ | = |\ \bar{v}' - \bar{v}_1'\ | = v_r \tag{4.30}$$

and the scattering angles are equal (or $d\sigma(\theta)/d\Omega$). Furthermore, $d^3\bar{v}'d^3\bar{v}_1' = d^3\bar{v}d^3\bar{v}_1$ as seems entirely reasonable and can be proved (see Kennard, 1938).

The collision integral $S = (\delta f/\delta t)_{coll}$ then equals the reverse process minus the forward process after the preceding terms are integrated over all values of $\Omega$ and all values of the velocity of the collision partner $\bar{v}_1$.

$$\frac{d}{dt}\,(f\ d^3\bar{v}\ d^3\bar{r})_{coll} = \text{flux scattered in} - \text{flux scattered out} \quad .$$

Therefore,

$$S = \left(\frac{df}{dt}\right)_{coll} \tag{4.31}$$

$$= \int d^3\bar{v}_1 \int d\Omega\ [f(\bar{r},\bar{v}',t)\ F(\bar{r},\bar{v}_1',t) - f(\bar{r},\bar{v},t)\ F(\bar{r},\bar{v}_1,t)] \times |\ \bar{v} - \bar{v}_1\ |\ d\sigma(\theta)/d\Omega \quad .$$

The total form of the Boltzmann equation is therefore

$$\frac{\partial f}{\partial t} + \bar{v}\cdot\nabla_r f + \frac{\bar{F}}{m}\cdot\nabla_v f = \int d^3\bar{v}_1 \int d\Omega\ [f'F_1' - f\ F_1] \times v_r\ d\sigma/d\Omega \quad . \tag{4.32}$$

Implicit in the form of the collision integral is the assumption that only binary collisions are important and the mean distance between collisions is much larger than the range of interparticle forces. In most cases this is valid, but for electron-electron collisions it is necessary to use a different form of the collision integral.

Now that we have an explicit form for the Boltzmann equation, we can solve for the distribution under the influence of an external field. This is usually done by assuming that any external fields cause only a small perturbation in the form of the distribution function. In the next section we shall perform this perturbation theory solution for an externally applied electric field.

## 4.4 EXPANSION OF THE BOLTZMANN EQUATION FOR AN APPLIED ELECTRIC FIELD

In order to determine the response of a plasma to an external force, such as an electric field, it is necessary, in general, to determine the distribution function (of the electrons for now) by solving the Boltzmann equation.

We know that the directed velocity gained by the electrons from an external field is, in most cases of interest, much less than the thermal or random velocity. Therefore the change in the distribution function, f, due to an external field constitutes only a small perturbation of the distribution function from its equilibrium value. The normal method of solution is therefore to express the distribution function as a sum of an isotropic (equilibrium) part plus small perturbations. See Chapman and Cowling (1952) for a classic treatment of this procedure. The technique which is most applicable to our case is the expansion of the distribution function in terms of spherical harmonics in velocity space. This approach has been presented by Allis (1956) and has the great advantage that often only two terms in the expansion are sufficient for obtaining an accurate approximation to the distribution function.

As shown in Fig. 4.11, the coordinate system is selected so that the preferred direction of orientation is the direction of the applied electric field. In this case we can expand the distribution function in terms of Legendre polynomials

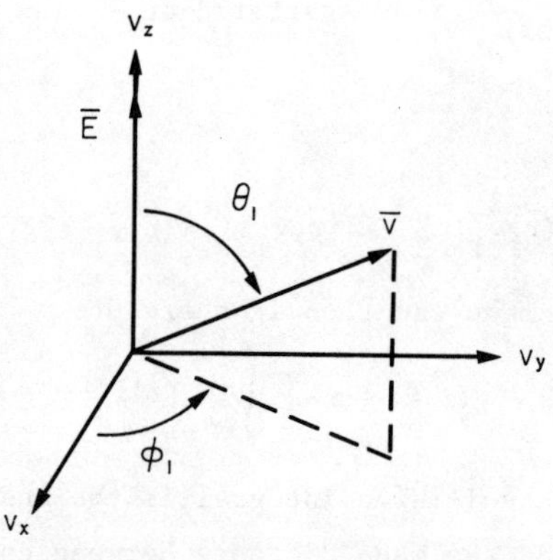


*Fig. 4.11. Coordinate system for the expansion of the distribution function*

$$f(\bar{v},\bar{r},t) = \sum_{k=0}^{\infty} P_k\,(\cos\,\theta_1)\, f_k\,(v,\bar{r},t) \qquad (4.33)$$

where $P_k\,(\cos\,\theta_1)$ are the Legendre polynomials and the $f_k$ depend only on the magnitude of the velocity. If we were to include magnetic fields, we would need the associated Legendre polynomials.

The Boltzmann equation now becomes

$$\frac{d}{dt}\sum_k P_k f_k + \bar{v}\cdot\nabla_r \sum_k P_k f_k - \frac{e\bar{E}}{m}\cdot\nabla_v \sum_k P_k f_k = S \quad . \qquad (4.34)$$

(Since we are considering only electrons, we have chosen the sign of the force term in equation (4.34) accordingly).

Now, by the use of the orthogonality properties of the Legendre polynomials, as well as the appropriate recursion relations, it is possible to convert this one equation with an infinite number of terms into an infinite set of coupled equations with a finite number of terms (see Appendix A). The resulting master equation is:

$$\frac{\partial f_k}{\partial t} + v\left(\frac{k}{2k-1}\frac{\partial f_{k-1}}{\partial Z} + \frac{k+1}{2k+3}\frac{\partial f_{k+1}}{\partial Z}\right) - \frac{eE_z}{m}\left(\frac{k}{2k-1}v^{k-1}\frac{\partial}{\partial v}\frac{f_{k-1}}{v^{k-1}}\right.$$
$$\left. + \frac{k+1}{2k+3}\frac{1}{v^{k+2}}\frac{\partial}{\partial v}v^{k+2}f_{k+1}\right) = \frac{2k+1}{4\pi}\int P_k\, S\, d\Omega_1 \quad . \tag{4.35}$$

From this equation we can obtain the following set of coupled equations:

$$\frac{\partial f_o}{\partial t} + \frac{v}{3}\frac{\partial f_1}{\partial Z} - \frac{eE_z}{m}\frac{1}{3}\frac{1}{v^2}\frac{\partial}{\partial v}(v^2 f_1) = S_o \quad , \tag{4.36}$$

$$\frac{\partial f_1}{\partial t} + v\left[\frac{\partial f_o}{\partial Z} + \frac{2}{5}\frac{\partial f_2}{\partial Z}\right] - \frac{eE_z}{m}\left[\frac{\partial f_o}{\partial v} + \frac{2}{5v^3}\frac{\partial}{\partial v}(v^3 f_2)\right] = S_1 \quad , \tag{4.37}$$

$$\frac{\partial f_2}{\partial t} + v\left[\frac{2}{3}\frac{\partial f_1}{\partial Z} + \frac{3}{7}\frac{\partial f_3}{\partial Z}\right] - \frac{eE_z}{m}\left[\frac{2}{3}v\frac{\partial}{\partial v}\left(\frac{f_1}{v}\right) + \frac{3}{7}\frac{1}{v^4}\frac{\partial}{\partial v}(v^4 f_3)\right] = S_2 \quad , \tag{4.38}$$

etc.

### 4.4.1 Expansion of the Collision Integral

Now that we have expanded the distribution function to obtain a set of coupled equations for f, we must also obtain the specific form of the collision integral to be used with each equation.

$$S_k = \frac{2k+1}{4\pi}\int P_k\, S\, d\Omega_1 \quad . \tag{4.39}$$

Now, you will remember that the form of the collision integral is

$$S = \iint d\bar{v}_1\, d\Omega\, \frac{d\sigma}{d\Omega}\, v_r\, \{f(\bar{v}')\, F(\bar{v}_1') - f(\bar{v})\, F(\bar{v}_1)\} \quad . \tag{4.40}$$

In this equation we have to substitute the expansion for $f(\bar{v})$ and $f(\bar{v}')$

$$f(\bar{v}) = \sum_{\ell=0}^{\infty} P_\ell\,(\cos\theta_1)\, f_\ell(v, \bar{r}, t) \tag{4.41}$$

and

$$f(\bar{v}') = \sum_{\ell=0}^{\infty} P_\ell\,(\cos\theta_1')\, f_\ell(v', \bar{r}, t) \quad . \tag{4.42}$$

Where the angle $\theta_1'$ is defined in Fig. 4.12 as the angle between $\bar{v}'$ and $\bar{E}$. Therefore

$$S_k = \frac{2k+1}{4\pi}\int d\bar{v}_1 \int d\Omega \int d\Omega_1\, \frac{d\sigma}{d\Omega}\, v_r\, P_k\,(\cos\theta_1)$$
$$\times \{F(\bar{v}_1') \cdot \sum_\ell P_\ell\,(\cos\theta_1')\, f_\ell' - F(\bar{v}_1) \cdot \sum_\ell P_\ell\,(\cos\theta_1)\, f_\ell\} \quad . \tag{4.43}$$

In order to integrate through on $d\Omega_1$, it is necessary to expand $\cos\theta_1'$ in terms of the angles $\theta$, $\theta_1$ and $\phi_1$.

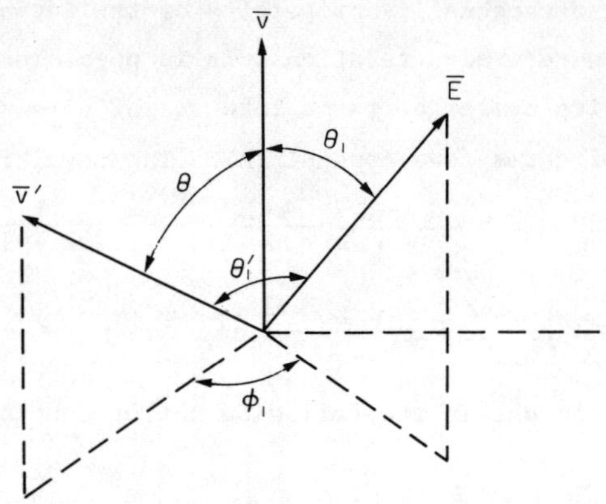


*Fig. 4.12. Coordinate system for the expansion of the collision integral*

From Fig. 4.12

$$\cos\theta_1' = \cos\theta\cos\theta_1 + \sin\theta\sin\theta_1\cos\phi_1 \tag{4.44}$$

and from the addition theorem for Legendre polynomials (see Jahnke and Emde, 1945),

$$P_\ell(\cos\theta_1') = P_\ell(\cos\theta)\,P_\ell(\cos\theta_1) + 2\sum_{m=1}^{\ell}\frac{(\ell-m)!}{(\ell+m)!}P_\ell^m(\cos\theta)$$
$$\times\, P_\ell^m(\cos\theta_1)\cos(m\,\phi_1)\quad . \tag{4.45}$$

Substituting this term in the equation for $S_k$ and integrating over $d\Omega_1$, we find that most of the terms drop out (by orthogonality of the Legendre polynomials)

$$\int_{-1}^{+1} P_\ell(x)\,P_k(x)\,dx = 0 \qquad (\text{for } \ell \neq k)$$

and we obtain

$$S_k = \int d\bar{v}_1 \int d\Omega\,\frac{d\sigma}{d\Omega}\,v_r\,\{F(\bar{v}_1')\,f_k(v')\,P_k(\cos\theta) - F(\bar{v}_1)\,f_k(v)\}\quad . \tag{4.46}$$

Explicit equations for the zero and first order collision integrals are:

$$S_o = \int d\bar{v}_1 \int d\Omega\,\frac{d\sigma}{d\Omega}\,v_r\,\{F(\bar{v}_1')\,f_o(v') - F(\bar{v}_1)\,f_o(v)\}\quad , \tag{4.47}$$

$$S_1 = \int d\bar{v}_1 \int d\Omega\,\frac{d\sigma}{d\Omega}\,v_r\,\{F(\bar{v}_1')\,f_1(v')\cos\theta - F(\bar{v}_1)\,f_1(v)\}\quad . \tag{4.48}$$

The solution of the equations for $S_0$ and $S_1$ can be quite complicated, as we shall see, but we can solve these equations using some simple approximations and go to more complicated models when these approximations appear to be unsatisfactory.

4.4.1.1 Solution of the collision integral---first approximation. In our first approximation, we shall assume that we are only considering elastic collisions between electrons and heavy collision partners, such as atoms, ions or molecules. We shall therefore make the simplest approximation that

$$v' = v \quad , \quad \bar{v}_1' = \bar{v}_1 \quad .$$

Therefore

$$S_o = \int d\bar{v}_1 \int d\Omega \, \frac{d\sigma}{d\Omega} \, v_r \, \{F(\bar{v}_1) \, f_o(v) - F(\bar{v}_1) \, f_o(v)\} \tag{4.49}$$

or

$$S_o = 0 \quad .$$

This is not satisfactory, so we will have to evaluate $S_o$ to the next order of approximation.

For $S_1$ we obtain

$$S_1 = \int d\bar{v}_1 \int d\Omega \, \frac{d\sigma}{d\Omega} \, v_r \, \{F(\bar{v}_1) \, f_1(v) \cos\theta - F(\bar{v}_1) \, f_1(v)\} \quad . \tag{4.50}$$

Therefore,

$$S_1 = - f_1(v) \int\int d\bar{v}_1 \, d\Omega \, \frac{d\sigma}{d\Omega} \, v_r \, F(\bar{v}_1) \, (1 - \cos\theta) \quad . \tag{4.51}$$

From the previous definitions, the collision frequency for momentum transfer

$$\nu_m(v) = N_1 \, v_r \int \frac{d\sigma}{d\Omega} \, (1-\cos\theta) \, d\Omega \tag{4.52}$$

where the density of the scattering centers

$$N_1 = \int F(\bar{v}_1) \, d\bar{v}_1 \quad . \tag{4.53}$$

Therefore,

$$S_1 = - \nu_m(v) \, f_1(v) \quad . \tag{4.54}$$

This is the form of the collision integral that will be used in subsequent solutions to the Boltzmann equation. Keep in mind, however, that energy and momentum conservation are not satisfied by this equation.

4.4.1.2 Solution of the collision integral--second approximation. In the previous section it was shown that the assumption $v' = v$ and $\bar{v}_1' = \bar{v}_1$ did not give a solution for $S_o$. If the electric field is weak, this is not too important and we can assume that the distribution function is Maxwellian (if we consider only elastic collisions). In a strong field, it is necessary to take into account the change in electron momentum and energy during a collision. In fact, a strong electric field can radically alter the shape of the isotropic part of the distribution function, therefore $S_o$ must be evaluated taking into account energy and momentum conservation. This sounds straightforward, but in reality it is quite complicated. A development is given by Ginzburg and Gurevich (1960) for those interested in the details.

The desired solution is

$$S_o = \frac{1}{2v^2}\frac{\partial}{\partial v}\left\{v^2\delta\nu_m\left[\frac{kT}{m}\frac{\partial f_o}{\partial v} + vf_o\right]\right\} \tag{4.55}$$

where $\delta = 2m/M$ is the fractional energy transfer in an elastic collision and the neutral particles are assumed to have a Maxwellian distribution characterized by the gas temperature, T. You will notice that if $f_o$ is also Maxwellian at the temperature T, $S_o = 0$ as expected.

## 4.5 DISTRIBUTION FUNCTIONS FOR AN APPLIED ELECTRIC FIELD - ELASTIC COLLISIONS ONLY

Now that $S_o$ and $S_1$ have been evaluated for elastic collisions, the coupled set of equations for $f_o$ and $f_1$ can be used to solve for the distribution function (ignoring $f_2$ and higher terms)

$$\frac{\partial f_o}{\partial t} + \frac{v}{3}\frac{\partial f_1}{\partial z} - \frac{eE_z}{3mv^2}\frac{\partial}{\partial v}(v^2 f_1) = \frac{1}{2v^2}\frac{\partial}{\partial v}\left\{v^2\delta\nu_m\left[\frac{kT}{m}\frac{\partial f_o}{\partial v} + vf_o\right]\right\} \tag{4.56}$$

and

$$\frac{\partial f_1}{\partial t} + v\frac{\partial f_o}{\partial z} - \frac{eE_z}{m}\frac{\partial f_o}{\partial v} = -\nu_m f_1 \quad . \tag{4.57}$$

From equation (4.37) it can be seen that the neglect of $f_2$ and higher order terms is only valid if $\partial f_o/\partial z >> \partial f_2/\partial z$ and if $\partial f_o/\partial v >> 1/v^2\ \partial/\partial v\ (v^3 f_2)$. For the case when elastic collisions dominate, Ginzburg and Gurevich (1960) have pointed out that these conditions are usually satisfied. Assuming a steady-state situation with no gradients of temperature or number density

$$f_1 = \frac{eE}{m\nu_m}\frac{\partial f_o}{\partial v} \quad . \tag{4.58}$$

Therefore, if $f_o$ is known, we can evaluate $f_1$, the anisotropic part of the distribution function. This is the foundation for the evaluation of the transport coefficients, as will be seen later, but now equation (4.58) can be substituted in equation (4.56) to give

$$-\frac{eE}{3mv^2}\frac{\partial}{\partial v}\left\{v^2\frac{eE}{m\nu_m}\frac{\partial f_o}{\partial v}\right\} = \frac{1}{2v^2}\frac{\partial}{\partial v}\left\{v^2\delta\nu_m\left[\frac{kT}{m}\frac{\partial f_o}{\partial v} + vf_o\right]\right\} \quad . \tag{4.59}$$

Equation (4.59) can be reexpressed in terms of the energy, $\varepsilon = mv^2/2$, and the distribution function for energy, $f_o(\varepsilon) = 4\pi\sqrt{2}\, f_o(v)/m^{3/2}$ where

$$\int \varepsilon^{1/2} f_o(\varepsilon)d\varepsilon = 1.$$

The resultant equation is

$$\frac{\partial}{\partial \varepsilon}\left[\frac{e^2E^2\varepsilon}{3N\sigma_m}\frac{\partial f}{\partial \varepsilon}\right] + \frac{2m}{M}\frac{d}{d\varepsilon}\left[\varepsilon^2 N\sigma_m[f_o + kT\frac{\partial f}{\partial \varepsilon}]\right] = 0 \tag{4.59a}$$

where $\nu_m$ has been re-expressed as $N\sigma_m v$ and $\delta = 2m/M$. This form of the solution will be seen again in equation (4.73). Referring back to equation (4.59), the derivative terms conveniently cancel out and we obtain

$$\frac{\partial f_o}{\partial v} + \frac{m\beta v f_o}{1 + \frac{2}{3}\frac{\beta}{\delta\nu_m^2}\frac{e^2E^2}{m}} = 0 \tag{4.60}$$

where $\beta = 1/kT$.

The general form of the solution is

$$f_o(\bar{v}) = C \exp - \int_o^v \frac{m\beta v dv}{1 + \frac{2}{3}\frac{\beta}{\delta\nu_m^2}\frac{e^2E^2}{m}} \,. \tag{4.61}$$

Note that $f_o(\bar{v})$ is a distribution function for <u>velocity</u> but that it depends only on the magnitude of the velocity just as specified in equation (4.33). If the electric field $\bar{E} = 0$, then the distribution function for the electrons is Maxwellian at the gas temperature, T, as expected. If $\bar{E}$ is not zero, then the specific form of the distribution function will depend on the velocity dependence of the collision frequency.

### 4.5.1 Constant Collision-Frequency Case

The simplest case to consider is when $\nu_m$ = constant. In this case, $f_o$ can be written as

$$f_o = C \exp - \int_o^v \frac{m v dv}{k\left[T + \frac{2}{3}\frac{e^2E^2}{km\delta\nu_m^2}\right]} \,. \tag{4.62}$$

Again, this can be recognized as a Maxwellian distribution with an electron temperature

$$T_e = T + \frac{2}{3}\frac{e^2E^2}{km\delta\nu_m^2} \,. \tag{4.63}$$

This is identical to equation (2.36), which was based on a simple energy balance from the Langevin equation. Since $\nu_m$ = constant is basically the Langevin approximation, this result is expected.

### 4.5.2 Constant Mean Free-path Case

The second simplest case to consider is when the mean free path, $\ell$, between collisions is independent of electron energy. In this case $\nu_m = v/\ell$ and

$$f_o(\bar{v}) = C \exp - \int_o^v \frac{m\beta v dv}{1 + \frac{2}{3}\frac{\beta}{\delta v^2}\frac{\ell^2e^2E^2}{m}} \,. \tag{4.64}$$

If the collision frequency is large or

$$\frac{2}{3} \frac{\beta}{\delta\nu^2} \frac{\ell^2 e^2 E^2}{m} \ll 1 ,$$

then, as before, the distribution function is essentially Maxwellian. On the other hand, for a strong electric field, this solution becomes the Druyvesteyn distribution

$$f_o(\bar{v}) = C \exp\left\{-\frac{3}{8}\frac{\delta m^2 v^4}{\ell^2 e^2 E^2}\right\} . \tag{4.65}$$

The point to notice here is that $f_o$ varies as $e^{-v^4}$ rather than as $e^{-v^2}$ as for a Maxwellian. We see that the Druyvesteyn distribution drops off much more rapidly than the Maxwellian at high energies. As we would imagine, this can have a significant effect on ionization and excitation rates so it is very important to evaluate $f_o$ if we wish to determine accurately the excitation and ionization processes in a gas discharge.

## 4.6 DISTRIBUTION FUNCTIONS WHEN ELECTRON-ELECTRON COLLISIONS ARE IMPORTANT

We have seen that an impressed electric field can have the effect of driving the electron distribution function away from a Maxwellian and towards a Druyvesteyn. Only the energy loss due to collisions ($\delta\nu_m$) tends to keep the distribution Maxwellian.

In the case of electron-electron collisions, the energy transfer factor $\delta \simeq 1$ so these collisions can play a strong part in equilibrating the electron energy and retaining the Maxwellian distribution.

In order to assess the role of electron-electron collisions, it is necessary to determine the appropriate terms in the collision integral. In this case, the previous form of the collision integral is of little use since the electron-electron interaction can be best approximated by multiple small angle collisions rather than few large angle collisions. The proper approach is then to employ the Fokker-Planck collision term. We will not specifically deal with this formulation since we are going to be primarily interested in weakly ionized gases. The reader is instead referred to the very thorough treatment of this subject by Shkarofsky, Johnston and Bachynski (1966). A very readable account is also given by W. P. Allis (1956). Ginzburg and Gurevich (1960), and Shkarofsky, Johnston and Bachynski (1966) have both given the zero order collision term that we need to insert in the Boltzmann equation:

$$S_{oe} = \frac{1}{v^2}\frac{\partial}{\partial v}\left\{v^2\left[A_1(f_o)vf_o + A_2(f_o)\frac{\partial f_o}{\partial v}\right]\right\} \tag{4.66}$$

where

$$A_1 = \frac{4\pi\nu_{ee}(v)}{n_e} \int_o^v x^2 f_o(x)dx \tag{4.67}$$

and

$$A_2 = \frac{4\pi\nu_{ee}(v)}{3n_e} \left\{ \int_o^v x^4 f_o(x)dx + v^3 \int_v^\infty xf_o(x)dx \right\} . \tag{4.68}$$

We take $\nu_{ee}$, the electron-electron collision frequency, to be

$$\nu_{ee}(v) = n_e v\sigma_{ee} = \frac{n_e ve^4}{4\pi\varepsilon_o^2 (mv^2)^2} \ell n\Lambda \tag{4.69}$$

(see equation (3.43)). We note that $S_{oe} = 0$ for a Maxwellian distribution. The form for $S_{1e}$ is considerably more complicated but not of such importance so refer to Ginzburg and Gurevich (1960) for the details.

If we consider a discharge in which elastic electron neutral and/or electron-ion collisions are important as well as electron-electron collisions, we have the following equation to solve:

$$\frac{\partial f_o}{\partial t} = \frac{1}{2v^2} \frac{\partial}{\partial v} \left\{ v^2 \left( [\delta\nu_m \frac{kT}{m} + \frac{2e^2E^2}{3m^2\nu_m} + 2A_2(f_o)] \frac{\partial f_o}{\partial v} + [\delta\nu_m + 2A_1(f_o)]vf_o \right) \right\} . \tag{4.70}$$

The solution is not straightforward since $A_1(f_o)$ and $A_2(f_o)$ are integrals depending on $f_o$. To obtain a solution we essentially have to iterate the equation. Ginzburg and Gurevich (1960) have shown that usually the first iteration (initially assuming $f_o$ to be Maxwellian) is sufficient.

Under stationary conditions (DC electric field or high frequency AC electric field) we can neglect $\partial f_o/\partial t$. Choosing the initial $f_o$ to be Maxwellian, we find to the next approximation

$$f_o = C \exp - \int_o^v \frac{vdv[\delta\nu_m + 2A_1(o)]}{\frac{kT}{m}\delta\nu_m + \frac{2e^2E^2}{3m^2\nu_m^2} + 2A_2(o)} . \tag{4.71}$$

This form of the equation shows that the balance between electron-electron collisions ($\nu_{ee}$), elastic electron-neutral or electron-ion collisions ($\delta\nu_m$) and the electron drift energy gained from the electric field will determine the relative effect of electron-electron collisions. Megill and Cahn (1964) and Rockwood (1974) have discussed the relative importance of these terms. Ginzburg and Gurevich (1960) have also pointed out the importance of the relationship between electron-electron and electron-neutral (or ion) elastic collisions in an excellent manner as shown in Fig. 4.13. In this figure, the factor

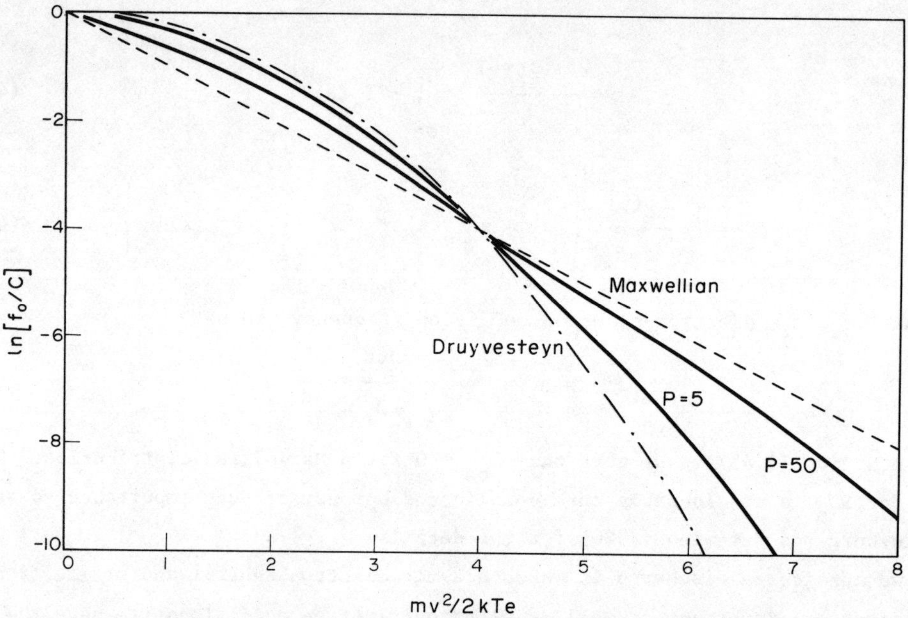


*Fig. 4.13. Effect of electron-electron collisions on the distribution function (from Ginzburg and Gurevich, 1960)*

$$p = \frac{\nu_{ee}(v_o)}{\delta\nu_m(v_o)} \qquad (4.72)$$

where $v_o = \sqrt{\frac{2kT_e}{m}}$ and $T_e = \frac{eE\ell}{\sqrt{6\delta}}$

($\ell$ is the assumed constant mean free path).

For p >> 5, the distribution tends towards Maxwellian. For p << 5, the distribution tends toward the Druyvesteyn. Note, however, that even a large relative number of electron-electron collisions (corresponding to a large fractional ionization) may not Maxwellianize the high energy tail that is responsible for excitation and ionization in the medium. In general, as Rockwood (1974) has pointed out, electron-electron collisions may be important in atomic gases even at relatively low fractional ionization ($n_e/N \approx 10^{-6}$), but electron-electron interactions are unlikely to play any significant role in molecular gas discharges (particularly molecular gas lasers) in the range of most practical interests.

## 4.7 DISTRIBUTION FUNCTIONS WHEN INELASTIC COLLISIONS DOMINATE

In the study of weakly ionized gases, it is usually the inelastic collisions, primarily excitation, that dominate in determining the shape of the distribution functions. Unfortunately, this is also the most difficult case to analyze. In an inelastic collision between electrons and neutrals, the electron can give up some of its energy into exciting rotational, vibrational or electronic degrees of freedom of the molecule or in ionizing the molecule. The inverse process can also occur when an electron gains energy in assisting an excited neutral in returning to a lower energy level. To do a complete analysis, we would need to know the cross sections for the forward and inverse processes as well as the populations of all of the excited states. Obviously this is a very complicated problem in general and we will examine a few specific examples in a later section.

### 4.7.1 The Boltzmann Equation Including Inelastic Processes

We can, in a general sense, formulate the Boltzmann equation by including the appropriate collision integral term for inelastic processes. If the distribution function for energy $f(\varepsilon)$ is used rather than the distribution function for velocity, and electron-electron collisions are assumed to have a negligible effect, then the Boltzmann equation is usually re-cast in the following form (see Engelhardt and Phelps, 1963, and Lowke, Phelps and Irwin, 1973):

$$\frac{d}{d\varepsilon}\left[\frac{e^2E^2\varepsilon}{3N\sigma_m}\frac{\partial f}{\partial \varepsilon}\right] + \frac{2m}{M}\frac{d}{d\varepsilon}\left[\varepsilon^2 N\sigma_m\left(f + kT\frac{\partial f}{\partial \varepsilon}\right)\right]$$

$$+ \sum_j [(\varepsilon + \varepsilon_j)f(\varepsilon + \varepsilon_j)N\sigma_j(\varepsilon + \varepsilon_j) - \varepsilon f(\varepsilon)N\sigma_j(\varepsilon)]$$

$$+ \sum_j [(\varepsilon - \varepsilon_j)f(\varepsilon - \varepsilon_j)N^*\sigma_{-j}(\varepsilon - \varepsilon_j) - \varepsilon f(\varepsilon)N^*\sigma_{-j}(\varepsilon)] = 0 \qquad (4.73)$$

where $N^*$ is the excited state density. In this case $\varepsilon = mv^2/2$ and $\sigma_m$ is the cross section for elastic collisions. It is also assumed that the distribution function is normalized so that

$$\int_0^\infty \varepsilon^{1/2} f(\varepsilon)d\varepsilon = 1 \ .$$

The first term in equation (4.73) represents the energy gain from the electric field and the second term represents the elastic energy losses, as treated before. The third and fourth terms represent the inelastic processes in a manner following Holstein (1946). Specifically, the third term accounts for the inelastic loss processes - the first factor represents electrons of energy $\varepsilon + \varepsilon_j$ undergoing a collision in which they lose an energy $\varepsilon_j$ (the cross section for this process being $\sigma_j$ $(\varepsilon + \varepsilon_j)$) and appear as electron of energy $\varepsilon$ - the second factor represents electrons of energy $\varepsilon$ undergoing an inelastic energy loss and appearing as electrons of some lower energy. The fourth term is basically the inverse of the third and represents electrons gaining energy due to superelastic collisions. Of course, the evaluation of this last term requires a knowledge of the densities of the excited species in the medium. Figure 4.14 indicates the processes accounted for in the inelastic collision terms.

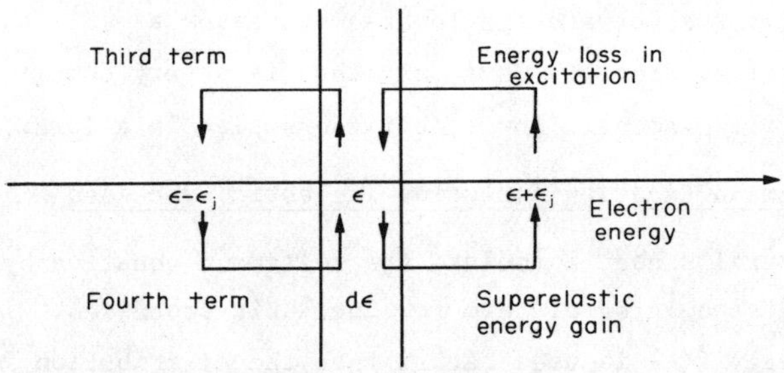


*Fig. 4.14. Schematic representation of the inelastic collision term*

Equation (4.73) is based upon the two term expansion of the distribution function. However, it should be kept in mind at this point that the two term expansion of the distribution function may not be valid when inelastic collisions occur. Ginzburg and Gurevich (1960) have discussed the criteria for validity (see section 4.5). Holstein (1946) has also indicated that the two term expansion breaks down at electron energies where the inelastic collision cross section becomes comparable to the elastic collision cross section.

### 4.7.2 The Distribution Function for Atomic Gases

In a few cases, the elastic and inelastic processes can be expressed in terms of a small number of cross sections or collision frequencies which can be reasonably accurately approximated by simple analytic functions. The distribution function can then be obtained analytically. A very early calculation of this sort was made for helium by Smit (1936). More recent calculations for helium have been performed by Abdelnabi and Massey (1953) and Reder and Brown (1954). The results of some of these calculations are shown in Fig. 4.15.

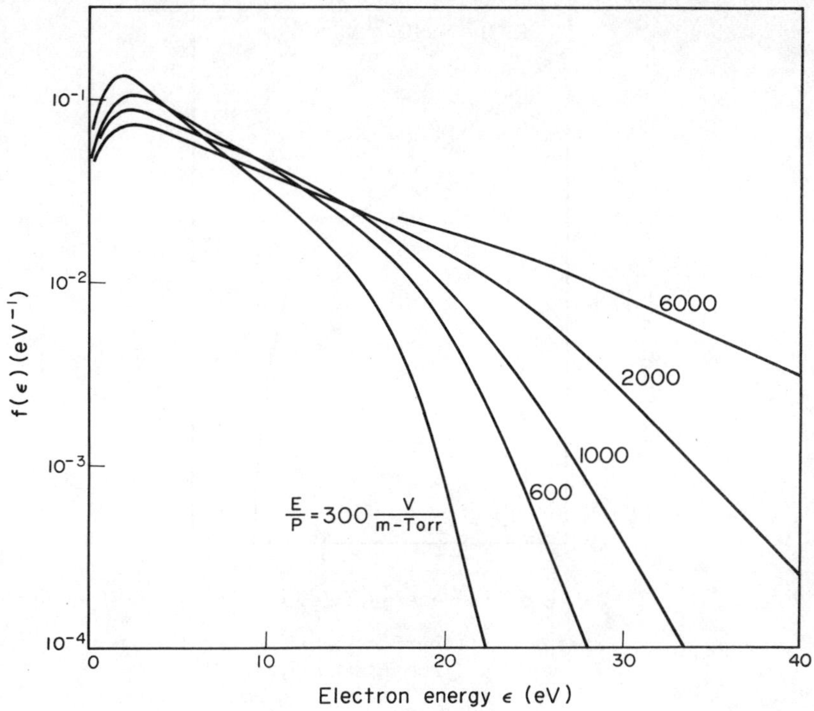


*Fig. 4.15. Energy distribution function for electrons in helium (from Smit, 1936, and Reder and Brown, 1954)*

The important point to notice is the rather sharp break in the distribution just above the threshold for inelastic collisions. Also notice that the distribution is essentially Maxwellian (straight line on the log scale) below the excitation threshold and is also essentially Maxwellian, but at a much lower "temperature", above the inelastic threshold. This approximation turns out to be useful for gases where only one inelastic process dominates the shape of the distribution function.

Of course, when the inelastic processes are more complicated, computer coding of the Boltzmann equation can be and has been used to solve for the distribution function in atomic gases. Figure 4.16 shows such a distribution for mercury as calculated by Rockwood (1973).

The obvious deduction from both Figs. 4.15 and 4.16 is that the calculation of excitation and ionization rates, even in the atomic gases, cannot be made with any degree of accuracy if a simple Maxwellian distribution is used. At the very least, a two temperature Maxwellian with a break point at the inelastic threshold should be used. The utility of such an approach will be seen in Section 4.8.1 when Vriens' analytic approach is considered.

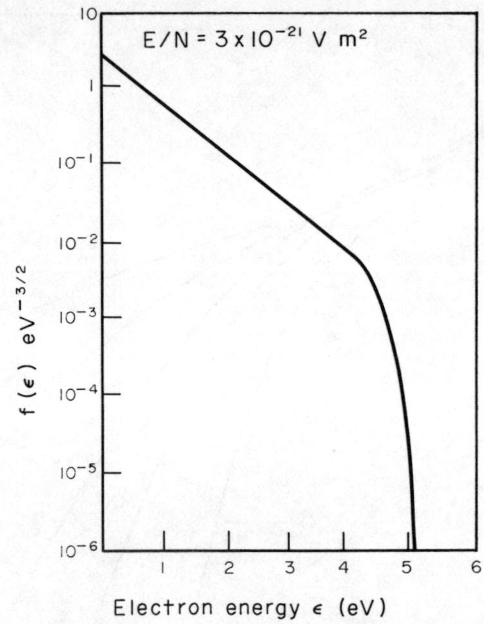


*Fig. 4.16. Energy distribution function for electrons in mercury (from Rockwood, 1973)*

### 4.7.3 The Distribution Function for Molecular Gases

Because of the variety and complexity of the inelastic processes occurring in molecular gases, analytic solutions to the Boltzmann equation are usually not attempted, rather computer techniques are used. Carleton and Megill (1962) used this approach to solve for the distribution function in air (for upper atmospheric studies). Figure 4.17 shows a typical example of the distribution functions.

The very rapid depletion of the high-energy tail by inelastic collisions is very evident. The well-known calculations of the distribution functions in $N_2$, CO and $CO_2$ by Nighan (1970) show this same behavior as well as the effect of varying E/N. Figure 4.18 shows Nighan's calculation for the distribution functions in $N_2$ as a function of E/N. These distributions are normalized so that

$$\int_0^{\infty} \varepsilon^{1/2} f(\varepsilon)\,d\varepsilon = 1 \quad . \tag{4.74}$$

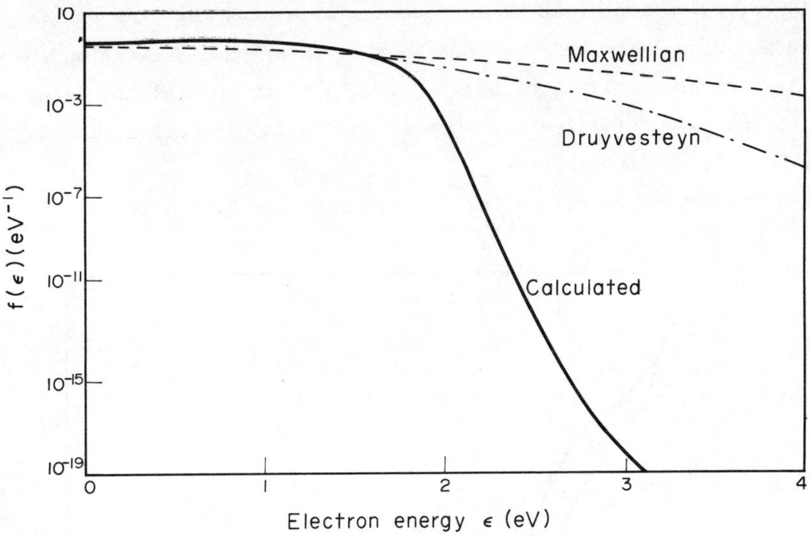


*Fig. 4.17. Electron energy distribution in a mixture of 60% $N_2$, 13% $O_2$ and 27% O for an E/P of 107 V/m - torr (from Carleton and Megill, 1962)*

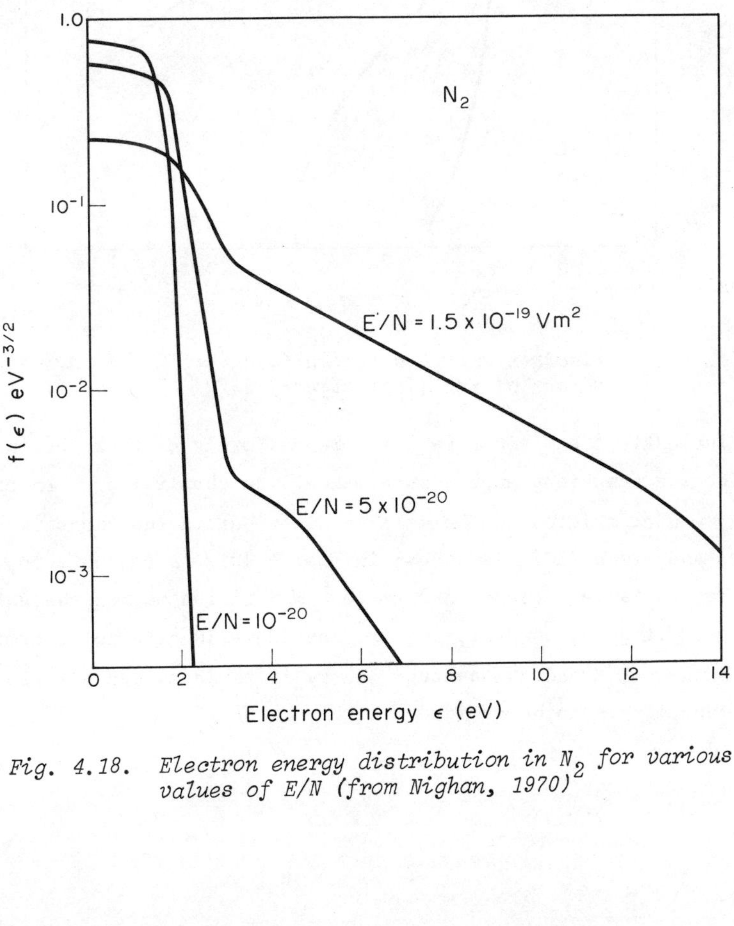


*Fig. 4.18. Electron energy distribution in $N_2$ for various values of E/N (from Nighan, 1970)*

A Maxwellian distribution would be represented by a straight line on Fig. 4.18, so the very non-Maxwellian behavior due to the vibrational excitation threshold near 2 eV is evident. Only at higher E/N values is it possible to accelerate electrons to energies sufficient to surpass the inelastic barrier and begin to approach a Maxwellian distribution.

Nighan's distribution functions for $CO_2$ are shown in Fig. 4.19.

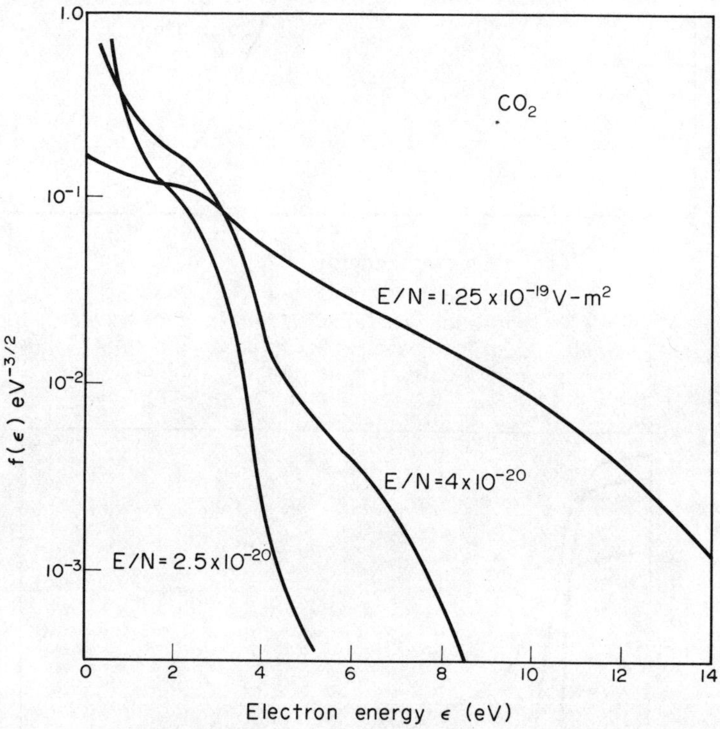


*Fig. 4.19. Electron energy distributions in $CO_2$ for various values of E/N (from Nighan, 1970)*

For $CO_2$, the inelastic cross sections are more uniformly distributed in energy, so the distribution functions appear more Maxwellian than for $N_2$. An excellent comparison between calculated and Maxwellian distribution functions is given by Lowke, Phelps and Irwin (1973) as shown in Fig. 4.20. As can be seen, the exact calculated solution is very close in form to a Maxwellian having the same average electron energy. It does, however, require considerable care to determine the relationship between E/N and the average energy if the complete solution for the distribution function is to be avoided.

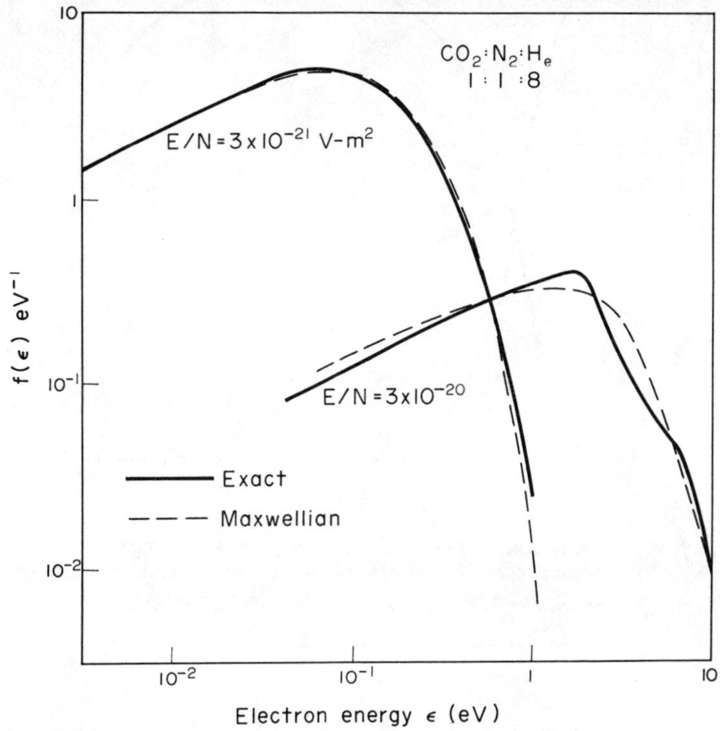


*Fig. 4.20. Electron energy distribution in $CO_2$:$N_2$:He in proportions 1:1:8 for two values of E/N (from Lowke, Phelps and Irwin, 1973)*

### 4.7.4 Rate-process Calculations

Once the distribution function is known, which requires knowledge of the elastic and inelastic cross sections, then it is possible to solve for transport coefficients, ionization rates, excitation rates and the associated power flows. Transport coefficients will be treated in the next chapter; however, we can show one important type of calculation that can be performed at this point - the fraction of the total power dissipated in a gas discharge that goes into each of the possible energy loss processes. From Rockwood's (1973) calculations on mercury, the fractional power partition is shown in Fig. 4.21. A very similar plot by Nighan (1970) for $N_2$ is shown in Fig. 4.22. Similar calculations for $CO_2$ and typical $CO_2$ laser mixtures will be discussed in the section on $CO_2$ lasers.

## 4.8 APPROXIMATE ANALYTIC TECHNIQUES FOR DETERMINING DISTRIBUTION FUNCTIONS AND RATES

Even though computer solutions are powerful techniques for determining distribution functions and rates, it is desirable to have analytic techniques, even if they are

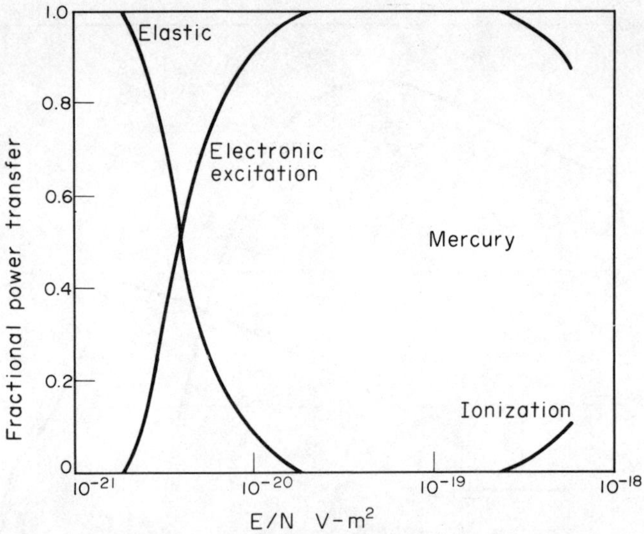


*Fig. 4.21. Fractional partition of total discharge power among elastic and inelastic processes in mercury as a function of E/N (from Rockwood, 1973)*

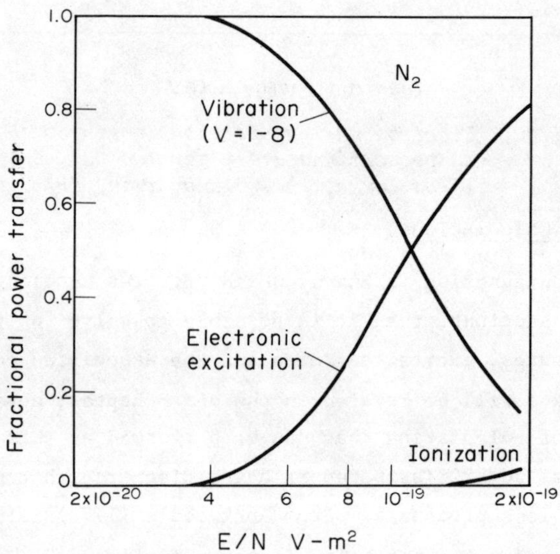


*Fig. 4.22. Fractional partition of total discharge power in $N_2$ as a function of E/N (from Nighan, 1970)*

approximate, to act as guides for experiment and more detailed theory. Two such approaches have been developed, one primarily for atomic systems and the other primarily for molecular systems.

### 4.8.1 Two and Three electron Group Models

L. Vriens (1973, 1974) and Vriens and Ligthart (1977) have developed a simple but effective model to deal primarily with rare gas or alkali vapor-rare gas mixtures (of particular importance in lighting applications). This model allows the calculation of collision rates, energy flow terms, excited state densities and radiation output under conditions where large deviations of the electron distribution function from Maxwellian are present without actually having to solve for the distribution functions.

In Vriens' approach, a distribution function is postulated in terms of very simple parameters, i.e. it is assumed that the distribution is made up of two, or possibly three, portions - each of which can be described by a Maxwellian distribution but each of which is defined by a different electron temperature. The previously described calculation of the distribution functions in helium shows that this is an excellent first approximation. In the simplest case, the two electron group model, all of the electrons below the first excitation level are called the "bulk" electrons and are characterized by a bulk electron temperature $T_b$. The major deviations from Maxwellian occur above the first excitation level where inelastic collisions rapidly deplete the tail of the distribution function. Therefore, a different temperature, $T_t$, is used to define the tail temperature and $T_t$ is usually significantly less than $T_b$ to indicate the degree of depletion.

Instead of using the Boltzmann equation to solve for the fraction of electrons in each energy interval (the continuum approach) it is now only necessary to specify $T_b$ and $T_t$. The approach is conceptually the same as in the solution of the Boltzmann equation. The energy flow terms are accounted for and the balance is found.

The energy conservation relation for all of the electrons in a gas discharge can be written as

$$P_E + P_{sel} + P_{rec} = P_{el} + P_{exc} + P_{ion} + P_{diff} \quad . \tag{4.75}$$

The left hand terms are energy sources: the electric field (E), superelastic collisions (sel) and electron-ion recombination (rec). The right hand side represents energy sinks: elastic collisions (el), excitation (exc), ionization (ion), and diffusion (diff). In most cases for low pressure discharges, recombination and diffusion terms can be neglected, so we write the energy balance equation for all electrons as

$$P_E + P_{sel} = P_{el} + P_{exc} + P_{ion} \quad . \tag{4.76}$$

Each term in equation (4.76) can be evaluated starting from the simple energy level diagram shown in Fig. 4.23

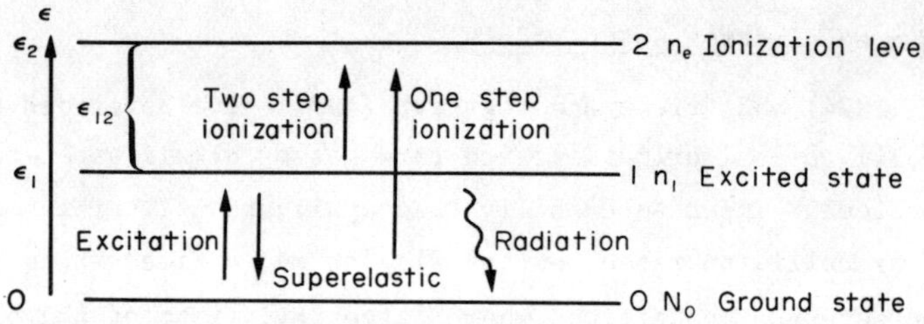


*Fig. 4.23. Energy-level diagram for the two-electron group model*

For purposes of discussion, we will now confine ourselves to the case $\varepsilon_1 < (\varepsilon_2 - \varepsilon_1)$ so two step ionizing collisions require electrons of energy $\varepsilon_e > \varepsilon_1$. This condition is not a necessary prerequisite for the use of this method and, indeed it is not satisfied for the rare gases, but it is a good approximation for the alkali vapors.

In order to evaluate equation (4.76), it is useful to notice that only the tail electrons have sufficient energy to excite or ionize so these processes represent energy losses for tail electrons only. Similarly, any superelastic collision will give an additional energy $\varepsilon_1$ to the electron, making it a tail electron regardless of where it originated, and so represents an energy gain for tail electrons only. Naturally, the energy gain due to the electric field and the energy loss due to elastic collisions affect both bulk and tail electrons.

The overall energy flow, from which we can derive various energy balance relationships, is shown in Fig. 4.24 as presented by Vriens (1974).

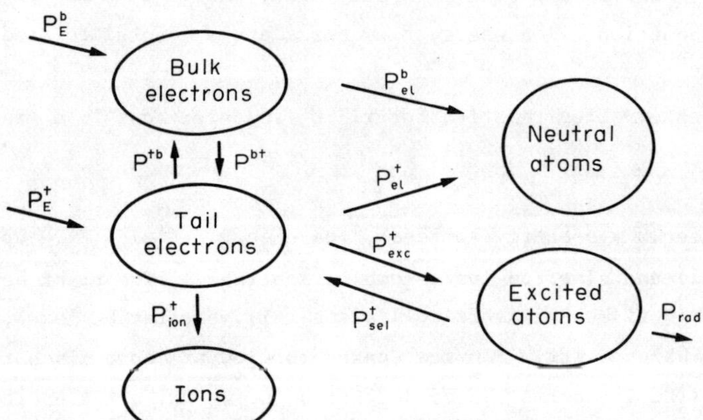


*Fig. 4.24. Energy flow in the two-electron group model (from Vriens, 1974).*

The superscripts "b" or "t" indicate whether or not the processes represent energy gain or loss terms for the bulk or tail electrons.

The only terms that do not appear in equation (4.76) are $P^{bt}$ and $P^{tb}$. These terms represent the processes by which energy is exchanged between bulk and tail electrons, such as Coulomb collisions.

From Fig. 4.24 it is possible to write separate energy balance equations for bulk and tail electrons. For the bulk electrons:

$$P_E{}^b + P^{tb} = P_{el}{}^b + P^{bt} \quad . \tag{4.77}$$

For the tail electrons:

$$P_E{}^t + P_{sel}{}^t + P^{bt} = P_{el}{}^t + P_{exc}{}^t + P_{ion}{}^t + P^{tb} \quad . \tag{4.78}$$

Actually, the sum of equations (4.77) and (4.78) equals equation (4.76). For the two electron group model, we now have two independent energy balance equations. This gives us one more equation and one more variable $T_t$ which is therefore not a free parameter. We can proceed to evaluate the terms in equations (4.77) and (4.78).

First, we note that the energy flow into the electrons comes from the applied electric field so

$$P_E = \sigma E^2 \tag{4.79}$$

where $\sigma$ is the electrical conductivity of the medium. The Langevin equation for the conductivity is less precise than would be desired so a more accurate equation for the conductivity (as will be derived in the next chapter) is used.

$$\sigma = \frac{-4\pi}{3}\frac{e^2}{m}\int_0^\infty \frac{v^3 \partial f_o/\partial v}{\nu_m(v) + j\omega}\, dv \ . \tag{4.80}$$

For a DC field and a Maxwellian $f_o$,

$$f_o = n_e\left(\frac{m}{2\pi k T_e}\right)^{3/2} e^{-mv^2/2kT_e} , \tag{4.81}$$

the conductivity becomes

$$\sigma = \frac{n_e e^2}{3kT_e}\left\langle \frac{v_e{}^2}{\nu_m} \right\rangle . \tag{4.82}$$

Therefore,

$$P_E = \frac{n_e e^2 E^2}{3kT_e} \left\langle \frac{v_e}{N_o \sigma_m(v_e)} \right\rangle \tag{4.83}$$

where $\sigma_m$ is the cross section for elastic collisions of electrons with the neutrals of density $N_o$ and < >, as before, represents an average over the distribution function. For the contribution to the bulk electrons, the integration implied by the bracket would only be over the energy range $o \leq \varepsilon_e \leq \varepsilon_1$. This is denoted by

$$< >^b \; .$$

Similarly, the contribution to the tail electrons would involve the integration over the energy range $\varepsilon_1 \leq \varepsilon_e < \infty$, and is signified by

$$< >^t \; .$$

Therefore,

$$P_E^{\,b} = \frac{n_e e^2 E^2}{3kT_b} \left\langle \frac{v_e}{N_o \sigma_m} \right\rangle^b \tag{4.84}$$

and

$$P_E^{\,t} = \frac{n_e e^2 E^2}{3kT_t} \left\langle \frac{v_e}{N_o \sigma_m} \right\rangle^t \; . \tag{4.85}$$

$P_{el}$ represents the energy loss in elastic collisions between electrons and neutral atoms on a per unit volume per second basis (watts/m$^3$) and is represented simply as

$$P_{el}^{\,b} = n_e N_o (2m/M) < v_e \sigma_m \varepsilon_e >^b \tag{4.86}$$

and

$$P_{el}^{\,t} = n_e N_o (2m/M) < v_e \sigma_m \varepsilon_e >^t \tag{4.87}$$

where $\delta\varepsilon_e = (2m/M)\varepsilon_e$ is the energy lost by an electron of energy $\varepsilon_e$ when it undergoes an elastic collision with a neutral of mass M. The energy loss in excitation is strictly a tail process

$$P_{exc}^{\,t} = n_e N_o \varepsilon_1 < v_e \sigma_{01} >^t \tag{4.88}$$

where $\sigma_{01}$ is the cross section for excitation to the first excited state at energy $\varepsilon_1$.

The ionization energy loss is also a tail process

$$P_{ion}^{\,t} = n_e N_o \varepsilon_2 < v_e \sigma_{02} >^t + n_e n_1 \varepsilon_{12} < v_e \sigma_{12} >^t \; . \tag{4.89}$$

The first term represents one step ionization (ground state to ion), but in the rare gases and particularly in the alkalis the second term which is two step ionization (excited state to ion) usually dominates.

The superelastic term represents an energy gain for the tail electrons but, in this case, it is primarily the bulk electrons that take part in this process. First of all, the bulk electrons are by far the most numerous in the discharge, secondly, the cross section for superelastic collisions, $\sigma_{10}$, is much higher for low energy electrons than for high energy electrons (see equation (4.23)). For this reason, the predominant integral is that over the bulk electron energies and

$$P_{sel}^{\ t} = n_e n_1 \varepsilon_1 < v_e \sigma_{10} >^b \quad . \qquad (4.90)$$

It should be pointed out here that most solutions of the Boltzmann equation do <u>not</u> take into account two-step ionization or superelastic collisions. In essence, the density of excited states is assumed to be negligible. For rare gases and alkali vapors in particular, these collisional processes play a very important role, in fact they usually have a dominant role, and must be taken into account if any realistic assessments of the distribution functions and collisional rates are to be made. The two step and superelastic terms are explicit in the Vriens approach, so more realistic answers can be expected from such calculations.

The $P^{tb}$ and $P^{bt}$ terms representing the energy flow between bulk and tail electrons consist of two important classes. The first is Coulomb, or electron-electron collisions by which electrons equilibrate their energy. In this process a slow (bulk) electron and a fast (tail) electron share their energy resulting in two faster bulk electrons. This process takes energy from the tail to the bulk. The inverse process is also possible, so that energy can be transferred from the bulk to the tail. The relative importance of this "Maxwellianizing" process can be determined by comparing the electron-electron collision frequency to the inelastic collision frequencies.

$$\nu_{ee} : \nu_{inel} \quad .$$

If $\nu_{ee} >> \nu_{inel}$, in all probability the distribution will be Maxwellian.

The Coulomb terms are somewhat difficult to derive, so they will simply be presented (from Vriens and Ligthart, 1977) in equation (4.91). It should be pointed out here that continued development of the two electron group model and especially the Coulomb terms is continuing. Recent improvements by F.A.S. Ligthart and R.A.J. Keijser will be published in the near future.

$$P_c^{\ tb} = n_e^{\ 2} e^4 (\varepsilon_1 + kT_t)(8\pi/m(kT_t)^3)^{1/2} \ \ell n\Lambda_1 \exp(-\varepsilon_1/kT_t),$$

$$P_c^{\ bt} = n_e^{\ 2} e^4 (\varepsilon_1 + kT_b)(8\pi/m(kT_b)^3)^{1/2} \ \ell n\Lambda_1 \exp(-\varepsilon_1/kT_b) \qquad (4.91)$$

where

$$\Lambda_1 = (\varepsilon_1{}^2/R - 3kT_b)(\varepsilon_1 - \tfrac{3}{2}kT_b)(\hbar\omega_p)^{-2}$$

for

$$\varepsilon_1 > 2(kT_b R)^{1/2}$$

and

$$\Lambda_1 = kT_b(\varepsilon_1 - \tfrac{3}{2}kT_b)(\hbar\omega_p)^{-2} \quad (4.92)$$

for

$$2kT_b < \varepsilon_1 \leq 2(kT_b R)^{1/2}$$

$R = 13.6$ eV and $\omega_p$ is the plasma frequency. The other class of flow terms involve the energy transferred in excitation or ionization as shown in Fig. 4.25.

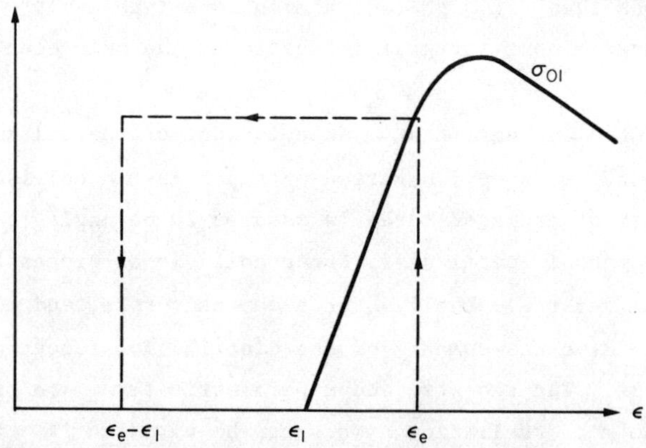


*Fig. 4.25. Energy flow in inelastic collisions*

An electron of energy $\varepsilon_e$ will lose an energy $\varepsilon_1$ in excitation and will be transferred to the bulk with an energy $\varepsilon_e - \varepsilon_1$. We therefore have an energy transfer term due to excitation.

$$P_{exc}^{tb} = n_e N_o < v_e \sigma_{01}(\varepsilon_e - \varepsilon_1) >^t \quad (4.93)$$

and a similar term due to ionization, which is assumed to be two step.

$$P_{ion}^{tb} = n_e n_1 < v_e \sigma_{12}(\varepsilon_e - \varepsilon_{12}) >^t \quad . \quad (4.94)$$

In superelastic collisions, an electron of energy $\varepsilon_e$ gains an additional energy $\varepsilon_1$ and becomes a tail electron. Since these are primarily bulk electrons, the bulk electrons lose an energy $\varepsilon_e$ in such a collision and this energy is transferred to the tail.

$$P_{sel}^{bt} = n_e n_1 < v_e \sigma_{10} \varepsilon_e >^b \quad . \quad (4.95)$$

The overall transfer terms can then be identified in terms of the previous five equations as

$$P^{bt} = P_c^{bt} + P_{sel}^{bt} \quad , \quad (4.96)$$

$$P^{tb} = P_c^{tb} + P_{exc}^{tb} + P_{ion}^{tb} \quad . \quad (4.97)$$

The identification of the terms in equations (4.77) and (4.78) is now complete.

Keep in mind that these equations cannot yet be solved since the evaluation of the superelastic terms and two step ionization terms require knowledge of the density of excited species $n_1$. In fact, $n_1$ is one of the important quantities that we want to obtain. (see Fig. 4.26).

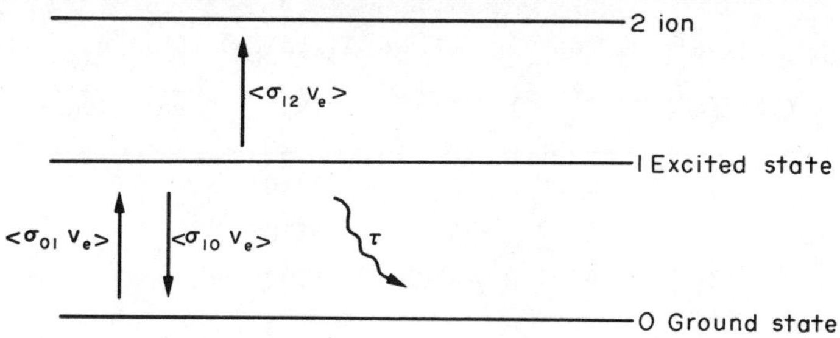


*Fig. 4.26. Important excitation and de-excitation processes for the excited state*

It is possible to write a continuity equation for $n_1$ provided that the processes shown represent all of the important excitation and de-excitation processes.

$$\frac{\partial n_1}{\partial t} = n_e N_o < \sigma_{01} v_e >^t - n_e n_1 < \sigma_{10} v_e >^b - n_e n_1 < \sigma_{12} v_e >^t - n_1/\tau \quad . \tag{4.98}$$

$\tau$ is the effective radiative lifetime for the excited states.

The evaluation of $\tau$ is not a straightforward process, as will be discussed in chapter nine, since the radiation from the excited states is usually heavily trapped within the medium and rather complicated calculations are necessary in order to determine $\tau$. From equation (4.98) in steady state

$$n_1 = \frac{n_e N_o < \sigma_{01} v_e >^t}{n_e < \sigma_{10} v_e >^b + n_e < \sigma_{12} v_e >^t + 1/\tau} \quad . \tag{4.99}$$

Equation (4.99) plus the energy-conservation equations (4.76) and (4.78) (or (4.77) and (4.78)) then give us three equations from which three unknowns can be obtained.

The parameters of the discharge as given in the preceding equations are E, $n_e$, $N_o$, $n_1$, $T_b$ and $T_t$. Generally then we can specify the neutral density $N_o$, and electron density $n_e$ (which is related to the current) and the bulk (or average) electron temperature $T_b$ and solve for E, $T_t$ and $n_1$ as well as all the power flow terms and rates. If we added one additional independent equation, the ionization balance, we could also solve for the bulk temperature, $T_b$.

The utility of the two electron group model has been clearly shown by Vriens and Ligthart (1977) when they compared the calculated and measured excited state densities in a Cs-Ar discharge. From the experimental data, E, $n_e$ and $T_b$ were specified so only the tail electron energy balance equation (4.78) and the continuity equation (4.99) were needed to solve for $T_t$ and $n_1$. Table 4-1 shows the comparison. The second last column shows the results of the two electron

*Table 4.1. Experimental and Calculated Parameters in a Cs-Ar Discharge. Tube radius = 1.5 cm, Ar pressure = 5 torr, wall temperature = 378 K (from Vriens and Ligthart, 1977).*

| I | τ | $N_o$(Cs) | $n_e$ | $T_b$ | $T_t$(calc.) | $n_1$ $(10^{17}/m^3)$ | | |
|---|---|---|---|---|---|---|---|---|
| (A) | $(10^{-6}$S) | $(10^{18}/m^3)$ | $(10^{17}/m^3)$ | (K) | (K) | experiment | two temp. | one temp. |
| .1 | 9.8 | 19 | 1.14 | 8000 | 5910 | 6 | 7.9 | 19 |
| .3 | 8.8 | 15 | 4.4 | 5720 | 5510 | 8 | 11.6 | 13 |
| .5 | 7.5 | 13 | 10 | 4820 | 4790 | 7 | 8.5 | 8.7 |

group model calculation. As can be seen, the calculated results are in quite good agreement with the experimental values. The last column (one temp.) shows the value of the excited state density that would be calculated if the electron distribution function was assumed to be a single temperature Maxwellian with an electron temperature equal to the bulk temperature. A significant deviation from the actual values is evident, particularly at lower electron densities. Clearly, then, rate calculations based on a single temperature Maxwellian are inadequate and give rise to serious error unless the electron density is sufficiently high so that Coulomb collisions make the tail temperature and the bulk temperature approximately equal.

Tabel 4-2 shows a similar comparison between calculated and experimental values for a discharge in argon. The difference between the calculations based on a two temperature model and a one temperature model is even more striking for this case. The two temperature model calculation is a vast improvement and is in much better agreement.

*Table 4.2. Experimental and Calculated Parameters in an Ar Discharge. Discharge tube radius = 1.2 cm. Ar pressure = 5 torr. τ = 1.12 × $10^{-5}$S (from Vriens and Ligthart, 1977).*

| I | $n_e$ | $T_b$ | $T_t$(calc.) | $n_1$ $(10^{16}/m^3)$ | | |
|---|---|---|---|---|---|---|
| (A) | $(10^{18}/m^3)$ | $(10^3$K) | $(10^3$K) | experiment | two temp. | one temp. |
| .025 | .27 | 25 | 13.6 | 3.6 | 9.6 | 1240 |
| .05 | .6 | 22 | 13.8 | 6.2 | 24 | 1180 |
| .1 | 1.2 | 18 | 13.2 | 6.5 | 30 | 541 |
| .2 | 2.2 | 16 | 12.9 | 7.0 | 41 | 355 |
| .4 | 4.6 | 13 | 11.7 | 4.4 | 27 | 96 |

Tables 4.1 and 4.2 show an additional important gas discharge phenomenon that can be accurately predicted by the two electron group model. As the current increases, the excited state density does not increase proportionately, but saturates and then decreases. This saturation phenomenon represents an important limitation to lighting sources and lasers and can be directly related to superelastic and two step ionizing collisions.

4.8.2 The "Upflux" Approach

W.P. Allis and H.A. Haus (1974) and W.P. Allis (1974) presented an analytic approach to distribution function calculations which is based upon the energy-flow processes in the electron gas, consisting of acceleration by the electric field, deceleration by elastic collisions and discontinuous energy loss processes such as excitation and ionization. This approach is closely related to the Boltzmann equation approach, and can be characterized by the sketch of electron energy flow in energy space given in Fig. 4.27. $G_E$ represents the energy gain, or "upflux" from the electric field. There is a more or less continuous energy loss due to elastic collisions, $\delta\nu_{elastic}$, which tends to flow in the opposite direction to $G_E$. If the electron has an energy $\Delta\varepsilon$ above the excitation threshold, there is a probability of an inelastic collision in which it will lose an energy $\varepsilon_x$ and discontinuously move down to an energy $\Delta\varepsilon$.

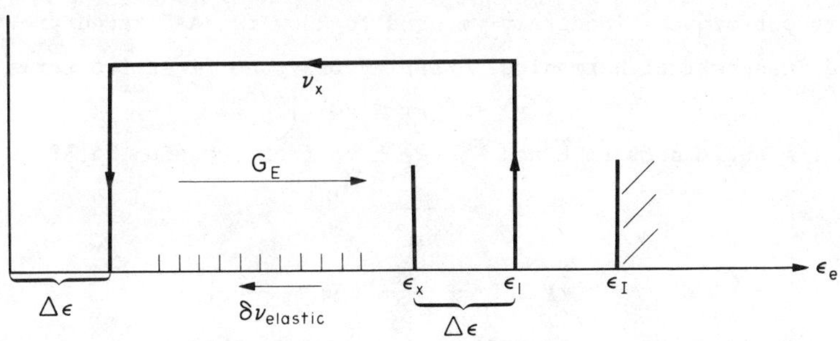


*Fig. 4.27. Energy flow in the upflux model*

The upflux analysis is based on the balance of these processes at steady state and leads to a formulation of the approximate shape of the distribution function. First of all, let us established the upflux term $G_E$. This term is derived from a flow vector in velocity space. In other words, the distribution function is more or less spherically symmetric in velocity space if the electric field is moderate, and the electrons tend to move outwards in velocity space (increasing energy) under the influence of the electric field.

We, therefore, define a flow vector in velocity space which is

$$\bar{g} = \bar{a}f \quad . \tag{4.100}$$

$\bar{a}$ is the acceleration of the electrons due to the electric field (or "velocity" in velocity space) and is

$$\bar{a} = -e\bar{E}/m \quad .$$

The upflux, $G_E$, is defined as the surface integral of this flow vector dotted onto the velocity over a sphere of radius v in velocity space.

$$G_E = \oint \bar{g}.\bar{v}\ v\ \sin\theta_1 d\theta_1 d\phi \quad . \tag{4.101}$$

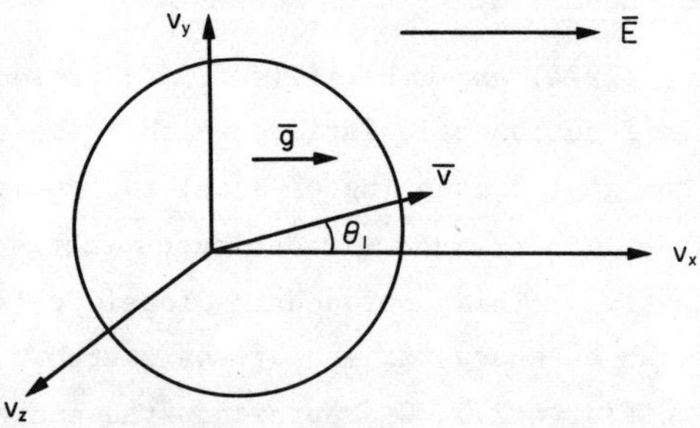


*Fig. 4.28. Relationship of flows and velocities in velocity space*

$G_E$ is, therefore, a measure of the total number of particles flowing out of the sphere per second and so is a measure of the energy gained, or $G_E$ is an energy flux.

In order to perform the integral, we need to know f. As stated previously, f is expressed in spherical harmonics. Keeping only the first two terms

$$f(\bar{v}) = f_o + f_1 \cos\theta_1 \tag{4.102}$$

where $\theta_1$ is the angle between $\bar{E}$ and $\bar{v}$. We have from equation (4.58)

$$f_1 = \frac{eE}{m\nu_m} \frac{\partial f_o}{\partial v} \quad .$$

Therefore,

$$f(\bar{v}) = f_o + \frac{eE}{m\nu_m} \cos\theta_1 \frac{\partial f_o}{\partial v} \quad . \tag{4.103}$$

To follow Allis' notation, we re-express in terms of energy

$$e\varepsilon = \frac{1}{2} mv^2$$

($\varepsilon$ is in volts).

Therefore,

$$f(\bar{v}) = f_o + \frac{Ev}{\nu_m} \cos\theta_1 \frac{\partial f_o}{\partial \varepsilon} \quad . \tag{4.104}$$

Introducing this term into the upflux equation

$$G_E = \int -\frac{e}{m} \bar{E}.\bar{v} \left[\frac{Ev}{\nu_m} \cos\theta_1 \frac{\partial f_o}{\partial \varepsilon}\right] v \sin\theta_1 d\theta_1 d\phi \quad . \tag{4.105}$$

Obviously, the $f_o$ term drops out since it is spherically symmetric and the resulting integral would go to zero. Now $\bar{E} \cdot \bar{v} = Ev \cos\theta_1$ so

$$G_E = -\frac{e}{m}\frac{E^2v^3}{\nu_m}\int_0^{2\pi}\int_0^{\pi}\frac{\partial f_o}{\partial \varepsilon}\cos^2\theta_1\sin\theta_1 d\theta_1 d\phi \tag{4.106}$$

and

$$G_E = -\frac{4\pi}{3}\frac{e}{m}\frac{E^2}{\nu_m}v^3\frac{\partial f_o(v)}{\partial \varepsilon} \quad . \tag{4.107}$$

Renormalizing f so that

$$dn = f_o(v)4\pi v^2 dv = f(\varepsilon)\sqrt{\varepsilon}\,d\varepsilon$$

we find that $G_E$ can be re-expressed as

$$G_E = -\frac{4}{3}\varepsilon_d\nu_m\varepsilon^{3/2}\frac{\partial f(\varepsilon)}{\partial \varepsilon} \tag{4.108}$$

where

$$\varepsilon_d = \frac{eE^2}{2m\nu_m^2} = \frac{mu^2}{2e} \quad .$$

Now u is the electron drift velocity, so $\varepsilon_d$ is the <u>drift energy</u> in volts. ($\varepsilon_d$ is a direct measure of E/N or E/P).

$G_E$ is the essential term since it represents the energy gained by the electrons per second as a function of the electron energy and the electric field. The next step is to equate that to the energy loss. Consider the simplest case first.

4.8.2.1 <u>Elastic losses.</u> In this case, the only losses are by continuous elastic collisions, where $\delta = 2m/M$ of the electron energy is lost per collision. Therefore, there is a negative energy flow term (see Fig. 4.27).

$$G_c = -\frac{2m}{M}\varepsilon\nu_m\sqrt{\varepsilon}\,f \quad . \tag{4.109}$$

The extra $\sqrt{\varepsilon}$ is to preserve the new normalization. If the gas has a temperature $T_g$, then energy can flow from the gas to the electrons if the electron energy is less than the gas energy. In that case

$$G_c = -\frac{2m}{M}\nu_m\varepsilon^{3/2}\left[f + T_g\frac{\partial f}{\partial \varepsilon}\right] \quad . \tag{4.110}$$

$T_g$ is expressed in volts. (As a general rule, energy losses enter as terms with an f and energy gains enter as a term with $\partial f/\partial\varepsilon$.) In the steady state

$$G = G_E + G_c = 0$$

or

$$G = -2\nu_m\varepsilon^{3/2}\left[\frac{2}{3}\varepsilon_d\frac{\partial f}{\partial \varepsilon} + \frac{m}{M}\left(f + T_g\frac{\partial f}{\partial \varepsilon}\right)\right] = 0 \ . \tag{4.111}$$

In this case

$$\frac{\partial f}{\partial \varepsilon} + \frac{f}{T_g + \frac{2}{3}\frac{M}{m}\varepsilon_d} = 0 \quad . \tag{4.112}$$

Therefore,

$$f(\varepsilon) = C\exp - \int_0^{\varepsilon}\frac{d\varepsilon}{T_g + \frac{2M}{3m}\varepsilon_d} \quad . \tag{4.113}$$

If

$$\varepsilon_d = \frac{eE^2}{2m\nu_m^2}$$

is constant, which implies that $\nu_m$ is independent of electron energy, f is Maxwellian with an electron temperature

$$T_e(\text{volts}) = T_g + \frac{2}{3}\frac{eE^2}{\delta m\nu_m^2} \quad . \tag{4.114}$$

The electron temperature (and hence the distribution function) depends on E/P or E/N.

Equation (4.114) is the same result as obtained in equation (4.63), however, the present result was obtained in a somewhat simpler manner.

The more important case by far is when the distribution function is influenced by inelastic collisions.

4.8.2.2 Inelastic losses. In this case, the net upflux $G = G_E + G_C$ is balanced by the loss of energy in excitations which bring the electron back in with an energy $\Delta\varepsilon$. If $n_e\nu_x$ represents the total excitation rate per unit volume, then the rate at which electrons increase their energy, G, must equal the rate at which they decrease their energy $\nu_x$ and must be constant between $\Delta\varepsilon$ and $\varepsilon_1$.

Therefore

$$G = -2\nu_m \varepsilon^{3/2}\left[\frac{2}{3}\varepsilon_d\frac{\partial f}{\partial\varepsilon} + \frac{m}{M}\left(f + T_g\frac{\partial f}{\partial\varepsilon}\right)\right] = \nu_x \quad . \tag{4.115}$$

This is a non-homogeneous equation which can be solved using the previous homogeneous solution (equation (4.113)).

For the case of particular interest here, the elastic collisions can be ignored (assume $m/M \to 0$ in equation (4.115)) and equation (4.115) can be solved to give:

$$f = \frac{\nu_x}{\nu_m}\frac{3}{2\varepsilon_d}\left(\frac{1}{\sqrt{\varepsilon}} - \frac{1}{\sqrt{\varepsilon_1}}\right) \quad \Delta\varepsilon < \varepsilon < \varepsilon_1 \tag{4.116}$$

for a constant $\nu_m$. From Fig. 4.27 it can be seen that all of the excitations are assumed to occur at $\varepsilon_1$ (if other inelastic processes are neglected) so the boundary condition $f(\varepsilon_1) = 0$ has been used to obtain equation (4.116). This implies an impenetrable barrier at $\varepsilon = \varepsilon_1$, an assumption which can be relaxed.

In order to plot the distribution function simply, we define

$$F = \frac{2\nu_m\varepsilon_d}{3}\sqrt{\varepsilon}\, f \tag{4.117}$$

which becomes

$$F = \nu_x\left(1 - \sqrt{\varepsilon/\varepsilon_1}\right) \tag{4.118}$$

as shown in Fig. 4.29.

Equation (4.118) is valid between $\varepsilon_1 - \varepsilon_x$ and $\varepsilon_1$. This corresponds to a "G" plot as shown in Fig. 4.30.

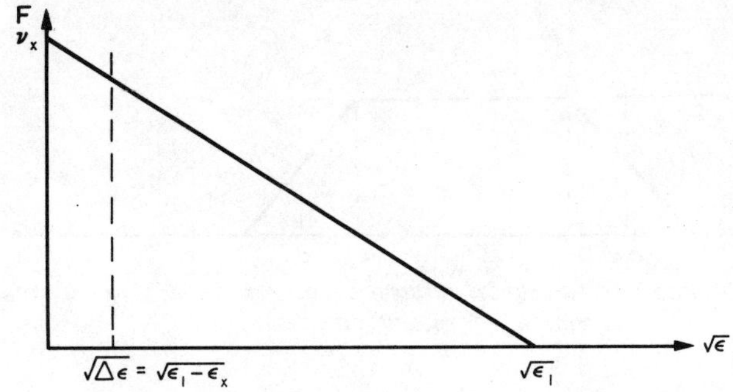


*Fig. 4.29. F vs $\sqrt{\varepsilon}$*

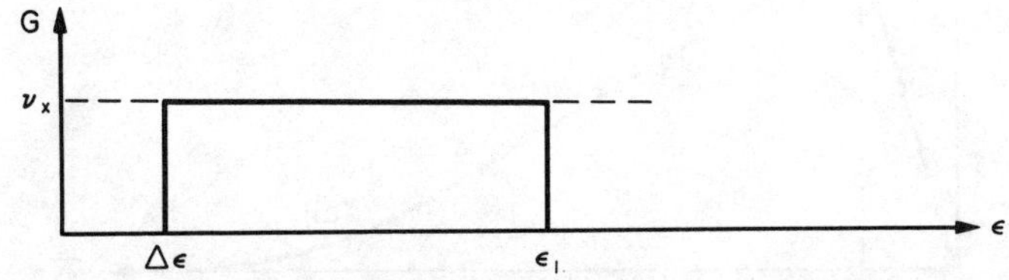


*Fig. 4.30. "G" plot corresponding to Fig. 4.29*

For the region below $\Delta\varepsilon = \varepsilon_1 - \varepsilon_x$, $G = 0$.

To the same approximation used here(i.e., $m/M \to 0$) if $G = 0$, $\partial f/\partial\varepsilon = 0$ so f is a constant and $F \sim \sqrt{\varepsilon}$. The composite distribution function would then be as shown in Fig. 4.31.

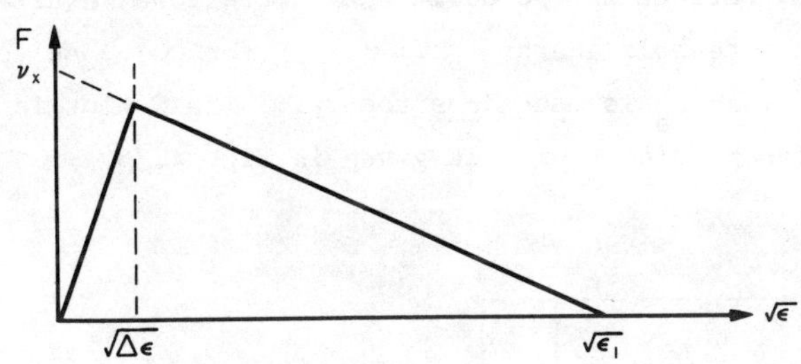


*Fig. 4.31. Simplified distribution function*

Figure 4.31 has been drawn under the assumption that all excitations occur at some value $\varepsilon_1$ above the excitation threshold $\varepsilon_x$. In general, however, the excitation rate will vary as a function of energy above $\varepsilon_x$.

If the excitation rate has an energy dependence as shown in Fig. 4.32 then the energy of the electrons translated back to the low energy range will show a similar dependence.

The resulting distributions are not quite rectilinear, but have some curvature as shown in Fig. 4.33.

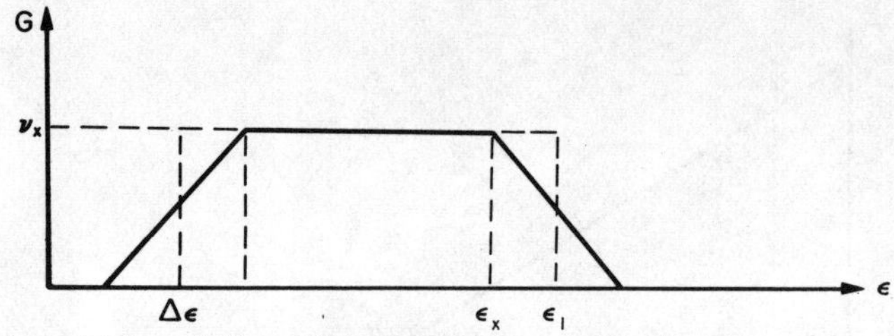


*Fig. 4.32. G function assuming excitation occurs over a range of energies*

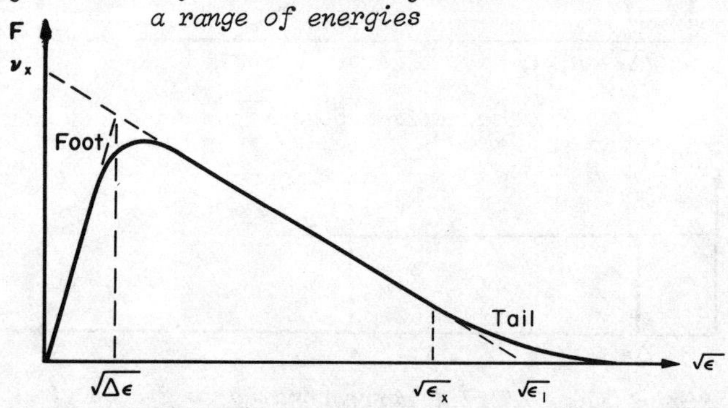


*Fig. 4.33. Distribution function corresponding to Fig. 4.32*

Now consider the applicability of this approach to molecular gases, in particular CO or $N_2$ - $CO_2$ as treated by Allis and Haus (1974). As before, we consider electronic excitation with a frequency $\nu_x$, a threshold at $\varepsilon_1$ and an energy loss $\varepsilon_x$. An important difference is the additional vibrational excitation with a frequency $\nu_a$ with a threshold energy $\varepsilon_2$ ($\sim$ 1.8 eV for CO). As you will remember, however, the energy loss $\varepsilon_a$ is <u>much</u> <u>less</u> than $\varepsilon_2$. The G plot is shown in Fig. 4.34. The corresponding F plot is given in Fig. 4.35.

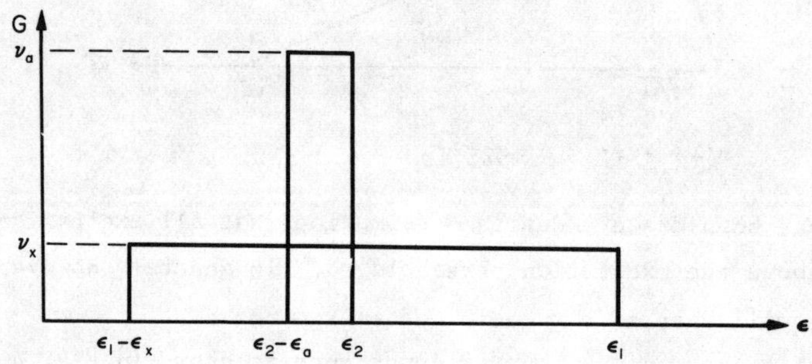


*Fig. 4.34. G plot for CO*

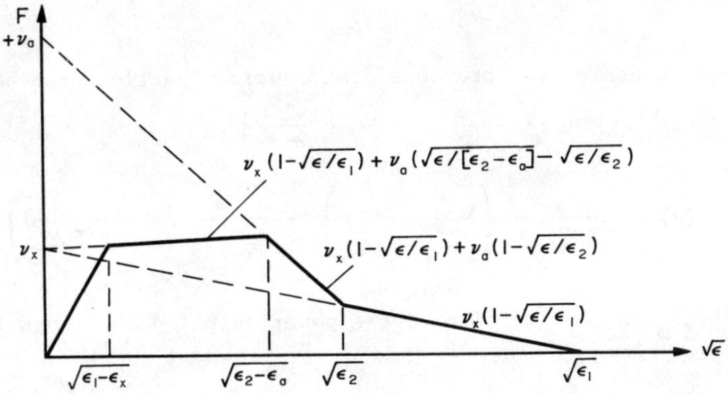


*Fig. 4.35. F plot for CO based on Fig. 4.34*

Since

$$f = \frac{3}{2\nu_m \varepsilon_d} \frac{F}{\sqrt{\varepsilon}} \tag{4.119}$$

then

$$f = \frac{\nu_x}{\nu_m}\frac{3}{2\varepsilon_d}\left(\frac{1}{\sqrt{\varepsilon}} - \frac{1}{\sqrt{\varepsilon_1}}\right) \qquad \varepsilon_2 < \varepsilon < \varepsilon_1 , \tag{4.120}$$

$$f = \frac{\nu_x}{\nu_m}\frac{3}{2\varepsilon_d}\left[\frac{1}{\sqrt{\varepsilon}} - \frac{1}{\sqrt{\varepsilon_1}} + \frac{\nu_a}{\nu_x}\left(\frac{1}{\sqrt{\varepsilon}} - \frac{1}{\sqrt{\varepsilon_2}}\right)\right] \qquad \varepsilon_2 - \varepsilon_a < \varepsilon < \varepsilon_2 , \tag{4.121}$$

$$f = \frac{\nu_x}{\nu_m}\frac{3}{2\varepsilon_d}\left[\frac{1}{\sqrt{\varepsilon}} - \frac{1}{\sqrt{\varepsilon_1}} + \frac{\nu_a}{\nu_x}\left(\frac{1}{\sqrt{\varepsilon_2 - \varepsilon_a}} - \frac{1}{\sqrt{\varepsilon_2}}\right)\right] \qquad \varepsilon_1 - \varepsilon_x < \varepsilon < \varepsilon_2 - \varepsilon_a . \tag{4.122}$$

Now, since $\varepsilon_a \ll \varepsilon_2$,

$$\frac{1}{\sqrt{\varepsilon_2 - \varepsilon_a}} - \frac{1}{\sqrt{\varepsilon_2}} \simeq \frac{\varepsilon_a}{2\varepsilon_2^{3/2}} .$$

Therefore,

$$f = \frac{\nu_x}{\nu_m}\frac{3}{2\varepsilon_d}\left[\frac{1}{\sqrt{\varepsilon}} - \frac{1}{\sqrt{\varepsilon_1}} + \frac{\nu_a}{\nu_x}\frac{\varepsilon_a}{2\varepsilon_2^{3/2}}\right] \qquad \varepsilon_1 - \varepsilon_x < \varepsilon < \varepsilon_2 - \varepsilon_a \tag{4.123}$$

and

$$f = f(\varepsilon) \qquad \varepsilon < \varepsilon_1 - \varepsilon_x . \tag{4.124}$$

These equations can be further simplified by invoking the power balance.

The power absorbed from the field (ignoring elastic losses) is

$$W_E = \int_0^\infty G_E d\varepsilon = -\frac{4}{3}\varepsilon_d \nu_m \int_0^\infty \varepsilon^{3/2} \frac{\partial f(\varepsilon)}{\partial \varepsilon} d\varepsilon = 2\nu_m \varepsilon_d . \tag{4.125}$$

This must balance the total energy lost in inelastic collisions

$$W_x = \nu_x \varepsilon_x + \nu_a \varepsilon_a . \tag{4.126}$$

Therefore,

$$2\nu_m\varepsilon_d = \nu_x\varepsilon_x + \nu_a\varepsilon_a \quad . \tag{4.127}$$

Using this relationship, the previous distribution functions can be expressed as

$$f(\varepsilon) = \frac{\varepsilon_x\nu_x}{2\varepsilon_d\nu_m}\frac{3}{\varepsilon_x}\left(\frac{1}{\sqrt{\varepsilon}} - \frac{1}{\sqrt{\varepsilon_1}}\right) \qquad \varepsilon_2 < \varepsilon < \varepsilon_1 \tag{4.128}$$

and

$$f(\varepsilon) = \frac{\varepsilon_x\nu_x}{2\varepsilon_d\nu_m}\frac{3}{\varepsilon_x}\left(\frac{1}{\sqrt{\varepsilon}} - \frac{1}{\sqrt{\varepsilon_1}}\right) + \frac{3}{2\varepsilon_2^{3/2}}\left(1 - \frac{\varepsilon_x\nu_x}{2\varepsilon_d\nu_m}\right) \quad . \tag{4.129}$$

$$\varepsilon_1 - \varepsilon_x < \varepsilon < \varepsilon_2 - \varepsilon_a$$

The term $\varepsilon_x\nu_x/2\varepsilon_d\nu_m$ is the fraction of the power supplied that goes into electronic excitation.

A plot of the distribution function expressed in equations (4.128) and (4.129) is given in Fig. 4.36 for various values of the excitation parameter $\mathrm{EXC} = \dfrac{\nu_x\varepsilon_x}{2\varepsilon_d\nu_m}$ .

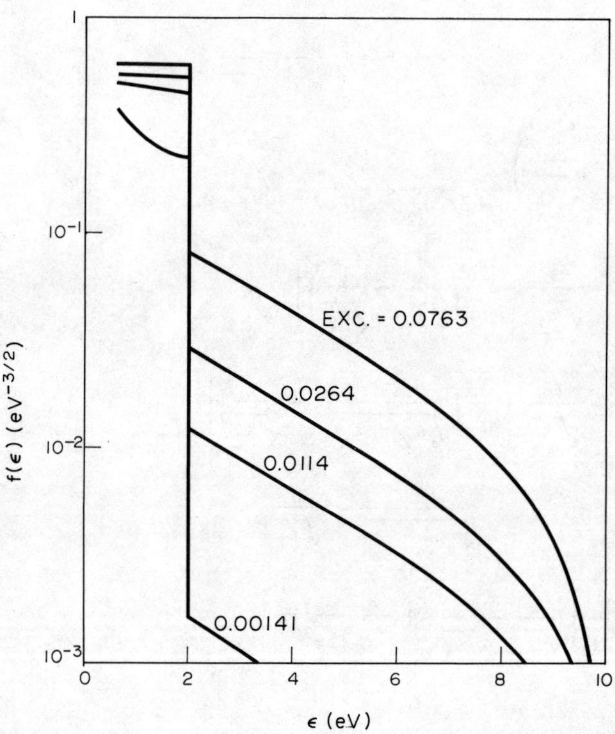


*Fig. 4.36. Distribution functions in CO as obtained from the upflux model (from Allis and Haus, 1974)*

Referring to the Boltzmann equation solutions for CO as performed by Nighan (1970), it can be seen that this very simple approach to distribution function calculations gives very good results. Furthermore, a comparison of the rates and excitation efficiencies shows quite good agreement of the "upflux" calculation with the computer solutions. The reader is referred to the Allis and Haus (1974) paper for a much more extensive discussion as well as the application of this technique to $CO_2$.

## 4.9 REFERENCES

Abdelnabi, I. and Massey, H. S. W. (1953), "Inelastic Collisions of Electrons in Helium and Townsend's Ionization Coefficient," Proc. Phys. Soc. A, 66, 288-296.

Allis, W. P. (1956), "Motions of Ions and Electrons," MIT Research Laboratory of Electronics, Technical Report 299 (1956); also Handbuch der Physik, Vol. XXI.

Allis, W. P. (1974), "Calcul Simple des Fonctions de Distribution D'électrons Faisant des Collisions Inélastiques," Note Technique No. 234, ONERA, Chatillon, FRANCE.

Allis, W. P. and Haus, H. A. (1974), "Electron Distributions in Gas Lasers," J. Appl. Phys. 45, 781-791.

Carleton, N. P. and Megill, L. R. (1962), "Electron Energy Distributions in Slightly Ionized Air under the Influence of Electric and Magnetic Fields," Phys. Rev. 126, 2089-2099.

Chapman, S. and Cowling, T. G. (1952), The Mathematical Theory of Non-Uniform Gases, Cambridge.

Engelhardt, A. G. and Phelps, A. V. (1963), "Elastic and Inelastic Collision Cross Sections in Hydrogen and Deuterium from Transport Coefficients," Phys. Rev. 131, 2115-2128.

Ginzburg, V. L. and Gurevich, A. V. (1960), "Nonlinear Phenomena in a Plasma Located in an Alternating Electromagnetic Field," Sov. Phys. Uspekhi 3, 115-146.

Hemenway, C. L., Henry, R. W. and Caulton, M. (1962), Physical Electronics, John Wiley & Sons, New York.

Holstein, T. (1946), "Energy Distributions of Electrons in High Frequency Gas Discharges," Phys. Rev. 70, 367-384.

Jahnke, E. and Emde, F. (1945), Tables of Functions with Formulae and Curves, Dover Publications, New York.

Kennard, E. H. (1938), Kinetic Theory of Gases, McGraw-Hill, New York.

Lowke, J. J., Phelps, A. V. and Irwin, B. W. (1973), "Predicted Electron Transport Coefficients and Operating Characteristics of $CO_2$-$N_2$-He Laser Mixtures," J. Appl. Phys. 44, 4664-4671.

Megill, L. R. and Cahn, J. (1964), "The Calculation of Electron Distribution Functions in the Ionosphere," J. Geophys. Res. 69, 5041-5048.

Nighan, W. L. (1970), "Electron Energy Distributions and Collision Rates in Electrically Excited $N_2$, CO and $CO_2$," Phys. Rev. A2, 1989-2000.

Reder, F. H. and Brown, S. C. (1954), "Energy Distribution Function of Electrons in Pure Helium," Phys. Rev. 95, 885-889.

Rockwood, S. D. (1973), "Elastic and Inelastic Cross Sections for Electron-Hg Scattering from Hg Transport Data," Phys. Rev. A8, 2348-2358.

Rockwood, S. D. (1974), "Effect of Electron-electron and Electron-ion Collisions in Hg, $CO_2/N_2/He$ and $CO/N_2$ Discharges," J. Appl. Phys. 45, 5229-5234.

Shkarofsky, I. P., Johnston, T. W. and Bachynski, M. P. (1966), The Particle Kinetics of Plasma, Addison-Wesley, Reading, Mass.

Smit, J. A. (1936), "Berechnung der Geschwindigkeitsverteilung der elektronen bei Gasentladungen in Helium" Physica 3, 543-560.

Vriens, L. (1973), "Energy Balance in Low-Pressure Gas Discharges," J. Appl. Phys. 44, 3980-3989.

Vriens, L. (1974), "Two and Three Electron Group Models for Low-Pressure Gas Discharges," J. Appl. Phys. 45, 1191-1195.

Vriens, L. and Ligthart, F. A. S. (1977), "Energy Balance and Coulomb Relaxation in Low-Pressure Gas Discharges," Philips Res. Repts. 32, 1-7.

# Chapter 5

# TRANSPORT COEFFICIENTS

The determination of the isotropic part of the distribution function as presented in the previous chapter is crucial to the determination of the important collisional excitation, de-excitation, ionization and elastic loss rates. For transport processes, however, the isotropic part of the distribution function by definition contributes no net or "drift" motion of the electrons and so contributes only secondarily to the transport properties. In this chapter, we concentrate on the anisotropic part of the distribution function from which we can derive the appropriate net motion or drift of electrons and hence the transport coefficients.

A good place to start is to use the Boltzmann equation to solve for the electrical conductivity of an ionized medium under the influence of an externally applied AC electric field.

## 5.1 ELECTRICAL CONDUCTIVITY

To determine the electrical conductivity, we will use only the first two terms in the expansion of the distribution function.

$$f = f_o + f_1 \cos\theta_1 \tag{5.1}$$

where $\theta_1$ is the angle between $\bar{v}$ and the electric field. This two term expansion is usually sufficiently accurate for cases when elastic collisions dominate. When inelastic collisions play an important role, the two term expansion may break down and the inclusion of the $f_2$ and $f_3$ terms may be necessary if accurate calculations of transport coefficients are desired. A discussion of the conditions of validity has been given by Ginzburg and Gurevich (1960). We will assume that $\bar{E} \sim e^{j\omega t}$ and solve for $f_1$ from equation (4.57).

$$\frac{\partial f_1}{\partial t} + v\frac{\partial f_o}{\partial z} - \frac{eE_z}{m}\frac{\partial f_o}{\partial v} = -\nu_m f_1 \quad . \tag{5.2}$$

In this equation we will assume that $f_o$ is isotropic and uniform, but not necessarily Maxwellian. For the present we shall also assume that $f_1$ is uniform in space ($\lambda >>$ dimensions), an assumption that is valid in all but unusual cases. Now, since $f_1$ arises in response to the externally applied electric field, we can also take $f_1 \sim e^{j\omega t}$.

Solving for $f_1$, we then obtain

$$(\nu_m + j\omega)f_1 = \frac{eE_z}{m}\frac{\partial f_o}{\partial v} \tag{5.3}$$

or

$$f_1 = \frac{eE_z}{m}\frac{\partial f_o/\partial v}{(\nu_m + j\omega)} \ . \tag{5.4}$$

Now we can solve for the current density due to the external field, and the conductivity.

$$J_z = -n_e e < v_z > = -e \int v_z f(\bar{v}) d^3v$$

$$= -e \int_o^\infty \int_o^\pi (v \cos\theta_1)[f_o + f_1\cos\theta_1]2\pi v^2 \sin\theta_1 d\theta_1 dv \ . \tag{5.5}$$

$f_o$ is isotropic and so does not contribute, therefore

$$J_z = \frac{-e^2E_z}{m} \cdot 2\pi \int_o^\infty \int_o^\pi \frac{\partial f_o/\partial v}{\nu_m + j\omega} v^3 \cos^2\theta_1 \sin\theta_1 d\theta_1 dv \tag{5.6}$$

and

$$J_z = \frac{-4\pi}{3}\frac{e^2}{m} E_z \int_o^\infty \frac{v^3 \partial f_o/\partial v}{\nu_m + j\omega} dv \ . \tag{5.7}$$

(Keep in mind that $f_o$ is the distribution function for velocity.) The <u>conductivity</u>, therefore, is

$$\sigma = \frac{-4\pi}{3}\frac{e^2}{m} \int_o^\infty \frac{v^3 \partial f_o/\partial v}{\nu_m(v) + j\omega} dv \ . \tag{5.8}$$

If $\nu_m$ is independent of velocity,

$$\sigma = \frac{ne^2}{m} \frac{1}{\nu_m + j\omega} \tag{5.9}$$

regardless of the shape of the distribution function. This is exactly the same answer as was found by using the Langevin equation.

If the distribution function is Maxwellian,

$$f_o = n_e \left( \frac{m}{2\pi kT_e} \right)^{3/2} e^{-mv^2/2kT_e} , \tag{5.10}$$

then the DC conductivity becomes

$$\sigma = \frac{n_e e^2}{3kT_e} \left\langle \frac{v_e^2}{\nu_m} \right\rangle \tag{5.11}$$

as was given in equation (4.82).

## 5.2 MOBILITY

Now, of course, there is a direct conversion possible between conductivity and mobility.

$$\bar{J} = \sigma\bar{E} = -ne\bar{u} = ne\mu_e\bar{E} . \tag{5.12}$$

Therefore, the mobility $\mu$ can be directly determined from the conductivity

$$\mu_e = \frac{\sigma}{ne} \tag{5.13}$$

or

$$\mu_e = \frac{-4\pi}{3} \frac{e}{mn} \int_o^\infty \frac{v^3 \partial f_o/\partial v}{\nu_m + j\omega} dv . \tag{5.14}$$

Integrating by parts, we can also write

$$\mu_e = \frac{e}{mn} \int f_o \frac{d}{dv} \left[ \frac{4\pi v^3}{3(\nu_m + j\omega)} \right] dv . \tag{5.15}$$

If $\omega = 0$ (DC) and $\nu$ is independent of velocity we can find

$$\mu_e = \frac{e}{m\nu_m} , \tag{5.16}$$

the same result as obtained in equation (2.6).

Equation (5.16) implies a constant mobility, or a drift velocity that varies linearly with E/N. In general, of course, this is not quite true and equation (5.14) should be solved to determine the drift velocity. An example of such a calculation is shown in Fig. 5.1 for $CO_2 : N_2 : H_e$ mixtures as obtained by Lowke, Phelps and Irwin (1973). A considerable amount of similar data is contained in the reference by Massey and Burhop (Vol. 1, 1969).

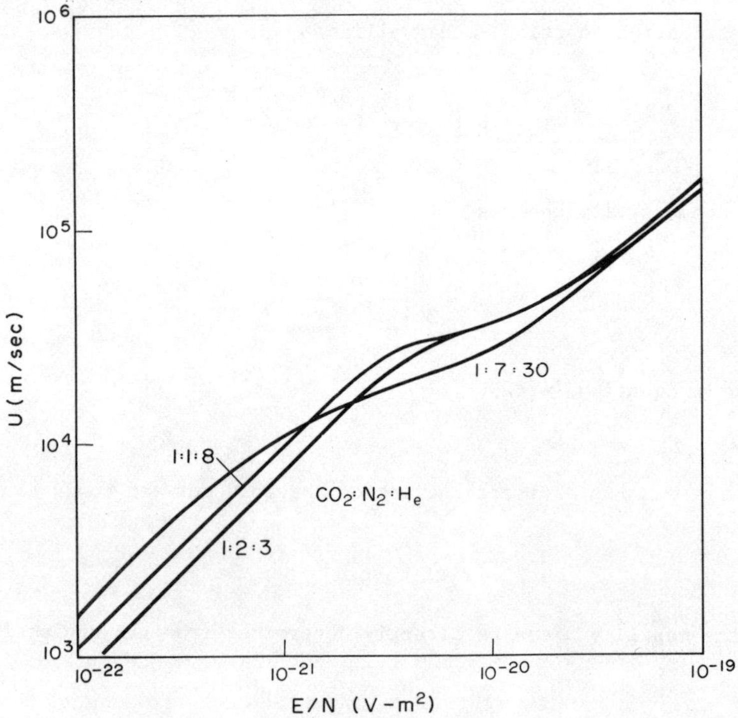


*Fig. 5.1. Calculated drift velocities of electrons for various gas mixtures of $CO_2$, $N_2$ and $H_e$ (from Lowke, Phelps and Irwin, 1973)*

Now, the same equations can be applied to calculating ion mobilities. The ion-molecule interaction is usually simpler than that for electrons since at low energies the polarization interaction potential applies (see Section 3.2.5) and the collision frequency and mobility are independent of electron energy. Correspondingly, at higher energies the ion-molecule interaction becomes a hard sphere interaction and the collision frequency increases with ion velocity, so the mobility decreases. Frost (1957) expressed the ion mobility in the following manner:

$$\mu_+ = \mu_o[1 + a(E/P)]^{-1/2} \tag{5.17}$$

where $\mu_o$ and a are constants depending on the ion and gas through which it moves. The variation of mobility with E/P is shown in Fig. 5.2.

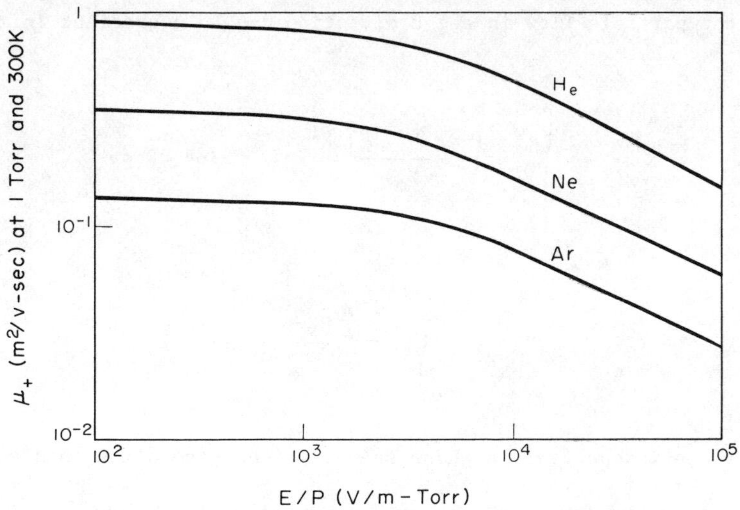


*Fig. 5.2. Variation of ion mobility for atomic ions in their parent gases as a function of E/P at 1 torr, 300 K, (from Frost, 1957)*

## 5.3 DIFFUSION

If a plasma is nonuniform in space, then there will be a diffusion of particles from regions of high density to areas of low density. This process can be characterized by a diffusion coefficient and we can determine the diffusion coefficient from the Boltzmann equation expansion.

If we assume a steady state solution and no externally applied field, from Equation (5.2)

$$v \frac{\partial f_o}{\partial z} = -\nu_m f_1 \tag{5.18}$$

and

$$f_1 = \frac{-v\partial f_o/\partial z}{\nu_m} \quad . \tag{5.19}$$

The particle flux is given by

$$\Gamma_z = n < v_z > = \int v_z f d^3v = \int v_z f_1 \cos\theta_1 d^3v$$

or

$$\Gamma_z = - \int d^3v \; v_z \frac{v\partial f_o/\partial z}{\nu_m} \cos\theta_1 \quad . \tag{5.20}$$

Considering the particle flow in the $\hat{z}$ direction due to a gradient in the $\hat{z}$ direction:

$$\Gamma_z = - \int_0^\infty \int_0^\pi \frac{v^2 \cos^2\theta_1}{\nu_m} \frac{\partial f_o}{\partial z} 2\pi v^2 \sin\theta_1 d\theta_1 dv$$

or

$$\Gamma_z = - \int_0^\infty 4\pi v^2 \frac{v^2}{3\nu_m} \frac{\partial f_o}{\partial z} dv \quad . \tag{5.21}$$

Since $\bar{r}$ and $\bar{v}$ are independent in phase space, we can generalize this equation to

$$\bar{\Gamma} = - \nabla \int_0^\infty 4\pi v^2 \frac{v^2}{3\nu_m} f_o dv = - \nabla(Dn) \tag{5.22}$$

where D is the diffusion coefficient. In integral form

$$D = \frac{1}{n} \int_0^\infty \frac{v^2}{3\nu_m} f_o 4\pi v^2 dv \ . \tag{5.23}$$

We can also express D as

$$D = < \frac{v^2}{3\nu} >. \tag{5.24}$$

If $\nu$ is independent of velocity

$$D = \frac{<v^2>}{3\nu} \tag{5.25}$$

and if $f_o$ is Maxwellian,

$$D = \frac{kT}{m\nu} \ . \tag{5.26}$$

The most general equation, equation (5.23), was derived assuming the absence of an electric field. If an electric field is present, then a distinction can be made between the diffusion coefficient transverse to the electric field, $D_T$, which is the same as the diffusion coefficient just derived, and the diffusion coefficient parallel to the electric field $D_L$. The distinction between these two cases has been carefully pointed out by Parker and Lowke (1969) and Lowke and Parker (1969) and is important to note in reducing data from time of flight experiments (analogous to the Haynes-Shockley experiment in semiconductors).

## 5.4 THE EINSTEIN RELATION AND CHARACTERISTIC ENERGY

From equations (5.14) and (5.23) it is possible to show that if $f_o$ is Maxwellian, then

$$\frac{D}{\mu} = \frac{kTe}{e} , \qquad (5.27)$$

the Einstein relation. This relationship is valid regardless of the velocity dependence of $\nu_m$.

Even if the distribution function is not Maxwellian, the Einstein relationship is useful in defining a "characteristic" energy

$$\varepsilon_k = \frac{eD}{\mu} \qquad (5.28)$$

which can be experimentally measured and used to define an equivalent Maxwellian temperature (or distribution). In fact, since two diffusion coefficients, $D_T$ and $D_L$ have been defined, both $D_T/\mu$ and $D_L/\mu$ can be measured. Two examples are given in Fig. 5.3 and 5.4 for $H_2$ (Engelhardt and Phelps, 1963) and $CO_2$:$N_2$:$H_e$ (Lowke, Phelps and Irwin, 1973).

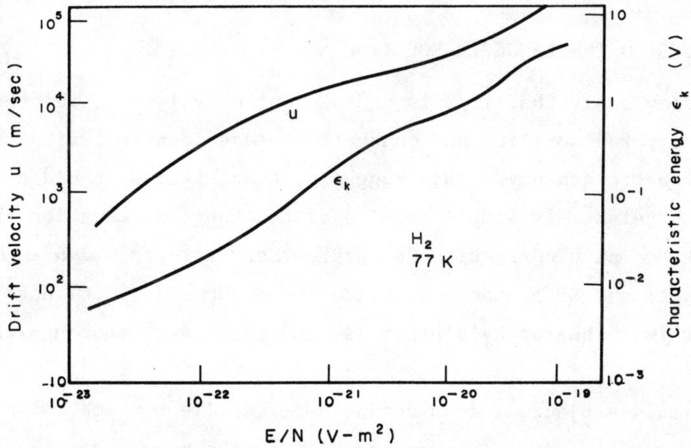


*Fig. 5.3. Drift velocity, u, and characteristic energy $\varepsilon_k$ for $H_2$ at 77 K as a function of E/N (from Engelhardt and Phelps, 1963)*

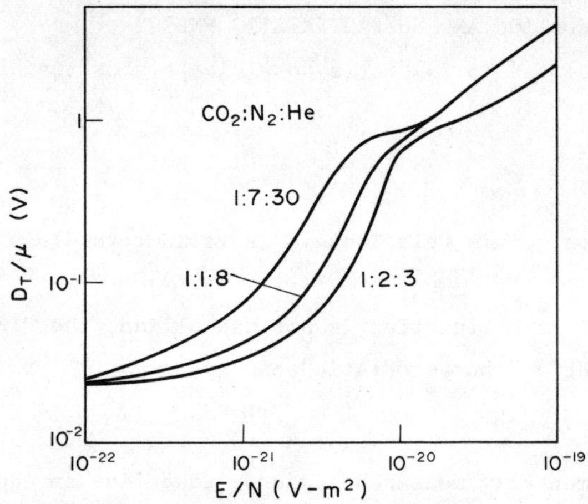


*Fig. 5.4. Calculated values of* $D_T/\mu$ *for various gas mixtures of* $CO_2$, $N_2$ *and* $H_e$ *(from Lowke, Phelps and Irwin, 1973)*

Extensive data on $D/\mu$ for rare and molecular gases have been presented by Lowke and Parker (1969) and are also contained in the reference work by Massey and Burhop (1969).

## 5.5 CORRECTIONS TO THE LANGEVIN EQUATION

We have seen previously that if $\nu$ is independent of velocity, the results of the more exact theory for mobility and conductivity are identical with the results of the Langevin equation theory. This suggests, then, that we could continue to make use of the relatively simple results of the Langevin equation if we use the exact theory to compute correction factors. Many authors, such as Molmud (1959), Shkarofsky (1961) and Shkarofsky, Johnston and Bachynski (1966) have successfully accomplished this. Shkarofsky's paper is, indeed, a very thorough treatment of this subject.

To approach this subject, let us define an effective $\nu$ which, when inserted in the Langevin equation, gives the same results as the more exact theory.

$$\sigma = \frac{ne^2}{m}\frac{1}{\nu_{eff} + j\omega} = \frac{-4\pi}{3}\frac{e^2}{m}\int_0^\infty \frac{v^3\partial f_o/\partial v}{\nu_m + j\omega}\,dv \quad . \tag{5.29}$$

Separating real and imaginary parts, we can write

$$\frac{ne^2}{m}\frac{\nu_{eff}}{\nu_{eff}^2 + \omega^2} = \frac{-4\pi}{3}\frac{e^2}{m}\int_0^\infty v^3\partial f_o/\partial v\,\frac{\nu_m}{\nu_m^2 + \omega^2}\,dv \tag{5.30}$$

and

$$\frac{ne^2}{m}\frac{1}{\nu_{eff}^2+\omega^2}=\frac{-4\pi}{3}\frac{e^2}{m}\int_o^\infty v^3\partial f_o/\partial v\,\frac{1}{\nu_m^2+\omega^2}\,dv\ . \tag{5.31}$$

These equations are quite general and not very useful. If we assume a specific form for $f_o$ we can simplify much further. In order to do this, we will assume $f_o$ is Maxwellian. This is about the best assumption you can make and we discussed previously the conditions of validity (Section 4.5). The previous equations then become

$$\frac{ne^2}{m}\frac{\nu_{eff}}{\nu_{eff}^2+\omega^2}=\frac{e^2}{m}\left(\frac{m}{3kT}\right)\int_o^\infty 4\pi v^2\left(\frac{v^2\nu_m}{\nu_m^2+\omega^2}\right)f_o dv. \tag{5.32}$$

and

$$\frac{ne^2}{m}\frac{1}{\nu_{eff}^2+\omega^2}=\frac{e^2}{m}\left(\frac{m}{3kT}\right)\int_o^\infty 4\pi v^2\left(\frac{v^2}{\nu_m^2+\omega^2}\right)f_o dv\ . \tag{5.33}$$

Therefore

$$\frac{\nu_{eff}}{\nu_{eff}^2+\omega^2}=\frac{\left\langle v^2\frac{\nu_m}{\nu_m^2+\omega^2}\right\rangle}{\langle v^2\rangle} \tag{5.34}$$

and

$$\frac{1}{\nu_{eff}^2+\omega^2}=\frac{\left\langle\frac{v^2}{\nu_m^2+\omega^2}\right\rangle}{\langle v^2\rangle} \tag{5.35}$$

where < > indicates an average over the Maxwellian distribution.

In this case

$$\nu_{eff}=\frac{\left\langle v^2\frac{\nu_m}{\nu_m^2+\omega^2}\right\rangle}{\left\langle\frac{v^2}{\nu_m^2+\omega^2}\right\rangle}\ . \tag{5.36}$$

Unfortunately, $\nu_{eff}$ in general depends on $\omega$ as well as the electron velocities. However, in the limiting cases, relatively simple equations can be obtained. When $\omega << \nu$ (DC limit)

$$\nu_{eff} = \frac{\langle v^2/\nu_m \rangle}{\langle v^2/\nu_m^2 \rangle} \quad . \tag{5.37}$$

When $\omega >> \nu$ (high frequency limit)

$$\nu_{eff} = \frac{\langle v^2 \nu_m \rangle}{\langle v^2 \rangle} \quad . \tag{5.38}$$

In both cases, if $\nu(v)$ is known then, for a given temperature ($f_o$), a unique value of $\nu_{eff}$ can be computed.

In terms of electromagnetic wave propagation through a plasma the high frequency limit is appropriate and we can examine this further as was demonstrated by Heald and Wharton (1965).

$$\nu_{eff} = \left(\frac{m}{3kT}\right)\frac{1}{n}\int_o^\infty 4\pi v^2 (v^2 \nu_m) f_o dv \quad . \tag{5.39}$$

If

$$f_o = n\left(\frac{m}{2\pi kT}\right)^{3/2} e^{-mv^2/2kT}$$

and if we take $\nu(v)$ to obey a simple power law

$$\nu(v) = Cv^\ell \quad ,$$

then

$$\nu_{eff} = \frac{8}{3\sqrt{\pi}}\left(\frac{m}{2kT}\right)^{5/2} C \int_o^\infty v^{\ell+4} e^{-mv^2/2kT} dv \quad . \tag{5.40}$$

We can express this in terms of Gamma functions

$$\nu_{eff} = C\left(\frac{2kT}{m}\right)^{\ell/2} \frac{\Gamma\left(\frac{5+\ell}{2}\right)}{\Gamma(5/2)} \quad . \tag{5.41}$$

If we define

$$\langle\nu\rangle = \frac{1}{n}\int \nu(v) f(v) 4\pi v^2 dv \quad , \tag{5.42}$$

for the Maxwellian distribution

$$\langle\nu\rangle = C\left(\frac{2kT}{m}\right)^{\ell/2} \frac{\Gamma\left(\frac{3+\ell}{2}\right)}{\Gamma(3/2)} \tag{5.43}$$

and

$$\nu_{eff} = \frac{3+\ell}{3}\langle\nu\rangle \quad . \tag{5.44}$$

So, as can be seen, $\nu_{eff} \simeq \langle\nu\rangle$. Remember, if $\nu$ is independent of v, then

$$\nu_{eff} = \nu = N\sigma_m v$$

evaluated for any value of v.

Consider specifically what happens when $\sigma_m$ is a constant (hard sphere). Then

$$\nu_{eff} = \frac{m}{3kT}\frac{N\sigma_m}{n}\int_0^\infty 4\pi v^2 (v^3) f_o dv \quad , \tag{5.45}$$

$$\nu_{eff} = \frac{m}{3kT} N\sigma_m 4\pi\left(\frac{m}{2\pi kT}\right)^{3/2}\int_0^\infty v^5 e^{-mv^2/2kT} dv \quad , \tag{5.46}$$

$$\nu_{eff} = \frac{4N\sigma_m}{3}\sqrt{\frac{8kT}{\pi m}} = \frac{4}{3} N\sigma_m v_{average} \quad . \tag{5.47}$$

Therefore

$$\nu_{eff} = \frac{4}{3}\nu_{average} \quad . \tag{5.48}$$

P. Molmud (1959) has calculated $\nu_{eff}$ for several cases of interest and his results are summarized in Table 5.1. The important thing to observe is that $\nu_{eff}$ to be used in the Langevin equation is almost the same as the "average" $\nu$ that you would use as a first approximation.

*Table 5.1. Comparison of effective and average collision frequencies for high frequency and DC cases (from Molmud, 1959)*

| Gas | $\nu_{eff}(\nu/\omega \ll 1)$ | $\nu_{eff}(\nu/\omega > 1)$ | $\nu$ |
|---|---|---|---|
| $\sigma \sim v$ | $\frac{5}{3}\nu(1+4.22\ \nu^2/\omega^2)$ | $\nu[1+4.57(\omega/\nu)^{3/2}]$ | $\nu = \frac{3\sigma}{v} N \frac{kT}{m}$ |
| air, $N_2$ | | | |
| $\sigma$ constant | $\frac{4}{3}\nu(1-0.22\ \nu^2/\omega^2)$ | $\frac{3\pi\nu}{8}(1+0.28\ \omega^2/\nu^2)$ | $\nu = 2\sigma N(\frac{2kT}{\pi m})^{1/2}$ |
| $\sigma \sim 1/v$ | $\nu$ | $\nu$ | $\nu = N\sigma v$ |
| $\sigma \sim 1/v^2$ | $\frac{2}{3}\nu(1-0.18\ \nu^2/\omega^2)$ | $\frac{3\pi\nu}{16}(1+0.30\ \omega^2/\nu^2)$ | $\nu = 2\sigma v^2 N(\frac{m}{2\pi kT})^{1/2}$ |

Shkarofsky (1961) and Shkarofsky, Johnston and Bachynski (1966) have done a very thorough job of determining these coefficients for various power law variations of $\nu(v)$, including the effects of magnetic fields, electron-ion and electron-electron collisions and have expressed the conductivity tensor as

$$\sigma = \frac{ne^2}{m}\frac{1}{g_o\langle\nu\rangle + jh_o\omega} \quad , \tag{5.49}$$

$$\sigma_{\ell,r} = \frac{ne^2}{m}\frac{1}{g^{\pm}\langle\nu\rangle + jh^{\pm}(\omega \pm \omega_c)} \quad . \tag{5.50}$$

The g's and h's are multiplicative factors which have been tabulated (as functions of $|\omega \pm \omega_c|/\langle\nu\rangle$) for power law variations of $\nu$ with velocity of $v^{-3}$, $v^{-2}$, ..., $v^{+3}$. Shkarofsky has also extended these results to non-Maxwellian distributions.

## 5.6 REFERENCES


Engelhardt, A. G. and Phelps, A. V. (1963), "Elastic and Inelastic Collision Cross Sections in Hydrogen and Deuterium from Transport Coefficients," Phys. Rev. 131, 2115-2128.

Frost, L. S. (1957), "Effect of Variable Ionic Mobility on Ambipolar Diffusion," Phys. Rev. 105, 354-356.

Ginzburg, V. L. and Gurevich, A. V. (1960), "Nonlinear Phenomena in a Plasma Located in an Alternating Electromagnetic Field," Sov. Phys. Uspekhi 3, 115-146.

Heald, M. A. and Wharton, C. B. (1965), Plasma Diagnostics with Microwaves, Wiley & Sons.

Lowke, John J. and Parker, James H., Jr. (1969), "Theory of Electron Diffusion Parallel to Electric Fields II Application to Real Gases," Phys. Rev. 181, 302-311.

Lowke, J. J., Phelps, A. V. and Irwin, B. W. (1973), "Predicted Electron Transport Coefficients and Operating Characteristics of $CO_2$-$N_2$-He Laser Mixtures," J. Appl. Phys. 44, 4664-4671.

Massey, H. S. W. and Burhop, E. H. S. (1969), Electronic and Ionic Impact Phenomena, Vol. 1, Oxford.

Molmud, P. (1959), "Langevin Equation and the AC Conductivity of Non-Maxwellian Plasmas," Phys. Rev. 114, 29-32.

Parker, James H., Jr. and Lowke, John J. (1969), "Theory of Electron Diffusion Parallel to Electric Fields. I. Theory," Phys. Rev. 181, 290-301.

Shkarofsky, I. P. (1961), "Values of the Transport Coefficients in a Plasma for any Degree of Ionization based on a Maxwellian Distribution," Can. J. Phys. 39, 1619-1703.

Shkarofsky, I. P., Johnston, T. W. and Bachynski, M. P. (1966), The Particle Kinetics of Plasmas, Addison-Wesley.

# Chapter 6

# THE FLUID EQUATIONS

In the previous chapters, a discussion of plasmas and gas discharges has been presented from a microscopic point of view based on distribution functions and the Boltzmann equation. This is rightly so since the form of the distribution function is vital in determining ionization and excitation rates and the other rate coefficients which play such an important part in gas discharges. In many cases we need not know the details of these processes but rather just the average values of quantities such as number density, velocity, and energy in order to explain most of the important processes occurring in gas discharges. The resulting equations are considerably simpler to use and are known as the fluid, or moment, equations. This is essentially the macroscopic point of view.

The use of the fluid equations suggests that we consider the plasma to act as a whole, or as a conducting fluid, rather than as individual particles. The condition for validity of this approximation is that the distance between particles be small with respect to the interparticle forces, i.e.

$$n\,\lambda_D^3 \gg 1 \,,$$

and the mean free path << scale of change of macroscopic quantities. This condition is usually well satisfied in most gas discharges and so we find the fluid equations to be extremely useful. In fact, the large majority of the basic processes occurring within gas discharges can be explained on the basis of these equations.

To obtain the fluid equations, we will start with the Boltzmann equation

$$\frac{\partial f}{\partial t} + \bar{v} \cdot \nabla_r f + \frac{\bar{F}}{m} \cdot \nabla_v f = \left(\frac{\delta f}{\delta t}\right)_c$$

and we will take the average, or moments of the quantity $\phi(\bar{v})$ over the Boltzmann equation

$$\int d^3v\phi(\bar{v})\,\frac{\partial f}{\partial t} + \int d^3v\phi(\bar{v})\bar{v}\cdot\nabla_r f + \int d^3v\phi(\bar{v})\,\frac{\bar{F}}{m}\cdot\nabla_v f = \int d^3v\phi(\bar{v})\left(\frac{\delta f}{\delta t}\right)_c \quad .(6.1)$$

We can simplify this equation one term at a time.

1. $$\int d^3v\phi(\bar{v})\,\frac{\partial f}{\partial t} = \frac{\partial}{\partial t}\int d^3v\phi(\bar{v})f = \frac{\partial}{\partial t}\,n < \phi(\bar{v}) > . \quad (6.2)$$

2. Using the vector identity

$$\nabla\cdot(\phi\bar{A}) = \bar{A}\cdot\nabla\phi + \phi\nabla\cdot\bar{A} \quad (6.3)$$

we can write

$$\int d^3v\phi(\bar{v})\bar{v}\cdot\nabla_r f = \int d^3v\nabla_r\cdot(\phi(\bar{v})\bar{v}f) - \int d^3vf\nabla_r\cdot(\phi(\bar{v})\bar{v})$$

or,

$$\int d^3v\phi(\bar{v})\bar{v}\cdot\nabla_r f = \nabla_r\cdot n < \phi(\bar{v})\bar{v} > . \quad (6.4)$$

3. If we assume $F_x$ is independent of $v_x$ (which still permits magnetic fields) we can write

$$\int d^3v\,\phi(\bar{v})\,\frac{\bar{F}}{m}\cdot\nabla_v f = \frac{\bar{F}}{m}\cdot\int d^3v\phi(\bar{v})\nabla_v f . \quad (6.5)$$

Upon integrating by parts

$$\int d^3v\phi(\bar{v})\,\frac{\bar{F}}{m}\cdot\nabla_v f = -\frac{\bar{F}}{m}\cdot\int d^3vf\nabla_v\phi(\bar{v}) = -\frac{\bar{F}}{m}\cdot n < \nabla_v\phi(\bar{v}) > , \quad (6.6)$$

therefore, we obtain

$$\frac{\partial}{\partial t}\,n < \phi(\bar{v}) > + \nabla_r\cdot n < \phi(\bar{v})\bar{v} > - \frac{\bar{F}}{m}\cdot n < \nabla_v\phi(\bar{v}) > = < \frac{\delta}{\delta t}\,n\phi(\bar{v}) >_c . \quad (6.7)$$

The collision term, $\int d^3v\phi(\bar{v})\left(\frac{\delta f}{\delta t}\right)_c$ can be treated in a number of ways. The simple Krook relaxation model, which is essentially the same as equation (4.54), can be used [see Tanenbaum (1967), chapter four or Ferraro and Plumpton (1961), chapter seven] but as was pointed out previously momentum and energy are not conserved in such a formalism. A phenomenological approach will be used in this section to determine the appropriate collision term for each of the fluid equations.

Now, in order to obtain the moment equations, all that is necessary is to substitute appropriate values of $\phi(\bar{v})$. In order to simplify these equations $\phi(\bar{v})$ should preferably represent quantities that are conserved in collisions between like particles. This, of course, allows us to set

$$< \frac{\delta}{\delta t}\,n\phi(\bar{v}) >_c = 0$$

in some cases.

## 6.1 THE CONTINUITY EQUATION

In the first case, <u>conservation of particles</u> we set <u>$\phi(\bar{v}) = 1$</u>, therefore,

$$\frac{\partial n}{\partial t} + \nabla_r\cdot n < \bar{v} > = 0 \quad (6.8)$$

if particles are conserved. By definition $< \bar{v} > = \bar{u}$, the average velocity, therefore

$$\frac{\partial n}{\partial t} + \nabla\cdot n\bar{u} = 0 \ , \quad (6.9)$$

the usual continuity equation. We have assumed no production or loss terms due to collisions with other species but this can be included as we shall show in equation (7.1).

## 6.2 THE MOMENTUM CONSERVATION EQUATION

In the second case, conservation of momentum, we set $\phi(\bar{v}) = m\bar{v}$, therefore,

$$\frac{\partial}{\partial t} nm\bar{u} + \nabla_r \cdot nm < \overline{vv} > - \bar{F}n = < \frac{\delta}{\delta t} nm\bar{v} >_c . \quad (6.10)$$

We know, for example, that electrons lose momentum by collisions with neutrals in a multi-component gas so we will retain our simple relaxation, or Langevin model, and let

$$< \frac{\delta}{\delta t} nm\bar{v} >_c = - nm\bar{u}\, \nu_m .$$

We then have

$$\frac{\partial}{\partial t} nm\bar{u} + \nabla \cdot nm < \overline{vv} > - \bar{F}n = - nm\bar{u}\, \nu_m . \quad (6.11)$$

We can rewrite

$$m \frac{\partial}{\partial t} n\bar{u} = m \left[ n \frac{\partial \bar{u}}{\partial t} + \bar{u} \frac{\partial n}{\partial t} \right]$$

and, from the continuity equation,

$$m \frac{\partial}{\partial t} n\bar{u} = m \left[ n \frac{\partial \bar{u}}{\partial t} - \bar{u}\nabla \cdot n\bar{u} \right] . \quad (6.12)$$

In the second term we have

$$< \overline{vv} > = < (\bar{u} + \bar{c})(\bar{u} + \bar{c}) > \quad (6.13)$$

where $\bar{v} = \bar{u} + \bar{c}$, $\bar{c}$ being the random (thermal) velocity. Therefore,

$$< \overline{vv} > = < \overline{cc} > + \overline{uu} \quad (6.14)$$

and, from

$$\nabla \cdot (\overline{ab}) = (\nabla \cdot \bar{a})\bar{b} + (\bar{a} \cdot \nabla)\bar{b} , \quad (6.15)$$

$$\begin{aligned} \nabla \cdot mn < \overline{vv} > &= \nabla \cdot mn < \overline{cc} > + \nabla \cdot mn\overline{uu} \\ &= \nabla \cdot mn < \overline{cc} > + m\bar{u}\, \nabla \cdot n\bar{u} + nm(\bar{u} \cdot \nabla)\bar{u} . \end{aligned} \quad (6.16)$$

If we define the pressure tensor $\bar{\bar{p}} = mn < \overline{cc} >$ we can write

$$mn[ \frac{\partial}{\partial t} + \bar{u} \cdot \nabla]\bar{u} + \nabla \cdot \bar{\bar{p}} - \bar{F}n = - nm\bar{u}\nu_m$$

or

$$\frac{d}{dt} m\bar{u} + \frac{\nabla \cdot \bar{\bar{p}}}{n} - \bar{F} = - m\bar{u}\nu_m \quad (6.17)$$

which is the momentum conservation equation. You will notice that this equation differs from the Langevin equation only by the pressure term. In the zero temperature, cold plasma model, we return precisely to the Langevin equation.

Return to the pressure tensor for a moment. In general, we can write the components of $\bar{c}$ as $\bar{c} = u\,\hat{x} + v\,\hat{y} + w\,\hat{z}$ and

$$\bar{\bar{p}} = nm \begin{bmatrix} < u^2 > & < uv > & < uw > \\ < vu > & < v^2 > & < vw > \\ < wu > & < wv > & < w^2 > \end{bmatrix} . \quad (6.18)$$

If the pressure is isotropic < uv > and similar terms disappear and

$$< u^2 > = < v^2 > = < w^2 > = \frac{1}{3} < u^2 + v^2 + w^2 > = \frac{1}{3} c^2 . \quad (6.19)$$

Therefore,

$$\bar{\bar{p}} = nm \begin{bmatrix} c^2/3 & 0 & 0 \\ 0 & c^2/3 & 0 \\ 0 & 0 & c^2/3 \end{bmatrix} . \quad (6.20)$$

Since $< mc^2/2 > = 3kT/2$ for thermodynamic equilibrium, we can write

$$p_{ij} = \delta_{ij} nkT \tag{6.21}$$

where $\delta_{ij}$ is the delta function and we can use the scalar pressure

$$p = nkT = nm < c^2 > /3 \tag{6.22}$$

to write

$$\frac{d}{dt} m\bar{u} = -\frac{\bar{\nabla} p}{n} + \bar{F} - m\bar{u}\nu_m \quad . \tag{6.23}$$

## 6.3 THE ENERGY-CONSERVATION EQUATION

In the third case, <u>conservation of energy</u>, we set

$$\phi(\bar{v}) = \frac{1}{2} mv^2/2 \quad , \tag{6.24}$$

$$\frac{\partial}{\partial t} n < mv^2/2 > + \nabla_r \cdot n < mv^2\bar{v}/2 > - \bar{F} \cdot n < \bar{v} > = < \frac{\delta}{\delta t} \frac{1}{2} nmv^2 >_c . \tag{6.25}$$

Now

$$< v^2 > = < c^2 > + u^2 \tag{6.26}$$

therefore,

$$n < mv^2/2 > = nmu^2/2 + 3p/2 \quad . \tag{6.27}$$

Also

$$\begin{aligned} < v^2\bar{v} > &= < (c^2 + u^2 + 2\bar{c} \cdot \bar{u})(\bar{c} + \bar{u}) > \\ &= < c^2\bar{c} + c^2\bar{u} + u^2\bar{c} + u^2\bar{u} + 2\bar{c}\bar{c} \cdot \bar{u} + 2\bar{u}\bar{c} \cdot \bar{u} > \quad . \end{aligned} \tag{6.28}$$

Since we have postulated an isotropic distribution in a drifting reference frame, the terms $c^2\bar{c}$, $u^2\bar{c}$ and $2\bar{u}\bar{c} \cdot \bar{u}$ should all drop out since they are odd in $\bar{c}$. For a distribution which is purely isotropic in velocity as well as space this is true. The more relevant case is when the temperature, or $\bar{c}$, changes as a function of position so that energy can be transported from one point to another. In this case there is a heat current defined as

$$\bar{q} = n < \frac{1}{2} mc^2\bar{c} > \tag{6.29}$$

which is the flux of random or thermal energy across a surface element moving with the mean velocity $\bar{u}$. Tanenbaum (1967) gives an excellent treatment of this concept in his chapter four. We will therefore assume that there is some slight deviation from a perfectly isotropic distribution and retain the term $c^2\bar{c}$ in equation (6.28). On the other hand, since $u << c$ for the case we are considering, the contribution of a slight anisotropy to the terms $u^2\bar{c}$ and $2\bar{u}\bar{c} \cdot \bar{u}$, both of which are of the order $u^2 < c^2 >^{1/2}$, is much less than the contribution to $c^2\bar{c}$, which is of the order $< c^2 >^{3/2}$, so the two terms $u^2\bar{c}$ and $2\bar{u}\bar{c} \cdot \bar{u}$ will be considered negligible in comparison to the other factors in equation (6.28).

Equation (6.28) can then be simplified to

$$< v^2\bar{v} > = < c^2\bar{c} > + < c^2 > \bar{u} + u^2\bar{u} + 2 < \bar{c}\bar{c} > \cdot \bar{u} \quad . \tag{6.30}$$

If the pressure is isotropic $< \bar{c}\bar{c} > \cdot \bar{u} = < c^2 > \bar{u}/3$

and

$$\begin{aligned} < v^2\bar{v} > &= < c^2\bar{c} > + u^2\bar{u} + 5 < c^2 > \bar{u}/3 \\ &= < c^2\bar{c} > + u^2\bar{u} + 5p\bar{u}/mn \quad . \end{aligned} \tag{6.31}$$

Then

$$\frac{\partial}{\partial t} n < mv^2/2 > = \frac{\partial}{\partial t} [nmu^2/2 + 3p/2] = nmu \frac{\partial u}{\partial t} + \frac{1}{2} mu^2 \frac{\partial n}{\partial t} + \frac{3}{2}\frac{\partial p}{\partial t} \qquad (6.32)$$

and

$$\nabla \cdot n < \frac{1}{2} mv^2\bar{v} > = \nabla_r \cdot \left[ mn < c^2\bar{c} > /2 + mnu^2\bar{u}/2 + 5p\bar{u}/2 \right] \qquad (6.33)$$

and

$$-\bar{F} \cdot n < \bar{v} > = - n\bar{F} \cdot \bar{u} \quad . \qquad (6.34)$$

By using the definition of the heat current $\bar{q}$ given in equation (6.29) and by invoking the continuity equation, equation (6.25) then becomes

$$\begin{aligned} & nmu \frac{\partial u}{\partial t} + \frac{1}{2} mu^2[- \nabla \cdot n\bar{u}] + \frac{3}{2}\frac{\partial p}{\partial t} + \nabla \cdot \bar{q} \\ & + \frac{1}{2} mu^2 \nabla \cdot n\bar{u} + \frac{1}{2} mn\bar{u} \cdot \nabla u^2 + \frac{5}{2} p \nabla \cdot \bar{u} \\ & + \frac{5}{2} \bar{u} \cdot \nabla p - n\bar{F} \cdot \bar{u} = < \frac{\delta}{\delta t} \frac{1}{2} nmv^2 >_c \quad . \end{aligned} \qquad (6.35)$$

We can simplify further by dotting the single species momentum conservation equation onto $\bar{u}$ to obtain

$$mn\bar{u} \cdot \left[ \frac{\partial}{\partial t} + \bar{u} \cdot \nabla \right] \bar{u} + \bar{u} \cdot \nabla p - n\bar{F} \cdot \bar{u} = 0. \qquad (6.36)$$

Substituting this result, equation (6.35) then simplifies to

$$\frac{3}{2}\left[ \frac{\partial p}{\partial t} + \bar{u} \cdot \nabla p \right] + \frac{5}{2} p\nabla \cdot \bar{u} + \nabla \cdot \bar{q} = < \frac{\delta}{\delta t} \frac{1}{2} nmv^2 >_c \quad , \qquad (6.37)$$

which can be written as

$$\frac{3}{2}\frac{dp}{dt} + \frac{5}{2} p\nabla \cdot \bar{u} + \nabla \cdot \bar{q} = < \frac{\delta}{\delta t} \frac{1}{2} nmv^2 >_c \cdot \qquad (6.38)$$

Now p = nkT, therefore equation (6.38) can be rewritten as

$$\frac{3}{2} nk \frac{dT}{dt} - kT\frac{dn}{dt} + \nabla \cdot \bar{q} = < \frac{\delta}{\delta t} \frac{1}{2} nmv^2 >_c \cdot \qquad (6.39)$$

Consider a single species so that average energy is conserved and the collision term goes to zero. If n is constant and if we define

$$\bar{q} = - K\nabla T \qquad (6.40)$$

where K is the thermal conductivity, then,

$$\frac{dT}{dt} = \frac{2K}{3kn} \nabla^2 T \ , \qquad (6.41)$$

the usual equation for heat transfer.

Equation (6.41) is perfectly suitable for the neutral gas by itself; however, in a gas discharge we are usually concerned with the electron gas and the neutral gas. This is essentially a two fluid (or three if you include the ions) system with strong collisional coupling and the collision term becomes important.

For the neutral gas, the main case of interest is the determination of the spatial temperature distribution when the source of energy is from collisions with the electron gas. If we assume that n is constant and that all of the energy deposited in the electron gas ($\bar{J} \cdot \bar{E}$) is lost by collisions with the neutral gas (Ecker and Zoller, 1964), then we can write for the <u>neutral gas</u> $< \frac{\delta}{\delta t} \frac{1}{2} nmv^2 >_c =$ $\bar{J} \cdot \bar{E}$ and

$$\frac{d}{dt}\left[ \frac{3}{2} nkT_g \right] - K\nabla^2 T_g = \bar{J} \cdot \bar{E} \quad . \qquad (6.42)$$

If inelastic collisions absorb a significant amount of the electron energy, then that amount should be subtracted from the $\bar{J} \cdot \bar{E}$ term.

Analogously, for the electron gas we can write

$$\frac{d}{dt}\left[\frac{3}{2} n_e k T_e\right] - K_e \nabla^2 T_e = \bar{J} \cdot \bar{E} - \frac{3n_e k(T_e - T_g)/2}{\tau} . \quad (6.43)$$

$\bar{J} \cdot \bar{E}$ is the energy source and the last term represents the energy loss by collisions with the neutral gas with an energy relaxation time constant $\tau$. Once again, the inelastic energy loss should be subtracted as well.

Some interesting experimental studies on electronic thermal conductivity have been performed by Goldstein and Sekiguchi (1958) and by Rostas, Bhattacharya and Cahn (1963). Certainly in cases where Coulomb collisions play an important role, the electronic thermal conductivity is a significant factor.

If the steady-state temperature distributions are important, then for the neutral gas equation (6.42) simplifies to

$$K\nabla^2 T + \bar{J} \cdot \bar{E} = 0 . \quad (6.44)$$

Actually, the thermal conductivity is generally a function of the gas temperature, so equation (6.44) should be written more correctly as

$$\nabla \cdot (K\nabla T) + \bar{J} \cdot \bar{E} = 0 . \quad (6.45)$$

An experimental study of the neutral gas density profiles in argon discharges has been performed by Eden and Cherrington (1973) and illustrates the solution to equation (6.45) quite well.

In the case of the electron gas in a DC discharge, equation (6.43) simplifies to

$$\bar{J} \cdot \bar{E} - \frac{3}{2} \frac{n_e k(T_e - T)}{\tau} + K_e \nabla^2 T_e = 0 . \quad (6.46)$$

You will notice that our previous simple derivation of the electron energy balance based on the Langevin equation (Section 2.3) differs from this equation only by the heat flow term, which in many cases is negligible with respect to the collisional energy loss. Once again, the Langevin approach to the determination of important discharge parameters is seen to give quite good results for the electron gas. For the neutral gas, of course, the heat flow terms usually represent the total heat loss and must be included.

## 6.4 REFERENCES

Ecker, G. and Zoller, O. (1964), "Thermally Inhomogeneous Plasma Column," Phys. Fluids 7, 1996-2000.

Eden, J. G. and Cherrington, B. E., (1973), "Radial Neutral Gas Temperature and Density Profiles in Low Pressure Argon Discharges," J. Appl. Phys. 44, 4920-4926.

Ferraro, V. C. A., Plumpton, C. (1961), An Introduction to Magneto-Fluid Mechanics, Oxford.

Goldstein, L. and Sekiguchi, T. (1958), "Electron-electron Interaction and Heat Conduction in Gaseous Plasmas," Phys. Rev. 109, 625-630.

Rostas, F., Bhattacharya, A.K. and Cahn, J. H. (1963), "Measurements of Electronic Thermal Conductivity in Magneto Plasmas," Phys. Rev. 129, 495-505.

Tanenbaum, B. Samuel (1967), Plasma Physics, McGraw-Hill.

# Chapter 7

# ELECTRON-DENSITY DECAY PROCESSES

Now that we have the three main fluid equations, the equations of continuity, momentum conservation and energy conservation, we can proceed to apply these equations to the examination of some of the fundamental processes occurring in gas discharges. The first application will deal with the electron density "rate processes" occurring within a discharge. These processes deal primarily with the continuity equation but now include important collisional processes such as ionization, recombination and attachment. We rewrite the continuity equation for electrons as

$$\frac{\partial n}{\partial t} + \nabla \cdot n\bar{u} = n\nu_I - \alpha n^2 - n\nu_a \tag{7.1}$$

where $\nu_I$ = electron collision frequency for ionization (assumed at this point to be a one step electron-atom or electron-molecule ionizing collision),

$\alpha$ = electron-ion recombination coefficient,

$\nu_a$ = electron-neutral attachment frequency (forming a negative ion).

There are, of course, other processes which could be included but equation 7.1 represents the major processes.

We could consider the steady-state solutions (DC discharges, for instance) and balance the source and loss terms, but for now we will talk about the solutions of these equations when the source term is absent (the afterglow of pulsed discharges). This approach gives particular insight into basic processes and is one of the major experimental approaches to evaluating the appropriate coefficients. What we will do initially, then, is to evaluate the electron loss processes in a plasma due to diffusion, recombination and attachment. Let us start with diffusion.

## 7.1 DIFFUSION

In this case, the electrons are lost by diffusion to the walls of the container and

$$\frac{\partial n}{\partial t} + \nabla\cdot n\bar{u} = 0 \,. \tag{7.2}$$

In order to evaluate the flux $\bar{\Gamma} = n\bar{u}$, we must turn to the momentum conservation equation

$$\frac{d}{dt}\, m\bar{u} = -\frac{\nabla p}{n} + \bar{F} - m\bar{u}\nu_m \,. \tag{7.3}$$

At this point, we assume that we can neglect the left hand side of the above equation. In essence, we assume that the electrons are inertialess, which is permissible if the time rate of change of the macroscopic variables, such as the decay time of the electron density, is much greater than the time between collisions. A good discussion of the necessary conditions is given by Tanenbaum (1967). If we then assume that $\bar{F}$ is due to an electric field acting on the electrons, we obtain

$$\bar{\Gamma} = n\bar{u} = -\frac{ne\bar{E}}{m\nu_m} - \frac{\nabla nkT}{m\nu_m} \,. \tag{7.4}$$

If we assume that the temperature T is constant, we can write

$$\bar{\Gamma} = -n\frac{e}{m\nu_m}\bar{E} - \frac{kT}{m\nu_m}\nabla n \tag{7.5}$$

or, the more familiar equation

$$\bar{\Gamma} = -n\mu\bar{E} - D\nabla n \,. \tag{7.6}$$

If it is necessary to retain the inertial term, then from equation (7.3) we can see that the appropriate equation is

$$\bar{\Gamma} = -n\mu\bar{E} - D\nabla n - \frac{1}{\nu_m}\frac{\partial\bar{\Gamma}}{\partial t} \,. \tag{7.7}$$

Examples of conditions under which this addition is necessary have been given by Shimony and Cahn (1965) and by Persson (1962).

Returning to equation (7.2), we can now write

$$\nabla\cdot n\bar{u} = \nabla\cdot\bar{\Gamma} = -\nabla\cdot n\mu\bar{E} - D\nabla^2 n \tag{7.8}$$

(assuming D is constant). For diffusion controlled decay, we assume that no electric field is present, and we obtain the diffusion equation

$$\frac{\partial n}{\partial t} = D\nabla^2 n \,. \tag{7.9}$$

The form of the solution to this equation will depend upon the boundary conditions applied. For instance, you can easily verify that

$$n(x,t) = \frac{C}{2(\pi Dt)^{1/2}}\, e^{-x^2/4Dt} \tag{7.10}$$

is a solution to the one dimensional diffusion equation and does, in fact, represent diffusion away from a point source in an infinite medium. Tanenbaum (1967) in his problem (3.5) shows that the density as a function of position and time in one dimension is

$$n(x,t) = \frac{1}{2(\pi D t)^{1/2}} \int_{-\infty}^{\infty} n_o(x') e^{-(x - x')^2/4Dt} dx' \qquad (7.11)$$

where $n_o(x')$ is the distribution at time $t = 0$. Solutions of the diffusion equation covering a wide range of conditions are given by Crank (1956), and I will refer you there for further details.

When we are dealing with gas discharges, we wish to obtain the solution to the diffusion equation in a bounded, rather than an infinite, medium. To obtain this solution, we assume that we can express

$$n(r,t) = R(r)T(t) \ . \qquad (7.12)$$

The diffusion equation then becomes

$$R \frac{\partial T}{\partial t} = TD\nabla^2 R \qquad (7.13)$$

and

$$\frac{1}{T}\frac{\partial T}{\partial t} = \frac{D}{R} \nabla^2 R = -\frac{1}{\tau} \ . \qquad (7.14)$$

Therefore,

$$n \sim e^{-t/\tau} , \qquad (7.15)$$

representing an exponential decay of the electron density with a diffusion decay time of $\tau$.

For the space dependence we have

$$\nabla^2 R = -\left(\frac{1}{D\tau}\right) R \qquad (7.16)$$

or

$$\nabla^2 n = -\left(\frac{1}{D\tau}\right) n = -\frac{n}{\Lambda^2} \qquad (7.17)$$

where $\Lambda = \sqrt{D\tau}$ is a characteristic diffusion length.

We can now solve the preceding equation for the geometries of interest and determine the spatial diffusion modes and the decay times for the modes. Let us begin with the simplest case.

### 7.1.1 Rectangular Geometry

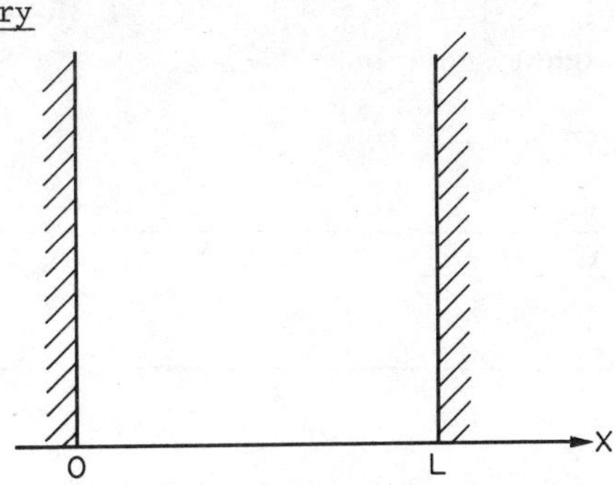


*Fig. 7.1. Boundary conditions for diffusion in one dimension*

Consider the one dimensional case shown in Fig. 7.1, with boundaries at x = 0, L. The appropriate boundary condition for charged particles is that n = 0 at x = 0 and x = L. In general, then,

$$n(x) = \sum_k n_k \sin\left(\frac{k\pi x}{L}\right)$$

for integer values of k. From the diffusion equation, we have

$$\nabla^2 n = \frac{\partial^2 n}{\partial x^2} = -\frac{n}{\Lambda^2}\,. \tag{7.18}$$

The appropriate solution to this equation is

$$n = \sum_k n_k \sin\left(\frac{x}{\Lambda_k}\right) \text{ where } \Lambda_k = \sqrt{D\tau_k}\,. \tag{7.19}$$

Equating the two solutions, we find

$$\frac{1}{\Lambda_k} = \frac{k\pi}{L} = \frac{1}{\sqrt{D\tau_k}}\,. \tag{7.20}$$

For each mode, k, we therefore have a decay time

$$\tau_k = \frac{1}{k^2 D}\left(\frac{L}{\pi}\right)^2. \tag{7.21}$$

Therefore $\dfrac{\tau_1}{\tau_2} = 4,\ \dfrac{\tau_1}{\tau_3} = 9$, etc.

In this case, we can see that the higher order modes will diffuse out quite rapidly leaving essentially only the lowest order mode

$$n(x) = n_1 \sin\left(\frac{\pi x}{L}\right) \tag{7.22}$$

with a decay time $\tau = \frac{1}{D}\left(\frac{L}{\pi}\right)^2$ and a characteristic diffusion length given by

$$\frac{1}{\Lambda^2} = \left(\frac{\pi}{L}\right)^2 = \frac{1}{D\tau}\,. \tag{7.23}$$

If we extend this analysis to a cube of sides, $L_1$, $L_2$ and $L_3$, we obtain, for the dominant diffusion mode

$$\frac{1}{\Lambda^2} = \left(\frac{\pi}{L_1}\right)^2 + \left(\frac{\pi}{L_2}\right)^2 + \left(\frac{\pi}{L_3}\right)^2 = \frac{1}{D\tau}\,. \tag{7.24}$$

Let us now consider a slightly more useful case.

### 7.1.2 Cylindrical Geometry

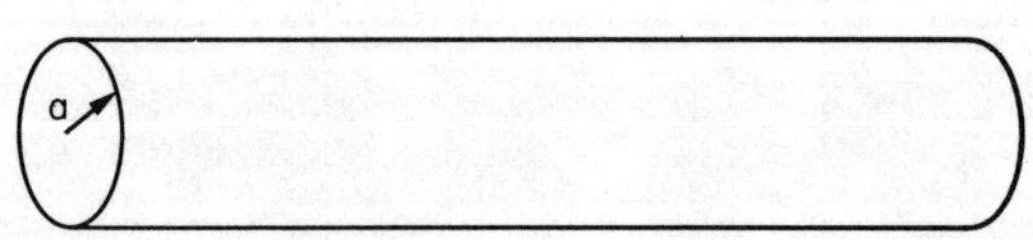


*Fig. 7.2. Boundary conditions for diffusion in cylindrical geometry*

Consider an infinitely long cylinder of radius a

$$\nabla^2 n = \frac{1}{r}\frac{\partial}{\partial r} r \frac{\partial n}{\partial r} + \frac{1}{r^2}\frac{\partial^2 n}{\partial \phi^2} + \frac{\partial^2 n}{\partial Z^2} . \tag{7.25}$$

If we assume no Z dependence and azimuthal symmetry, we have

$$\frac{1}{r}\frac{\partial}{\partial r} r \frac{\partial n}{\partial r} + \frac{n}{\Lambda^2} = 0 . \tag{7.26}$$

The equation

$$r \frac{\partial}{\partial r} r \frac{\partial n}{\partial r} + (\beta_o r)^2 n = 0 \tag{7.27}$$

defines the Bessel function

$$n_r \sim J_o(\beta_o r) . \tag{7.28}$$

Therefore, the dominant diffusion mode is given by

$$n(r) = n_o J_o(\beta_o r) \tag{7.29}$$

where $n_o$ is the density along the axis of the cylinder and $\beta_o^2 = 1/D\tau$. From the boundary condition that n = 0 at r = a,

$$\beta_o a = 2.405 .$$

Therefore,

$$\frac{1}{\Lambda^2} = \frac{1}{D\tau} = \left(\frac{2.405}{a}\right)^2 . \tag{7.30}$$

If we extend this solution to the case of cylindrical discharge tube of radius a and length L (Fig. 7.3), we have

$$\frac{1}{\Lambda^2} = \frac{1}{D\tau} = \left(\frac{\pi}{L}\right)^2 + \left(\frac{2.405}{a}\right)^2 . \tag{7.31}$$

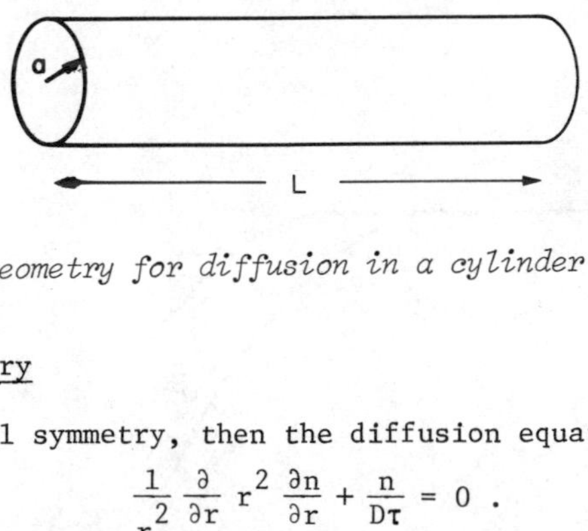


*Fig. 7.3. Geometry for diffusion in a cylinder of length L*

### 7.1.3 Spherical Geometry

If we consider spherical symmetry, then the diffusion equation becomes

$$\frac{1}{r^2}\frac{\partial}{\partial r} r^2 \frac{\partial n}{\partial r} + \frac{n}{D\tau} = 0 . \tag{7.32}$$

This is similar to the equation

$$\frac{d}{dr} r^2 \frac{\partial R}{\partial r} + [(kr)^2 - n(n+1)]R = 0 \tag{7.33}$$

which defines the spherical Bessel function $j_n(kr)$ where

$$R \sim j_n(kr) = \sqrt{\frac{\pi}{2kr}}\, J_{n+1/2}(kr) , \tag{7.34}$$

J being the ordinary Bessel function. In this case n = 0 and $k^2 = 1/D\tau = 1/\Lambda^2$.

The only solution which is finite at r = 0 is

$$n(r) = n_o j_o(kr) = n_o \sqrt{\frac{\pi}{2kr}}\, J_{1/2}(kr) . \tag{7.35}$$

Now

$$J_{1/2}(kr) = \sqrt{\frac{2}{\pi kr}} \sin kr \tag{7.36}$$

therefore,

$$n = n_o \frac{\sin kr}{kr} \qquad \text{where } k = \frac{1}{\sqrt{D\tau}} . \tag{7.37}$$

The appropriate boundary condition is that $n = 0$ at $r = a$, or, for the lowest order mode $k = \frac{\pi}{a}$ and

$$\frac{1}{\Lambda^2} = \frac{1}{D\tau} = \left(\frac{\pi}{a}\right)^2 . \tag{7.38}$$

### 7.1.4 Ambipolar Diffusion

In the previous section, we have made the implicit assumption that the diffusion coefficient, D, to be used in the previous equations is the diffusion coefficient for electrons. $D_e = kT_e/m\nu_e$. This, in fact, assumes that the electrons are free to diffuse through a uniform neutral background gas and that the ions have no substantial effect on this process. This "free diffusion", however, can only occur in very low density plasmas where the Coulomb forces can essentially be neglected. In higher density plasmas, such as commonly occur in gas discharges, this is definitely not the case and we must determine the diffusion rates by simultaneously solving the "two fluid" equations for electrons and ions keeping in mind that the very strong Coulomb forces will tend to maintain overall space charge neutrality ($n_e - n_+ << n_e$). To show what is meant, consider Fig. 7.4. In the presence of a

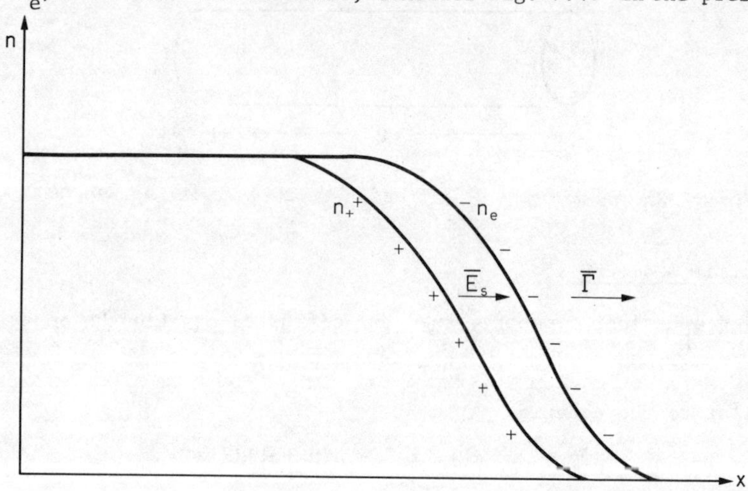


*Fig. 7.4. Ambipolar effects in a diffusing plasma*

gradient in plasma density, both the electrons and ions will tend to diffuse into the region of lower density. The electron diffusion coefficient, however, is usually considerably greater than the ion-diffusion coefficient.

$$\frac{kT_e}{m\nu_{eN}} >> \frac{kT_+}{M\nu_{iN}}$$

(You will notice that the interaction with the neutral background gas is expressed by the electron-neutral and ion-neutral collision frequencies $\nu_{eN}$ and $\nu_{iN}$.) The electrons will tend to diffuse more rapidly than the ions, setting up a space charge separation as shown. The resulting space charge field, $\bar{E}_s$, will then retard the electron diffusion and increase the ion diffusion so that space charge neutrality is maintained at all points in space. Under these conditions, the electrons and ions will diffuse at the same rate as determined by the ambipolar diffusion coefficient.

In order to determine the ambipolar diffusion coefficient, let us proceed as follows:

For the electrons

$$\bar{\Gamma}_e = -D_e \nabla n_e - n_e \mu_e \bar{E}_s \ . \tag{7.39}$$

For the ions

$$\bar{\Gamma}_+ = -D_+ \nabla n_+ + n_+ \mu_+ \bar{E}_s \ . \tag{7.40}$$

The space charge field $\bar{E}_s$, which is responsible for ambipolar diffusion, is included in these equations.

From the continuity equations, we have

$$\frac{\partial n_e}{\partial t} = - \nabla \cdot \bar{\Gamma}_e = D_e \nabla^2 n_e + n_e \mu_e \nabla \cdot \bar{E}_s + \mu_e \bar{E}_s \cdot \nabla n_e \tag{7.41}$$

and

$$\frac{\partial n_+}{\partial t} = -\nabla \cdot \bar{\Gamma}_+ = D_+ \nabla^2 n_+ - n_+ \mu_+ \nabla \cdot \bar{E}_s - \mu_+ \bar{E}_s \cdot \nabla n_+ \ . \tag{7.42}$$

We shall now assume that the space charge field is strong enough to maintain space charge neutrality, $n_e - n_+ << n_e$, and let

$$n_e \cong n_+ = n \ . \tag{7.43}$$

Multiplying the first equation by $\mu_+$ and the second equation by $\mu_e$ and adding, we obtain

$$(\mu_e + \mu_+) \frac{\partial n}{\partial t} = (D_e \mu_+ + D_+ \mu_e) \nabla^2 n \tag{7.44}$$

or

$$\frac{\partial n}{\partial t} = \frac{D_e \mu_+ + D_+ \mu_e}{\mu_e + \mu_+} \nabla^2 n = D_a \nabla^2 n \ . \tag{7.45}$$

Which defines the ambipolar diffusion coefficient

$$D_a = \frac{D_e \mu_+ + D_+ \mu_e}{\mu_e + \mu_+} \ . \tag{7.46}$$

Now, in the usual cases of interest

$$D_e >> D_+ \quad \text{and} \quad \mu_e >> \mu_+$$

so we can make some simplifications.

$$D_a = \frac{D_e \mu_+ + D_+ \mu_e}{\mu_e} = \frac{kT_e}{e} \mu_+ + D_+ \tag{7.47}$$

or

$$D_a = D_+(1 + T_e/T_+) \ . \tag{7.48}$$

If $T_e = T_+$, then $D_a = 2D_+$.

If we wish to obtain the space charge field $E_s$, we can return to the original equations

$$\Gamma = -D_e \nabla n - n\mu_e E_s \ , \tag{7.39}$$

$$\Gamma = -D_+ \nabla n + n\mu_+ E_s \ . \tag{7.40}$$

Subtracting, we obtain

$$\bar{E}_s = -\left(\frac{D_e - D_+}{\mu_e + \mu_+}\right) \frac{\nabla n}{n} \ . \tag{7.49}$$

We could repeat this same derivation for diffusion in a discharge when a magnetic field is present. This is a very important case related to plasma confinement and an excellent discussion is given by Brown (1966). The basic result is that the diffusion across the magnetic field (under suitable conditions) is reduced, with the effective diffusion coefficient being multiplied by $1/B^2$ or $1/B$ depending upon the discharge configuration. The interpretation can be subject to lengthy discussion, so refer to Brown for the details.

### 7.1.5 Transition Diffusion

We have covered both extremes: free diffusion of electrons when the charge density is so low that the Coulomb forces are ineffective and ambipolar diffusion when space-charge forces dominate. Obviously there must be a gradual transition between the two cases, and this transition region can be important in many discharge situations.

The first study of transition diffusion was performed by Allis and Rose (1954). Their approach is quite complicated, but Allis (1956) has presented a simpler approach which has also been covered by Brown (1966). To summarize this approach very briefly, we return to equations (7.39) and (7.40).

$$\Gamma_e = -D_e \nabla n_e - E_s \mu_e n_e \ , \tag{7.39}$$

$$\Gamma_+ = -D_+ \nabla n_+ + E_s \mu_+ n_+ \ . \tag{7.40}$$

In this case we still take the flow to be the same, $\Gamma_e = \Gamma_+$, but we no longer assume $n_e = n_+$. We do, however, assume that $\nabla n_e/n_e = \nabla n_+/n_+$ (the constant ratio approximation). We then have, by eliminating $E_s$,

$$\begin{aligned}(\mu_e n_e + \mu_+ n_+)\Gamma &= -\mu_+ n_+ D_e \nabla n_e - \mu_e n_e D_+ \nabla n_+ \\ &= -(\mu_+ D_e + \mu_e D_+) n_+ \nabla n_e\end{aligned} \tag{7.50}$$

and

$$\Gamma = -\frac{(\mu_+ D_e + \mu_e D_+) n_+}{\mu_e n_e + \mu_+ n_+} \nabla n_e \ . \tag{7.51}$$

If we use the usual definition of the ambipolar diffusion coefficient

$$D_a = \frac{\mu_e D_+ + \mu_+ D_e}{\mu_e + \mu_+}$$

we obtain

$$\Gamma = -D_s \nabla n_e = -D_a\left(1 + \mu_e \frac{\rho}{\sigma}\right)\nabla n_e \,. \tag{7.52}$$

$$D_s = D_a\left(1 + \mu_e \frac{\rho}{\sigma}\right) \tag{7.53}$$

is the effective diffusion coefficient for the transition case where $\rho = (n_+ - n_e)e$, the total charge density, and $\sigma = (\mu_+ n_+ + \mu_e n_e)e$, the conductivity. From equation (7.39) and $\Gamma = -D_s \nabla n_e$, the space charge field

$$E_s = \frac{D_s - D_e}{\mu_e}\frac{\nabla n_e}{n_e} \,. \tag{7.54}$$

From Poisson's equation

$$\nabla \cdot \varepsilon_o E_s = \rho = (n_+ - n_e)e = \varepsilon_o \frac{D_s - D_e}{\mu_e}\left[\frac{\nabla^2 n_e}{n_e} - \left(\frac{\nabla n_e}{n_e}\right)^2\right]. \tag{7.55}$$

Let us use the definition of diffusion length

$$\frac{\nabla^2 n}{n} = -\frac{1}{\Lambda^2}$$

and take the centre of the discharge (where $\nabla n_e = 0$). We then have

$$\rho_c = \varepsilon_o \frac{D_e - D_s}{\mu_e}\frac{1}{\Lambda^2} \tag{7.56}$$

and

$$D_s = D_a\left[1 + \frac{\mu_e \varepsilon_o (D_e - D_s)}{\sigma_c \mu_e \Lambda^2}\right] \tag{7.57}$$

or

$$D_s = D_a \frac{D_e + \Lambda^2 \sigma_c/\varepsilon_o}{D_a + \Lambda^2 \sigma_c/\varepsilon_o} \,. \tag{7.58}$$

A plot of this term as a function of $\Lambda^2\sigma_c/\varepsilon_o D_e$ (which is roughly proportional to the plasma density) is given in Fig. 7.5. As we can see, the transition is quite gradual over several orders of magnitude in plasma density.

This rather simple theory gives an approximate idea of how the transition from ambipolar to free diffusion occurs. Several more recent theoretical and experimental studies have shed further light on the subject and have sharpened up the definition of the beginning of transition diffusion. Kregel (1970) performed numerical calculations using Gauss' law to evaluate the space charge field and found that the Debye shielding length $\lambda_D$ is much smaller than the characteristic diffusion length $\Lambda$ during ambipolar diffusion. In the midst of the transition to free diffusion $\lambda_D$ and $\Lambda$ become comparable. By the very definition of the Debye length as being the distance over which substantial deviation from space charge neutrality is possible, it can be seen that the comparison of the diffusion length to the Debye length is an excellent manner in which to define the diffusion regime.

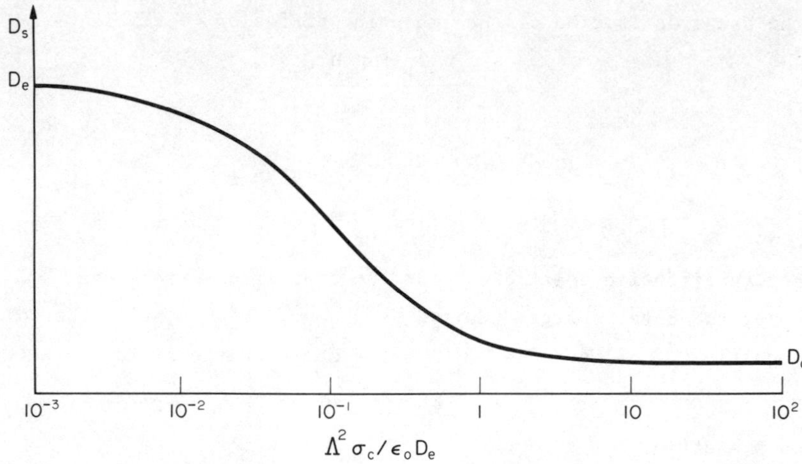


*Fig. 7.5. The transition from free to ambipolar diffusion based on the simplified theory*

Gerber and Gerardo (1973) performed detailed measurements of the decay of electrons and ions in a helium afterglow and concluded that the onset of transition diffusion occurs for ratios of the characteristic diffusion length to the Debye length as high as

$$\Lambda/\lambda_D = 86 \ . \tag{7.59}$$

Figure 7.6 shows experimentally determined electron density and helium ion density curves in units of $(\Lambda/\Lambda_D)^2$ as a function of normalized time as found by Gerber and Gerardo (1973).

Figure 7.7 gives a plot of the ratio of the transition diffusion coefficient to the ambipolar diffusion coefficient as a function of $(\Lambda/\lambda_D)^2$ as obtained by Gerber and Gerardo.

An important point to note is that the transition from ambipolar to free diffusion occurs over about five orders of magnitude of electron or ion density. Furthermore, since the Debye length increases as the square root of the electron temperature, high energy discharges (typically DC discharges at low pressures) can be in the transition diffusion regime even at moderately high electron densities. More will be said about this when DC discharges are considered.

### 7.1.6 Diffusion in Multi-species Discharges

We can add considerably to the difficulties of analysis if we assume that several different species of positive ions, or positive and negative ions, can exist in a discharge. When we consider ambipolar diffusion within such a medium, it is necessary to solve simultaneously the fluid equations for all species involved. This subject is discussed by Oskam (1958) and Brown (1966). This is far from an

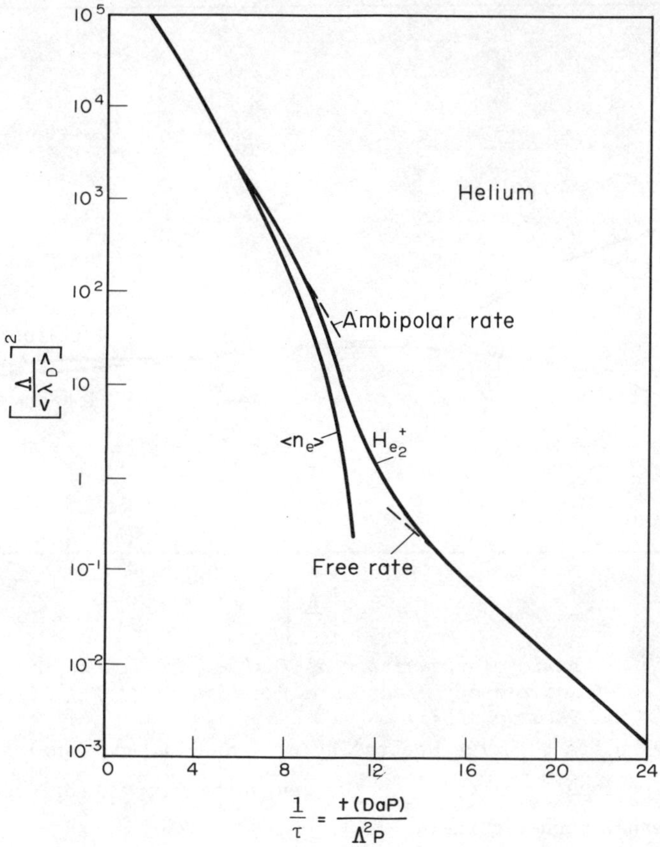


*Fig. 7.6. Experimentally obtained electron density decay curve in units of $(\Lambda/ < \lambda_D >)^2$ vs. normalized time. The scaled $He_2{}^+$ wall current is also shown (from Gerber and Gerardo, 1973)*

academic exercise since many important discharges, especially in molecular gases, fall into this category.

Let us consider the case of a plasma containing electrons and two types of positive ions. (This could, for instance, be helium with $He^+$ and $He_2^+$ present.) We write

$$\bar{\Gamma}_e = -D_e \nabla n_e - n_e \mu_e \bar{E} \ , \tag{7.60}$$

$$\bar{\Gamma}_1 = -D_1 \nabla n_1 + n_1 \mu_1 \bar{E} \ , \tag{7.61}$$

$$\bar{\Gamma}_2 = -D_2 \nabla n_2 + n_2 \mu_2 \bar{E} \ . \tag{7.62}$$

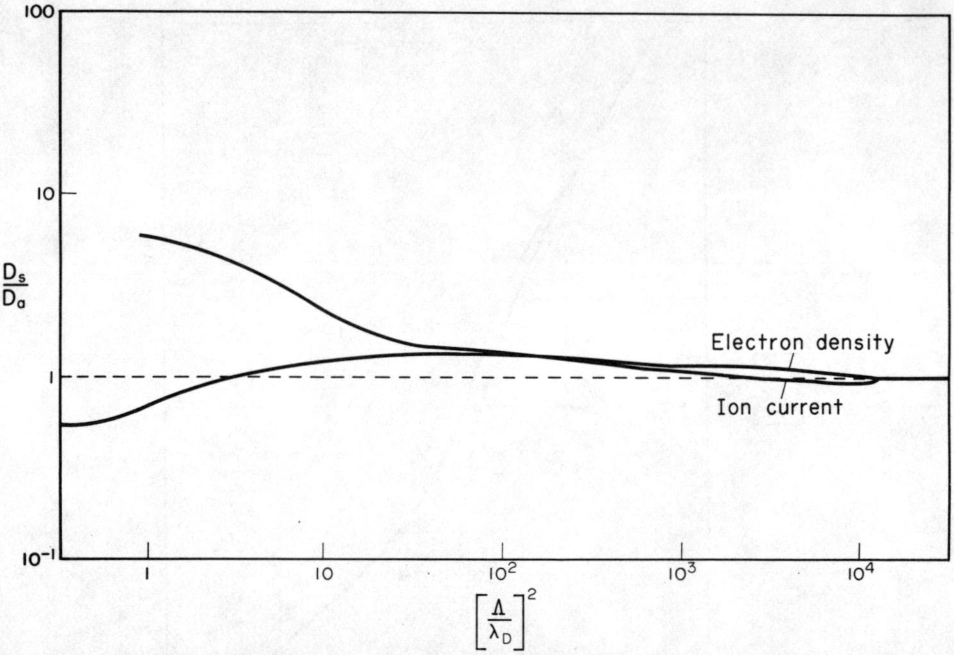


*Fig. 7.7. The ratio of transition diffusion coefficient to ambipolar diffusion coefficient as a function of $(\Lambda/\lambda_D)^2$ (from Gerber and Gerardo, 1973)*

As usual, we assume space charge neutrality $n_e = n_1 + n_2$ and equal flux of electrons and positive ions $\bar{\Gamma}_e = \bar{\Gamma}_1 + \bar{\Gamma}_2$. (We can write this vector equality only if there is no external magnetic field.)

If we make the further assumption that the density gradients are proportional,

$$\frac{\nabla n_e}{n_e} = \frac{\nabla n_1}{n_1} = \frac{\nabla n_2}{n_2} \tag{7.63}$$

then

$$D_{a,e} = \frac{n_1[\mu_e D_1 + \mu_1 D_e] + n_2[\mu_e D_2 + \mu_2 D_e]}{n_e\mu_e + n_1\mu_1 + n_2\mu_2}, \tag{7.64}$$

$$D_{a,1} = \frac{n_e[\mu_e D_1 + \mu_1 D_e] + n_2[\mu_2 D_1 - \mu_1 D_2]}{n_e\mu_e + n_1\mu_1 + n_2\mu_2}, \tag{7.65}$$

$$D_{a,2} = \frac{n_e[\mu_e D_2 + \mu_2 D_e] + n_1[\mu_1 D_2 - \mu_2 D_1]}{n_e\mu_e + n_1\mu_1 + n_2\mu_2} \tag{7.66}$$

and

$$\bar{E}_s = -\frac{n_e D_e - [n_1 D_1 + n_2 D_2]}{n_e\mu_e + n_1\mu_1 + n_2\mu_2}\,\frac{\nabla n_e}{n_e}. \tag{7.67}$$

As you see, we end up with three different diffusion coefficients for the three different species.

We can make some appropriate simplifications by assuming that the positive ions are in thermal equilibrium with the neutral gas at temperature T, $(D_1\mu_2 = D_2\mu_1)$ and by assuming that

$$D_e >> D_1, D_2 ,$$

$$\mu_e >> \mu_1, \mu_2 .$$

Then,

$$D_{a,e} \simeq \frac{n_1}{n_e} D_{a,1} + \frac{n_2}{n_e} D_{a,2} , \tag{7.68}$$

$$D_{a,1} \approx D_1(1 + T_e/T) , \tag{7.69}$$

and

$$D_{a,2} = D_2(1 + T_e/T) . \tag{7.70}$$

Therefore, the ions do not influence each other but both influence the electrons and vice versa. Note that $D_{a,1}$ and $D_{a,2}$ are independent of space and time but $D_{a,e}$, with the density ratios involved, shows a time dependence.

Keep in mind that the derivation of the diffusion coefficients involved a restrictive assumption, the constant ratio approximation (equation 7.63). If this approximation is not valid, then it is usually impossible to define diffusion coefficients to be independent of the charged particle concentrations and their gradients. The use of a diffusion coefficient to represent the ratio between a particle flux and a density gradient is of questionable validity under such circumstances and it is better to retain the space charge fields explicitly as in equations 7.60 through 7.62.

### 7.1.7 Diffusion Cooling

As one last topic in diffusion, I would like to briefly discuss diffusion cooling as presented in a short paper by Biondi (1954) and in a later paper by Bhattacharya and Ingold (1972). In order to understand this effect, it is necessary to remember that there is a distribution of electron energies within the plasma, a fact that tends to be overlooked when we use the fluid equations.

As was pointed out before, in ambipolar diffusion a space charge field is set up which tends to retard the electrons and accelerate the ions. If we integrate this space charge field, we find that a potential barrier exists. Now, of course, electrons with energies greater than the barrier energy, which are the faster electrons in the distribution, can easily escape to the walls. These escaping electrons carry away more than the average electron energy and so the remaining electrons tend to be cooled by this process. In a decaying plasma, we can then have electron cooling by collisions with the neutral gas and by diffusion cooling. Of course, if the electron temperature drops below the gas temperature then the reverse process occurs and the electrons can be heated by collisions with the neutrals. At higher neutral pressures and for gases such as helium which have a very good thermal contact with the electrons, diffusion cooling does not play an important role. At low pressures in gases that have a poor thermal contact with the electrons, such as neon, argon, krypton and xenon, the diffusion cooling can be clearly seen as illustrated by Biondi's measurements of the ambipolar diffusion coefficient.

Remember

$$D_a \simeq D_+ \left(1 + \frac{T_e}{T}\right).$$

If $T_e = T$ then $D_a \simeq 2D_+$. If, however, $T_e < T$ then $D_+ < D_a < 2D_+$.

Biondi has given the plot shown in Fig. 7.8 of $D_a P$ versus pressure for afterglows in low pressure neon where T = 300 K and the characteristic diffusion length of the container $\Lambda$ = 0.735 cm. We can clearly see the effect of diffusion cooling. At the lowest pressure for which measurements were made, the diffusion coefficient corresponded to an electron temperature of ~ 30 K.

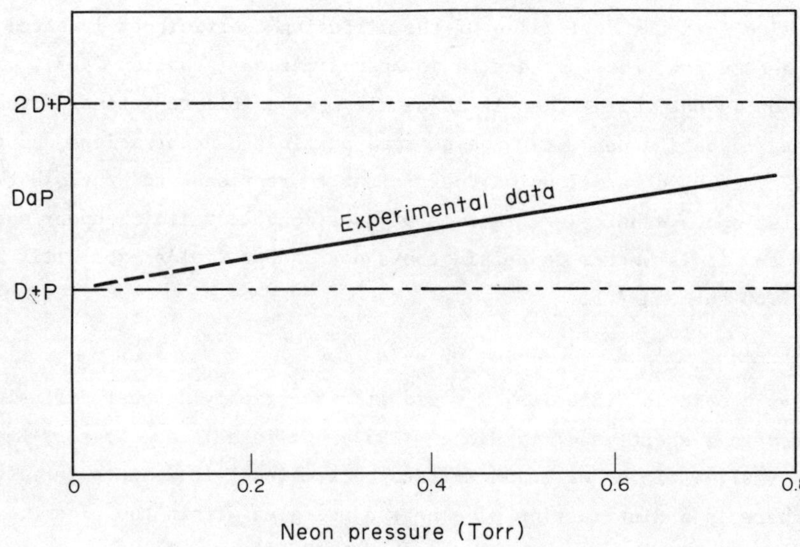


*Fig. 7.8. Normalized ambipolar diffusion coefficient vs. pressure in neon afterglows (from Biondi, 1954)*

## 7.2 RECOMBINATION

The second major electron loss process that we wish to discuss is recombination. In this process an electron and a positive ion unite to form a neutral molecule or atom in its ground state. The totality of microscopic processes involved in the capture of an electron to form an excited atom or molecule and the subsequent decay to the ground state neutral is known as recombination.

Recombination is essentially a volume, or chemical, process and the continuity equation when recombination dominates is written as

$$\frac{\partial n_+}{\partial t} = \frac{\partial n_e}{\partial t} = -\,\alpha n_e n_+ \tag{7.71}$$

where $\alpha$ is the recombination coefficient. In a quasi-neutral plasma containing electrons and one species of positive ions, we can write

$$n_e \simeq n_+$$

and

$$\frac{\partial n}{\partial t} = -\,\alpha n^2 . \tag{7.72}$$

The solution to this equation is

$$\frac{1}{n} = \frac{1}{n_o} + \alpha t . \tag{7.73}$$

If we plot $\frac{1}{n}$ vs t, we obtain a linear plot with an intercept at t = 0 of $\frac{1}{n_o}$ and a slope of $\alpha$ (Fig. 7.9)

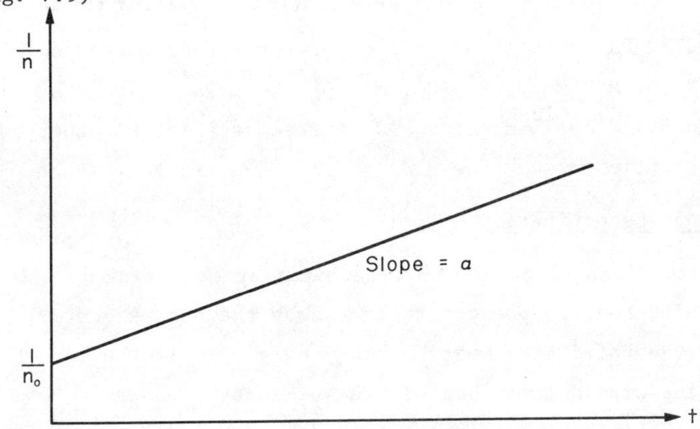


*Fig. 7.9. Inverse density vs. time for recombination decay*

This, of course, presents a convenient method of determining if the decay is recombination controlled and, if it is, the appropriate value of the recombination coefficient.

The time behaviour of the electron decay allows us to distinguish between diffusion and recombination decay. We can also distinguish between these processes by observing the spatial dependence of the electron density. Diffusion gives rise to spatial modes which are approximately sinusoidal in nature (depending on the geometry). Recombination, however, tends to produce a spatially uniform plasma. To understand this, remember that recombination is a volume loss process. The recombination rate depends on the local density and, therefore, recombination acts to flatten any non-uniformities in the plasma density.

The recombination process, being of a collisional microscopic nature is considerably more complicated in its details than diffusion. In a rather simple manner, however, we can appreciate that the probability that a recombination event will take place will depend upon the relative kinetic energies of the electron and the ion and, of more importance, how energy and momentum can be conserved in the recombination process. There are several different processes by which energy and momentum can be conserved and these give rise to different types of recombination. We shall outline some of them as follows:

### 7.2.1 Radiative Recombination

In this process, when an electron and ion recombine, the extra energy is given off in the form of radiation (photons).

$$X^+ + e \rightarrow X + h\nu \quad . \tag{7.74}$$

This process has a relatively low probability because of the difficulty of conservation of energy and momentum and so is usually important only at low densities where other types of recombination are very weak. As an example of the low probabilities, D. R. Bates and A. Dalgarno (1962) quote the total radiative recombination coefficients at 250 K for $H^+$, $He^+$, $Li^+$, $C^+$, $N^+$, $O^+$, $Ne^+$, $Na^+$ and $K^+$ to all be in the range of 3.0 to 4.8 × $10^{-18}$ $m^3$/sec. As we shall see later, other types of recombination give coefficients of the order of $10^{-12}$ $m^3$/sec.

### 7.2.2 Three-body Recombination

The probability of an electron-ion recombination occurring is substantially increased if a third body is present to take away the excess energy. Of course, the density of the third body must be reasonably high if the recombination event is going to occur with a third body in the vicinity.

Three body recombination can occur if the third body is a neutral, a positive ion or an electron. In general, there are roughly as many electrons available as positive ions, and since electrons are much more effective in removing excess energy, we can ignore the role of positive ions. We then have two important three body recombination processes to consider, electron collisional-radiative recombination

$$X^+ + e + e \rightarrow X + e \tag{7.75}$$

and neutral collisional-radiative recombination

$$X^+ + e + Y \rightarrow X + Y \ . \tag{7.76}$$

#### 7.2.2.1 Electron collisional radiative recombination

In this process, collisional recombination is not usually considered separately from radiative recombination since in a decaying plasma we tend to have collisional recombination at high densities and radiative at low densities. In a classic paper by D. R. Bates, A. E. Kingston and R. W. P. McWhirter (1962), the term Collisional Radiative Recombination was proposed to cover all of the rather complex processes involved in collisional and/or radiative capture and decay of the excited states to the ground state. Some of the individual processes that must be considered in order to obtain a collisional radiative recombination coefficient are

| | |
|---|---|
| $X^+ + e \rightarrow X'' + h\nu$ | (radiative capture) |
| $X^+ + e + e \rightarrow X'' + e$ | (collisional capture) |
| $X'' + e \rightarrow X^+ + e + e$ | (re-ionization) |
| $X'' + e \rightarrow X' + e$ | (collisional de-excitation) |
| $X' + e \rightarrow X'' + e$ | (excitation) |
| $X'' \rightarrow X' + h\nu$ | (line emission) |
| $X' + h\nu \rightarrow X''$ | (absorption) |

All of these processes have been considered in a statistical manner by Bates, Kingston and McWhirter. (Their approach has also been summarized by J. L. Delcroix, 1968).

Examples of collisional-radiative recombination coefficients for an optically thin $H^+$ ion plasma have been presented by Bates and Dalgarno (1962). They have shown that at 250 K, for example

$$\alpha = 8.8\text{x}10^{-17}\ \text{m}^3/\text{sec} \quad \text{for } n_e = 10^{14}/\text{m}^3$$

and increases to

$$\alpha = 2.6\text{x}10^{-11}\ \text{m}^3/\text{sec} \quad \text{for } n_e = 10^{20}/\text{m}^3 .$$

We can see quite a range in the recombination coefficient. Rather extensive tables and graphs of the electron density and temperature dependence of the recombination coefficients in hydrogenic plasmas ($H^+$, $He^+$ and pseudo-alkali ions) are contained in the paper by Bates, Kingston and McWhirter; the article by Bates and Dalgarno (1962) and the book by Massey and Gilbody (1974, Vol. IV).

An interesting comparison of calculated and measured electron collisional radiative recombination coefficients in cesium plasmas over a wide range of densities and temperatures is presented in Fig. 7.10 as obtained from Chen, Wu, Wu and Shaw (1973). A substantial increase in recombination coefficient with electron density and decrease with electron-temperature is evident. The theoretical predictions in Fig. 7.10 come from the work of Norcross and Stone (1968), who have shown that over limited ranges of the parameters, the recombination coefficient can be expressed as

$$\alpha = \alpha_o \exp\left[\frac{E_a}{(kT_e)^{1/2}}\right] \tag{7.77}$$

where $E_a$ and $\alpha_o$ are appropriate constants.

7.2.2.2 Neutral collisional radiative recombination

Neutral atoms or molecules are, of course, far less efficient in removing excess energy than are electrons, however, it is quite conceivable that in high pressure ionized media (multi-atmospheres), neutral collisional radiative recombination

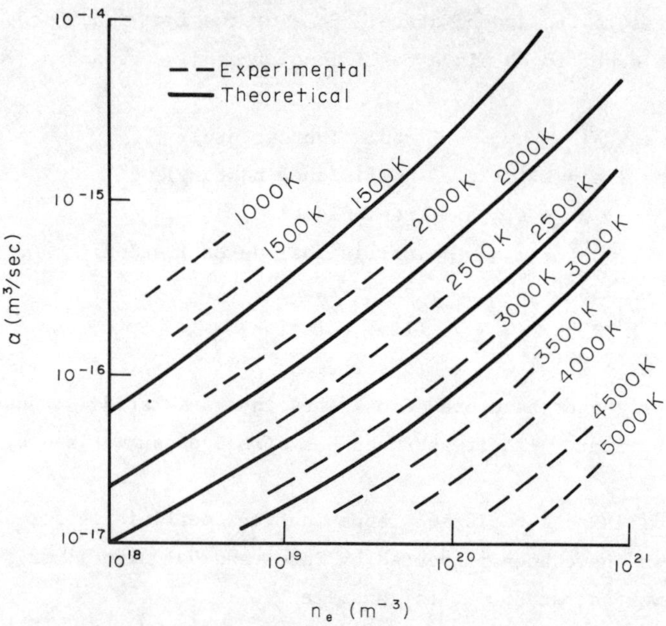


*Fig. 7.10. Measured and calculated electron-ion recombination coefficients in Cs (from Chen, Wu, Wu and Shaw, 1973), provided through the courtesy of Jet Propulsion Laboratory, California Institute of Technology*

could play a role. Bates and Khare (1965) analyzed this case in a manner similar to that for electron collisional radiative recombination. Bates, Malaviya and Young (1971) extended the analysis to molecular gases including the effects of molecular vibration and rotation.

Figure 7.11 shows the results for helium as calculated by Bates and Khare.

### 7.2.3 Dissociative Recombination

We now come to the predominant recombination process in gas discharges and plasmas (including the ionosphere). In this case a molecular ion plus an electron make a radiationless transition to an unstable state of the molecule which causes the atoms forming the molecule to move apart and gain kinetic energy due to their mutual repulsion.

$$XY^+ + e \rightleftarrows (XY)^* \rightarrow X + Y^* \quad . \qquad (7.78)$$

In this process X or Y could also be left in an excited state which could decay to the ground state. (For the rare gases, only one of the atoms can be in an excited state.) Potential energy curves illustrating this process are shown in Fig. 7.12. The intermediate state $(XY)^*$ can be formed by configuration interaction. In order to capture the electron (without transferring energy between electronic and nuclear motion) it is only necessary for the electron to have energy equal to the difference between the potential curves $XY^+ + e$ and $(XY)^*$. The circled point on

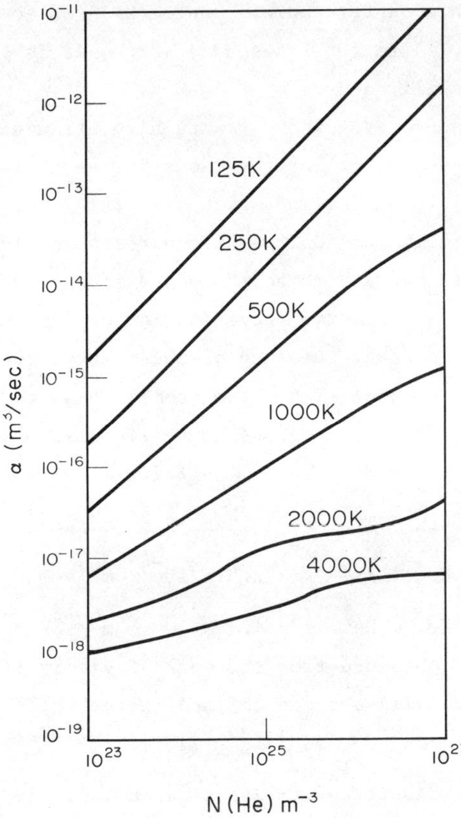


*Fig. 7.11.* *Neutral collisional radiative recombination coefficient for electrons with* $He^+$ *ions as a function of electron temperature and neutral helium density (from Bates and Khare, 1965, copyright The Institute of Physics)*

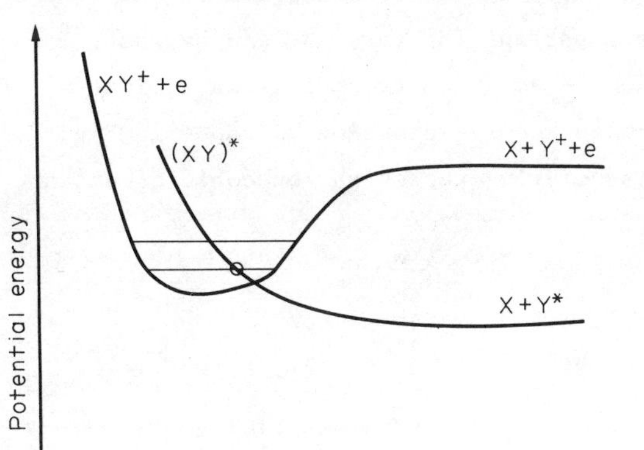


*Fig. 7.12.* *Potential energy curves related to dissociative recombination*

the curves indicate such a capture point. Once $(XY)^*$ is formed, the stabilization time is very short ($\sim 10^{-16}$ to $10^{-15}$ seconds) and so it is likely that dissociation into neutral atoms will occur.

The importance of this process is due to its high efficiency and large recombination coefficients. As we stated before, radiative recombination has a low probability and collisional recombination, although efficient, requires high densities. In dissociative recombination, we have a recombination event in which the positive ion provides its own "third body" to aid in energy and momentum conservation. Because of this, dissociative recombination coefficients are quite large, sometimes as much as $10^{-12}$ $m^3$/sec. Because of these large coefficients, this process can be important even in plasmas where atomic ions predominate since atomic ions can be converted into molecular ions by collisional processes. For instance

$$Ne^+ + 2Ne \rightarrow Ne_2^+ + Ne \qquad k = 4.4 - 7.7 \times 10^{-44}\ m^6/sec,$$

$$Ar^+ + 2Ar \rightarrow Ar_2^+ + Ar \qquad k = 1.9 \times 10^{-43}\ m^6/sec,$$

$$Kr^+ + 2Kr \rightarrow Kr_2^+ + Kr \qquad k = 2.3 \times 10^{-43}\ m^6/sec,$$

$$Xe^+ + 2Xe \rightarrow Xe_2^+ + Xe \qquad k = 1.8 \times 10^{-43}\ m^6/sec.$$

The conversion rates are obtained from the work of Vitols and Oskam (1972) and Bhattacharya (1970a,b,c, 1971) and are defined in the following sense

$$\frac{d}{dt}[X_2^+] = k[X^+]\,[X]^2 \qquad (7.79)$$

where [ ] represents the density of the species in $m^{-3}$. It is quite possible, therefore, to have sizeable molecular ion densities present. This is clearly shown in Fitzwilson and Chanin's (1973) measurements of the ratio of molecular to atomic ion densities in Ar, Kr and Xe discharges.

The presence of diatomic rare gas ions has led to the conclusion that dissociative recombination is important for rare gas discharges. The only exception is helium, where collisional-radiative recombination dominates. Presumably, this is due to the lack of a proper curve crossing as shown in Fig. 7.11.

Examples of some dissociative recombination coefficients are given in Table 7.1.

Table 7.1. Dissociative Recombination Coefficients ($m^3$/sec) @ 300 K

| Ion | $Ne_2^+$ | $Ar_2^+$ | $He_2^+$ | $Kr_2^+$ | $Xe_2^+$ | $N_2^+$ | $O_2^+$ | $NO^+$ | $CO_2^+$ |
|---|---|---|---|---|---|---|---|---|---|
| $\alpha$ | $1.8 \pm 0.1$ $\times 10^{-13}$ | $6.7 - 8.5$ $\times 10^{-13}$ | $< 10^{-15}$ | $1.2 \pm 0.1$ $\times 10^{-12}$ | $1.4 \pm 0.1$ $\times 10^{-12}$ | $2.7 \pm 0.3$ $\times 10^{-13}$ | 3 $\times 10^{-13}$ | $4.1 \pm 0.2$ $\times 10^{-13}$ | $3.8 \pm 0.5$ $\times 10^{-13}$ |

There are several excellent review articles on recombination in addition to the previously quoted references to Bates and Dalgarno (1962) and Massey and Gilbody

(1974). These are by H. J. Oskam (1969) and J. N. Bardsley and M. A. Biondi (1970).

7.2.3.1 Recombination at higher pressures

Very few measurements of recombination coefficients have been made at high pressures. Of those that have, the studies of R. E. Center (1973) on recombination in CO at pressures of up to 700 torr indicate the typical result. In this pressure range, the pressure itself seems to have only a small effect on the recombination coefficient. What seems to be more important is the production of more complex ions with larger recombination coefficients at high pressures. For instance, rather than $CO^+$, Center assumes that the dominant ion was $C_2O_2^+$ or possibly more complex. Similarly, nitrogen discharges at high pressures tend to have $N_4^+$ as the dominant ion.

7.2.4 Electron Temperature Dependence of Recombination

Recombination is very sensitive to electron temperature with a rather sharp decrease in $\alpha$ as the electron temperature increases. The theoretical calculations predict these variations and the experimental work of L. Goldstein (1955) has clearly shown this by demonstrating the sharp decrease ("quenching") of the recombination light emission in afterglow plasmas when the electron gas is momentarily heated by a microwave signal.

An example of the experimental observation of this effect is shown in Fig. 7.13. Immediately after the cessation of the discharge, the electron temperature is high so $\alpha$ is low. In the afterglow, the electron temperature drops rapidly (c) so that the recombination coefficient (e) rises rapidly to the steady state value corresponding to a roughly room temperature electron gas. The visible light emission (a) due to recombination is proportional to the net recombination rate, $\alpha\, n_e^2$, so it initially rises as $\alpha$ rises and then slowly decays as the electron density (d) decays. The light waveform shown in trace (a) is a characteristic signature of recombination light.

If at a later time a microwave heating pulse (b) is used to increase the electron temperature, the recombination coefficient will decrease correspondingly (e) and a decrease in the recombination light (a) will be seen. This effect shows the sensitivity of recombination to electron temperature.

The theoretical values of the electron temperature dependence of the recombination coefficient depends on the process under consideration. It is not always possible to give a broad prediction of the electron temperature dependence for recombination, however, as a first approximation, for radiative recombination

$\alpha \sim T_e^{-0.7}$,

for dissociative recombination $\alpha \sim T_e^{-1/2}$

and

for collisional-radiative recombination $\alpha \sim T_e^{-9/2}$.

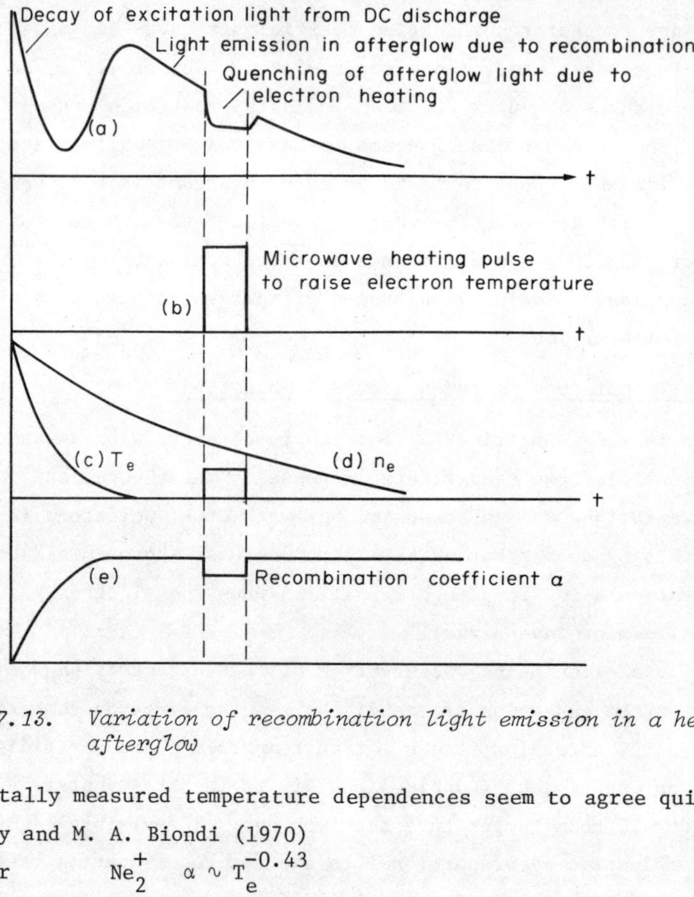


*Fig. 7.13. Variation of recombination light emission in a heated afterglow*

The experimentally measured temperature dependences seem to agree quite well. From J. N. Bardsley and M. A. Biondi (1970)

for $Ne_2^+$ $\alpha \sim T_e^{-0.43}$

$Ar_2^+$ $\alpha \sim T_e^{-0.67}$

$N_2^+$ $\alpha \sim T_e^{-0.39}$

$O_2^+$ $\alpha \sim T_e^{-0.70}$ (up to 1200 K) and $\sim T_e^{-0.56}$ (above 1200 K).

Now, $\alpha$ can also be decreased if the ion temperature is increased. For instance,

$Ne_2^+$ $\alpha \sim T_+^{-(0.42 \pm 0.4)}$ (see Bardsley and Biondi, 1970).

For an additional two references which show rather well the ion conversion and recombination processes that occur in discharges of interest, see Gerber, Sauter and Oskam (1966) and Sauter, Gerber and Oskam (1966).

### 7.2.5 Electron Density Decay in Plasmas with Diffusion and Recombination

If we consider the case when both diffusion and recombination are playing an important role in the plasma decay, the continuity equation becomes

$$\frac{\partial n_e}{\partial t} = - \alpha n_e^{\ 2} + Da\nabla^2 n_e \ . \tag{7.80}$$

This is a nonlinear equation and so no simple solution is available. In some cases, however, approximate solutions can be found. For instance, if the recombination is small with respect to diffusion, we can assume a fundamental diffusion mode with a decay time $\tau$ and rewrite

$$\frac{\partial n_e}{\partial t} = - \alpha n_e^{\ 2} - \frac{n_e}{\tau} \ . \tag{7.81}$$

An approximate solution to this equation is

$$\frac{n_e}{1 + \alpha\tau n_e} = n_e(o)\ e^{-\ (t/\tau)/(1+\alpha\tau n_e(o))} \ . \tag{7.82}$$

In the more general case, however, it is necessary to use a more sophisticated analysis. H. E. Wilhelm (1970a,b) has approached this problem from an analytical viewpoint. Similar forms of analysis have also been given by J. Crank (1956). Solutions to specific problems involving diffusion and recombination have been given by H. J. Oskam (1958) and by L. Frommhold and M. A. Biondi (1968). The reader may also be interested in referring to J. L. Delcroix (1968, Vol. 2) for an interesting evaluation of the relative roles of diffusion and recombination in plasmas for a range of plasma parameters.

## 7.3 ATTACHMENT

An important electron loss process in certain gases is electron attachment. In this process, electrons are captured by electro-negative gas atoms, or molecules, to form negative ions. As one would expect, this process can happen fairly easily in gases containing atoms that lack one (or two) outer electrons, such as the halogens. For this process, we can write

$$\frac{\partial n_e}{\partial t} = - \nu_a n_e = - h\nu_e n_e \tag{7.83}$$

where $\nu_a$ = collision frequency for attachment, $\nu_e$ is the electron neutral collision frequency and h is the attachment coefficient (attachments/collision). Obviously, this results in an exponential time decay of the electron density and since attachment, like recombination, is a volume process, the electron density distribution tends to be uniform in an attachment dominated plasma.

In order for the electron to be captured by a neutral atom or molecule, it is desirable that the total energy of the negative ion be <u>lower</u> than the energy of the neutral atom or molecule. This is related to the <u>electron affinity</u>, which is the difference in the ground state energies of the neutral atom and the negative atom. An example of a molecule forming a stable negative ion is $O_2$. As expected, larger electron affinites give rise to more stable negative ions. Several ions and their electron affinities are given as follows.

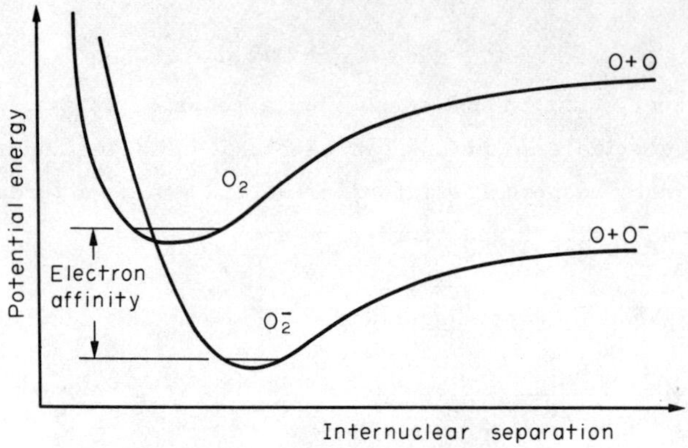


*Fig. 7.14. Potential energy curves for $O_2$ and $O_2^-$*

| Ion | Affinity |
|---|---|
| $NO^-$ | < .43 eV |
| $O_2^-$ | .43 eV |
| $OH^-$ | 1.83 eV |
| $NO_2^-$ | 2.8 eV |

A good listing is given in The Handbook of Chemistry and Physics.

It is also possible to have negative ions with ground state energies greater than the ground state energy of the neutral. An example of this is $H_2^-$ (see Branscomb, 1962). In this case the electron affinity is negative and this class of negative ions is very unstable and their lifetimes are very short. Similar unstable negative ions are $N_2^-$ and $N^-$.

Attachment is a collisional process and, like recombination, several different processes can occur which will result in the formation of a negative ion. The important attachment processes are: Radiative Attachment, Dissociative Attachment and Three-body Attachment.

### 7.3.1 Radiative Attachment

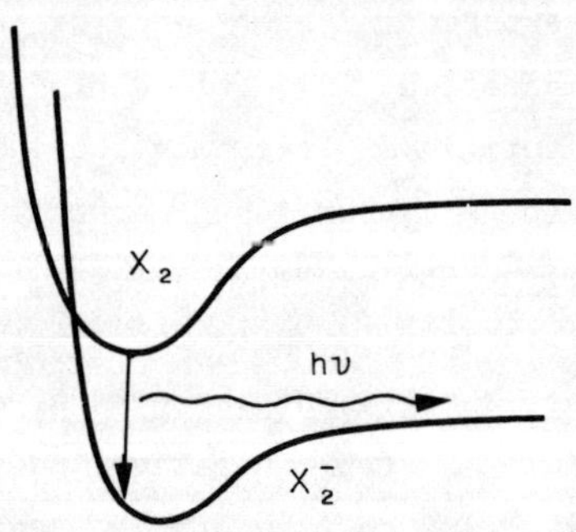


This is the inverse of photo-detachment.

*Fig. 7.15. Example of radiative attachment*

In this case, the energy difference results in the emission of radiation. An example is

$$O_2 + e \rightarrow O_2^- + h\nu . \tag{7.84}$$

Other negative ions that are formed by radiative attachment are

$$O^-, H^-, OH^-, NO_2^- .$$

7.3.2 Dissociative Attachment

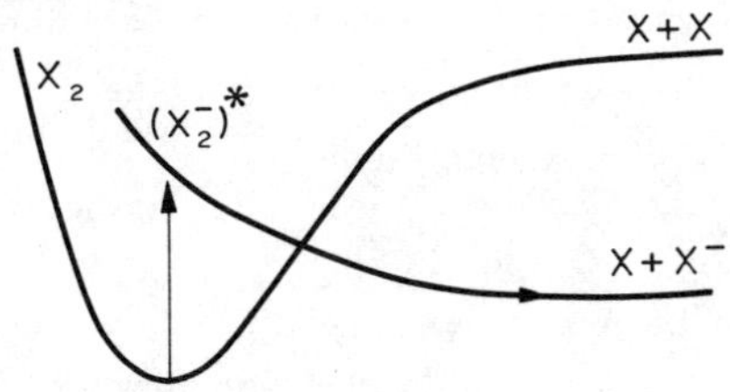


*Fig. 7.16. Example of dissociative attachment*

In this case, the electron energy is sufficient to excite the molecule to an unstable state which dissociates into a neutral and a negative atom.

Some examples are:

$$e + O_3 \rightarrow O^- + O_2 , \tag{7.85}$$

$$e + O_2 \rightarrow O^- + O , \tag{7.86}$$

$$e + CO_2 \rightarrow CO + O^- , \tag{7.87}$$

$$e + H_2O \rightarrow H + OH^- . \tag{7.88}$$

(Also see Section 3.3.5.) These last two processes (and their close equivalents) are very important in $CO_2$ lasers.

7.3.3 Three-body Attachment

In this case, the third body helps to conserve energy and momentum in the process. Examples are:

$$e + O_2 + N_2 \rightarrow O_2^- + N_2 , \tag{7.89}$$

$$e + O_2 + O_2 \rightarrow O_2^- + O_2 . \tag{7.90}$$

This last case may actually be a two step process (Bloch-Bradbury)

$$O_2 + e \rightarrow O_2^{-*} , \tag{7.91}$$

$$O_2^{-*} + O_2 \rightarrow O_2^- + O_2^* . \tag{7.92}$$

There are a host of other gases which attach electrons. Some of the atomic species are

$$C^-, F^-, S^-, Cl^-, Br^-, I^- .$$

Fairly complete tables of the binding energies of atomic negative ions are given by L. M. Branscomb (1962) and Bruce Steiner (1972).

Some more molecular species are:

HCl, HBr

$BCl_3$, $SF_6$, $CCl_4$, $CF_3I$, $CCl_2F_2$.

Some of these latter species, such as $SF_6$ and freon, are of considerable interest for use in suppressing the plasma sheath and the subsequent communication blackout experienced by re-entering space vehicles.

Let us now consider the attachment rates for various species of interest. One of the most important examples of attachment is oxygen. In this case we have two-body and three-body processes occurring and

$$\frac{\partial n_e}{\partial t} = - \nu_a n_e = - \beta n(O_2) n_e - K n^2(O_2) n_e \ . \tag{7.93}$$

The values of β and K have been measured and are shown in Figs. 7.17 and 7.18. The articles by Branscomb (1964) and Prasad and Craggs (1962) give a good review of attachment and detachment processes. The book by S. C. Brown (1966) and the article by Rapp and Briglia (1965) provide data on attachment cross sections for a variety of gases.

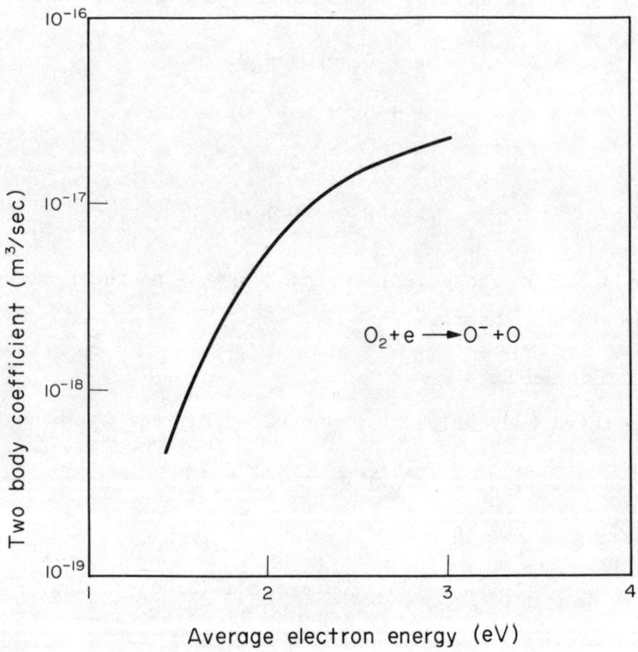


*Fig. 7.17. Two body attachment coefficient β, in $O_2$ (from Chanin Phelps and Biondi, 1959)*

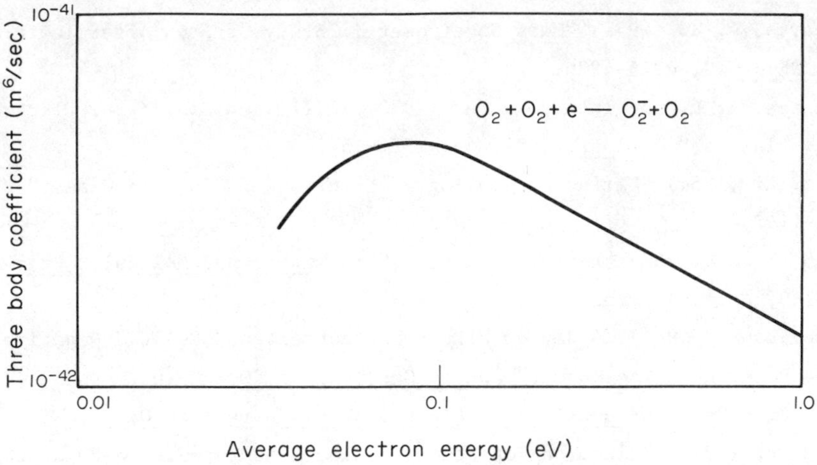


*Fig. 7.18. Three body attachment coefficient, K, in $O_2$ at 300 K (from Chanin, Phelps and Biondi, 1959)*

## 7.4 REFERENCES

Allis, W. P. (1956) "Motions of Ions and Electrons" MIT Research Laboratory of Electronics Technical Report 299.

Allis, W. P., and Rose, D. J. (1954) "Transition from Free to Ambipolar Diffusion," Phys. Rev. 93 84-93.

Bardsley, J. N., and Biondi, M. A. (1970) "Dissociative Recombination," in Advances in Atomic and Molecular Physics 6 pp. 1-57, Academic Press.

Bates, D. R. and Dalgarno, A. (1962) "Electronic Recombination," in Atomic and Molecular Processes, pp. 245-271, Academic Press.

Bates, D. R. and Khare, S. P. (1965) "Recombination of Positive Ions and Electrons in a Dense Neutral Gas," Proc. Phys. Soc. 85, 231-243.

Bates, D. R., Kingston, A. E., and McWhirter, R. W. (1962) "Recombination between Electrons and Atomic Ions. I. Optically Thin Plasmas," Proc. Roy. Soc. A 267, 297-312.

Bates, D. R., Malaviya, A. and Young, N. A. (1971) "Electron-ion Recombination in a Dense Molecular Gas," Proc. Roy Soc. A 320, 437-458.

Bhattacharya, A. K. (1970a) "Mass Spectrometric Study of Argon Afterglow Plasmas," J. Appl. Phys. 41, 1707-1710.

Bhattacharya, A. K. (1970b) "Ion Conversion in Krypton Afterglow Plasmas," Phys. Rev. A1, 1196-1200.

Bhattacharya, A. K. (1970c) "Ion Conversion in Xenon Afterglow Plasmas," Appl. Phys. Letters 17, 521-522.

Bhattacharya, A. K. (1971) "Mass Spectrometric Study of Neon Afterglow Plasmas," J. Appl. Phys. 42, 4216-4220.

Bhattacharya, A. K. and Ingold, J. H. (1972) "Diffusion Cooling of Electrons in Afterglow Plasmas," J. Appl. Phys. 43, 1535-1542.

Biondi, M. A. (1954) "Diffusion Cooling of Electrons in Ionized Gases," Phys. Rev. 93, 1136-1140.

Branscomb, L. M. (1962) "Photo detachment," in Atomic and Molecular Processes, pp. 100-140, Academic Press.

Branscomb, L. M. (1964) "A Review of Photodetachment and Related Negative Ion Processes Relevant to Aeronomy," Ann de Géophysique 20, 88-105.

Brown, S. C. (1966) Introduction to Electrical Discharges in Gases, Wiley & Sons.

Center, R. E. (1973) "Electron-ion Recombination Measurements in CO at High Pressures," J. Appl. Phys. 44, 3538-3542.

Chanin, L. M., Phelps, A. V. and Biondi, M. A. (1959) "Measurement of the Attachment of Slow Electrons in Oxygen," Phys. Rev. Letters 2, 344-346.

Chen, Che Jen, Wu, James T., Wu, Frank T., and Shaw, David T. (1973) "Measurement of Electron-ion Recombination Rate of a Dense High-temperature Cesium Plasma," J. Appl. Phys. 44, 3052-3054.

Crank, J. (1956) Mathematics of Diffusion, Oxford.

Delcroix, J. L. (1968) Plasma Physics, Vol. 2, Wiley & Sons.

Fitzwilson, R. L., and Chanin, L. M. (1973) "Positive Ion Ratio Measurements in Ar, Kr and Xe Glow Discharges," J. Appl. Phys. 44, 5337-5346.

Frommhold, Lothar and Biondi, M. A. (1968) "A Mathematical Study of the Electron Decay in Diffusion and Recombination Controlled Afterglows," Annals of Physics 48, 407-440.

Gerber, R. A. and Gerardo, J. B. (1973) "Ambipolar-to-Free Diffusion: The Temporal Behavior of the Electrons and Ions," Phys. Rev. A7, 781-790.

Gerber, R. A., Sauter, G. F., and Oskam, H. J. (1966) "Studies of Decaying Helium Plasmas," Physica 32, 2173-2191.

Goldstein, L. (1955) "Electrical Discharge in Gases and Modern Electronics," in Advances in Electronics and Electron Physics 7, pp. 339-503, Academic Press.

Kregel, Mark D. (1970) "Diffusion in Decaying Plasmas," J. Appl. Phys. 41, 1978-1983.

Massey, H. S. W., and Gilbody, H. B. (1974) Electronic and Ionic Impact Phenomena, 2nd ed., Vol. IV, Oxford.

Norcross, D. W., and Stone, P. M. (1968) "Recombination, Radiative Energy Loss and Level Populations in Non-equilibrium Cesium Discharges," J. Quant. Spectr. Radiat. Transfer 8, 655-684.

Oskam, H. J. (1958) "Microwave Investigation of Disintegrating Gaseous Discharge Plasmas," Philips Res. Repts. 13, 335-457.

Oskam, H. J. (1969) "Recombination of Rare Gas Ions with Electrons," Case Studies in Atomic Collision Physics 1, 463-523.

Persson, K. B. (1962) "Inertia Controlled Ambipolar Diffusion," Phys. Fluids 5, 1625-1632.

Prasad, A. N., and Craggs, J. D. (1962) "Attachment and Ionization Coefficients," in Atomic and Molecular Physics, pp. 206-244, Academic Press.

Rapp, D., and Briglia, D. D. (1965) "Total Cross Sections for Ionization and Attachment in Gases by Electron Impact. II. Negative Ion Formation," J. Chem. Phys. 43, 1480-1489.

Sauter, G. F., Gerber, R. A., and Oskam, H. J. (1966) "Studies of Decaying Plasmas Produced in Neon and Helium-Neon Mixtures," Physica 32, 1921-1937.

Shimony, Z., and Cahn, J. H. "Time Dependent Ambipolar Diffusion Waves," Phys. Fluids 8, 1704-1707.

Steiner, Bruce (1972) "Photodetachment: Cross Sections and Electron Affinities," Case Studies in Atomic Collision Physics 2, 483-545.

Tanenbaum, B. Samuel (1967) Plasma Physics, McGraw-Hill.

Vitols, A. P., and Oskam, H. J. (1972) "Reaction Rate Constant for $Ne^+ + 2Ne \rightarrow Ne_2^+ + Ne$," Phys. Rev. A5, 2618-2622.

Wilhelm, H. E. (1970a) "Analytical Solution of the Boundary-value problem for the Nonlinear Helmholtz Equation," J. Math. Phys. 11, 824-826.

Wilhelm, H. E. (1970b) "Analysis of Nonlinear Initial-Boundary-value problems in Recombination and Diffusion Kinetics," J. Chem Phys. 53, 1677-1682.

# Chapter 8

# DC DISCHARGES — THE POSITIVE COLUMN

Now that we have studied the important electron loss processes that determine the behavior of a plasma when there is no electron source, it is appropriate to consider the other case: when the source and loss terms balance. This is the steady state condition that applies to DC discharges and to high frequency RF discharges and we shall apply the analysis specifically to the positive column. This is the most common regime for gas discharges and gas lasers and so this analysis is of particular interest.

Depending on the pressure, current and gas medium that is under consideration, the dominant electron loss mechanism can be diffusion, recombination or attachment. The most common regime is when the loss of electrons by diffusion balances the production of electrons by collisional ionization. This case will be considered first (section 8.1) and then the situation where attachment or recombination dominates the electron loss will be considered (section 8.2).

## 8.1 DIFFUSION-DOMINATED DISCHARGES

To analyze this case, we use the fluid equations. The continuity equation becomes

$$\nabla \cdot n_e \bar{v}_e = n_e \nu_I = \nabla \cdot n_+ \bar{v}_+ \tag{8.1}$$

where $\nu_I$ is the electron collision frequency for ionization. We assume here that the source term is one-step ionization of a neutral by an electron.

For momentum conservation, the nonlinear terms (which are small) are ignored and

$$\bar{v}_e = -\mu_e \bar{E} - D_e \frac{\nabla n_e}{n_e}\,, \quad (8.2) \qquad \bar{v}_+ = \mu_+ \bar{E} - D_+ \frac{\nabla n_+}{n_+}\,. \quad (8.3)$$

In these equations we assume that the temperatures are constant and we do not use the equation for energy conservation. We will remove this restriction later.

We shall assume that ambipolar diffusion applies, that $\bar{\Gamma}_e = \bar{\Gamma}_+$ and $n_e = n_+ = n$, as usual. We then have

$$\nabla \cdot (n_e\bar{v}_e - n_+\bar{v}_+) = 0 = -\mu_e\nabla \cdot n\bar{E} - D_e\nabla^2 n - \mu_+\nabla \cdot n\bar{E} + D_+\nabla^2 n \ . \tag{8.4}$$

We also have

$$\nabla \cdot \ n_+\bar{v}_+ = \mu_+\nabla \cdot n\bar{E} - D_+\nabla^2 n = n\nu_I \tag{8.5}$$

from which

$$\nabla \cdot n\bar{E} = \frac{1}{\mu_+}\ [n\nu_I + D_+\nabla^2 n] \ . \tag{8.6}$$

Substituting back, we obtain

$$\nabla^2 n \left[\frac{\mu_+ D_e + \mu_e D_+}{\mu_e + \mu_+}\right] + n\nu_I = 0 \tag{8.7}$$

or

$$D_a\nabla^2 n + n\nu_I = 0 \ . \tag{8.8}$$

Equation 8.8 can be re-written as

$$\nabla^2 n + n\left(\frac{\nu_I}{D_a}\right) = 0 \ . \tag{8.9}$$

This equation allows us to solve for the electron density distribution in the positive column.

In cylindrical coordinates, with no $\phi$ or z variations

$$\frac{1}{r}\frac{\partial}{\partial r}\ r\ \frac{\partial n}{\partial r} + n\left(\frac{\nu_I}{D_a}\right) = 0 \ . \tag{8.10}$$

The solution of this equation, as we have seen before, is a zero order Bessel function

$$n = n_o J_o(\beta_o r) \ . \tag{8.11}$$

If one step ionization is the predominant electron source, then the zero order Bessel function describes the electron density spatial distribution across a cylindrical discharge tube.

We know that at the walls of the discharge tube, r = R, n = 0 (approximately) therefore

$$\beta_o^{\ 2} = \left(\frac{2.4}{R}\right)^2 = \frac{\nu_I}{D_a} \ . \tag{8.12}$$

The radial variation of n is therefore determined and, as you will note, so is $\nu_I$ if $D_a$ and R are specified. This is a statement of the balance condition. As many electrons must be produced per second as are lost to the walls by diffusion. As we shall see a little later this condition allows us to determine the electron energy (temperature) in the discharge. Before we take that step, let us determine the magnitude of the electron density now that we have determined its radial variation.

We determine the magnitude of n by solving for the current (an external variable).

$$I = \int_0^R 2\pi r\ n(r)e(v_{+z} - v_{ez})dr \ . \tag{8.13}$$

Now

$$v_{+z} = \mu_+ E_z = \frac{e}{M\nu_{iN}}\ E \tag{8.14}$$

and

$$v_{ez} = -\mu_e E_z = \frac{-e}{m\nu_{eN}} E . \quad (8.15)$$

Therefore,

$$I = 2\pi n_o \left( \frac{1}{M\nu_{iN}} + \frac{1}{m\nu_{eN}} \right) Ee^2 \int_o^R rJ_o(\beta_o r)dr . \quad (8.16)$$

The integral becomes

$$\frac{R}{\beta_o} J_1(\beta_o R) = 0.5191 \frac{R}{\beta_o} \quad (8.17)$$

so that, solving for $n_o$, we obtain

$$n_o = \frac{0.737 \; I}{e^2 ER^2(1/M\nu_{iN} + 1/m\nu_{eN})} . \quad (8.18)$$

If we can assume that the longitudinal electric field and the electron energy are reasonably constant for a fixed gas, pressure and discharge tube radius, then $n_o$ varies directly with I.

### 8.1.1 Electron Temperature in the Diffusion-dominated Discharge

As we said before, in order to balance the diffusion loss, $\nu_i$ is fixed by the gas, pressure and discharge tube radius

$$\nu_I = D_a \left( \frac{2.4}{R} \right)^2 . \quad (8.19)$$

We also know for a given gas and pressure, $\nu_I$ is determined only by the electron temperature (assuming a Maxwellian distribution) so we now can solve for $T_e$ by solving for $\nu_I$ as a function of $T_e$ and choosing the appropriate $\nu_I$. Before we begin, it should be pointed out that the expression for the electron temperature that will be derived here assumes that electrons have a Maxwellian electron energy distribution (which is usually only valid at high electron densities) and that one-step ionization is the dominant electron source (which is not the typical situation). A more accurate analysis of the positive column would require that we correct these assumptions, as will be discussed in section 8.1.3, so the analysis that follows should be considered simply as an illustration of a method and not as an accurate representation of the behavior of a positive column.

To obtain $\nu_I$, we must evaluate

$$\nu_I = N < \sigma_I v > \quad (8.20)$$

where $\sigma_I$ is the cross section for an ionizing collision. Fortunately, in most cases $\sigma_I$ can be approximated as a linear function of the electron energy for energies greater than some threshold value (see Brown, 1959).

So, for a given gas we can choose an ionizing threshold energy $\varepsilon_i$ and a slope, a, to describe $\sigma_I$. We then let

$$\sigma_I = a(\varepsilon - \varepsilon_i) \qquad \varepsilon > \varepsilon_i \quad (8.21)$$

and

$$\nu_I = Na < (\varepsilon - \varepsilon_i)v > \qquad \varepsilon > \varepsilon_i . \quad (8.22)$$

For a Maxwellian distribution

$$f(\varepsilon) = \frac{2}{\sqrt{\pi}(kT_e)^{3/2}} \sqrt{\varepsilon}\, e^{-\varepsilon/kT_e} . \tag{8.23}$$

The ionization frequency is

$$\nu_I = \int_{\varepsilon_i}^{\infty} \frac{2}{\sqrt{\pi}} \frac{\sqrt{\varepsilon}}{(kT_e)^{3/2}} e^{-\varepsilon/kT_e} Na(\varepsilon - \varepsilon_i) \sqrt{\frac{2\varepsilon}{m}}\, d\varepsilon . \tag{8.24}$$

Integrating, one obtains

$$\nu_I = \frac{4\sqrt{2}}{\sqrt{\pi}} Na \frac{(kT_e)^{3/2}}{m^{1/2}} e^{-\varepsilon_i/kT_e} \left[1 + \frac{\varepsilon_i}{2kT_e}\right] . \tag{8.25}$$

If it is assumed that $\varepsilon_i >> k\,T_e$, which is usually the case, and we take

$$D_a \simeq \frac{kT_e}{e} \mu_+ \quad (\text{i.e. } T_e >> T_+) . \tag{8.26}$$

Then equation 8.19 simplifies to

$$\frac{e^{\varepsilon_i/kT_e}}{(\varepsilon_i/kT_e)^{1/2}[1 + \varepsilon_i/2kT_e]} = \frac{0.5519\ e}{\sqrt{m}} \left[\frac{a\sqrt{\varepsilon_i}}{\mu_+ N}\right] N^2R^2 . \tag{8.27}$$

The factor in the bracket on the right hand side is a constant that depends only on the nature of the gas. The interpretation of equation (8.27) is that for a given gas, the electron temperature is a unique function of the density-radius product (N R) or the pressure-radius product (P R) and is <u>independent</u> of the electron density or current. A graphical solution of equation (8.27) is shown in Fig. 8.1.

In spite of the fact that experimental studies of electron temperature as a function of density-radius product in rare gas discharges have shown rather good agreement with this theory (see Brown, 1959, and Von Engel, 1965), it should be pointed out that the assumptions of a Maxwellian distribution and one step ionization are <u>not</u> typically met in these discharges, so the good agreement is mostly fortuitous and somewhat misleading (see Kagan and Lyagushchenko, 1962).

### 8.1.2 Longitudinal Electric Field in the Diffusion-dominated discharge

The simplest approach to determining the electric field along the positive column is to balance the energy gain from the electric field with the energy loss due to elastic collisions. Inelastic energy losses should also be included if they constitute a significant energy loss. That effect is usually very important, but since a general expression can't be easily formulated, inelastic processes won't be considered. Once again, consider this analysis simply as an illustration. Sections 4.7.4 and 12.4.3 will give a clearer picture of the role of inelastic collisions.

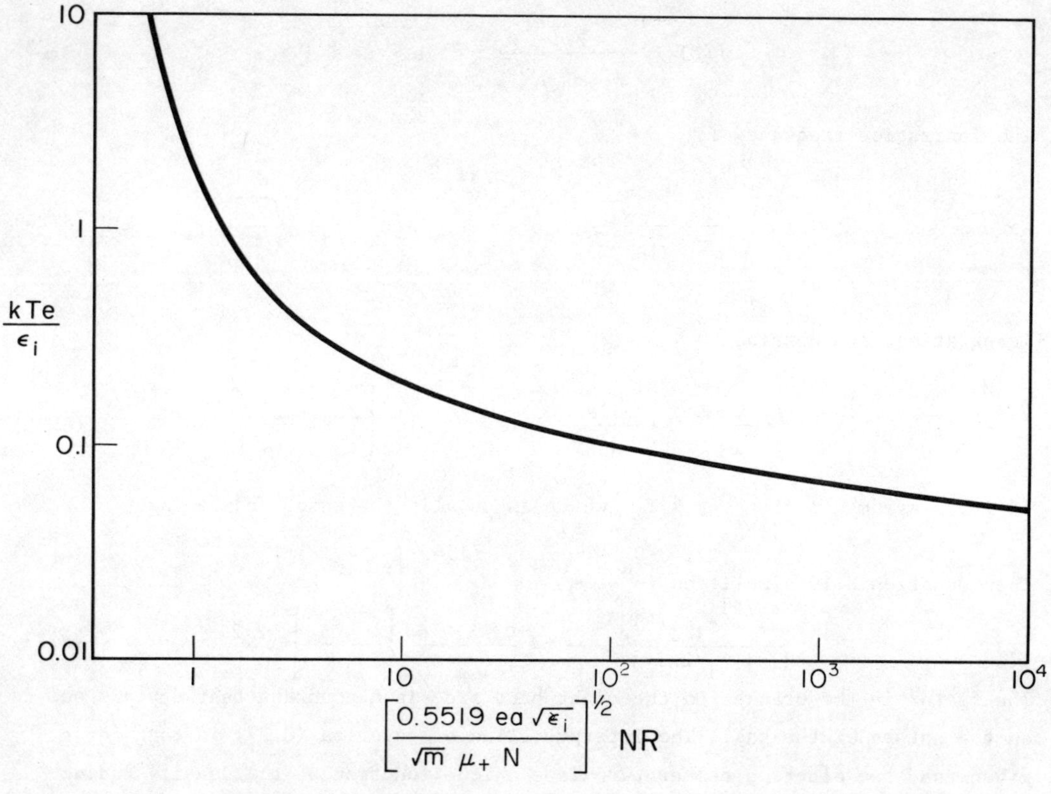


*Fig. 8.1. Normalized electron temperature vs normalized pressure density product*

Since uniform temperatures have been assumed, no heat flow terms need be included. The energy gained by the electrons per unit volume per second is

$$P = \bar{J} \cdot \bar{E} = \sigma E^2 . \tag{8.28}$$

On a per electron basis

$$dP_{gain} = \frac{\sigma E^2}{n_e} . \tag{8.29}$$

Assuming the energy loss is primarily due to elastic collisions, the energy loss per second per electron is

$$dP_{loss} = \delta < \nu_m \frac{mv^2}{2} > . \tag{8.30}$$

Balancing the energy gain and loss

$$\frac{\sigma E^2}{n_e} = \delta < \nu_m \frac{mv^2}{2} > . \tag{8.31}$$

Rewriting $\nu_m = N \sigma_m v$ and $\sigma = \sigma_o/N$, where $\sigma_o$ is independent of the neutral gas density, equation (8.31) can be written as

$$\left(\frac{E}{N}\right)^2 = \frac{\delta}{(\sigma_o/n_e)} < \sigma_m v \frac{mv^2}{2} > . \tag{8.32}$$

The right hand side of equation (8.32) is a function of the gas and the electron energy only. Within the previous assumption that the electron temperature (energy) is determined by the N R product, then we can write

$$\frac{E}{N} = f(NR) \ . \qquad (8.33)$$

For a fixed density-radius product, the longitudinal electric field, and hence the voltage drop along a discharge, is fixed and is <u>independent of the current through the discharge</u>. In spite of the oversimplified analysis, this conclusion is reasonably valid in many diffusion controlled positive column discharges.

Returning to the simplest example,

$$\nu_m = \text{constant} = \nu_o N \qquad (8.34)$$

then

$$\sigma_o = \frac{n_e e^2}{m\nu_o} \qquad (8.35)$$

and

$$< \frac{1}{2} mv^2 > = \frac{3}{2} kT_e \ . \qquad (8.36)$$

Therefore equation (8.32) becomes

$$\left(\frac{E}{N}\right)^2 = \frac{3}{2} \frac{\delta m\nu_o^2}{e^2} kT_e \ . \qquad (8.37)$$

We can also derive another interesting result from equation (8.31) under the assumption that $\nu_m$ is constant. Equation (8.31) then becomes

$$\frac{e^2E^2}{m\nu_m} = \delta\nu_m < \frac{mv^2}{2} > \qquad (8.38)$$

or

$$\frac{e^2}{m^2\nu_m^2} E^2 = \frac{\delta}{2} < v^2 > . \qquad (8.39)$$

Under these assumptions, the longitudinal drift velocity

$$\bar{u} = -\mu_e \bar{E} = -\frac{e}{m\nu_m} \bar{E} \ , \qquad (8.40)$$

therefore

$$u^2 = \frac{\delta}{2} < v^2 > \ . \qquad (8.41)$$

This is a simple relationship worth remembering when dealing with the positive column.

### 8.1.3 Deviations from the Simple Theory

Quite a number of simplifying assumptions, some usually unjustified, were made in the previous sections in order to obtain a simple and reasonably accurate picture of the diffusion dominated positive column discharge. We assumed:

1. One step ionization
2. Diffusion loss only
3. Ambipolar diffusion
4. Maxwellian electron distribution
5. Uniform temperature
6. Single species of positive ions
7. No contribution from excited species in the discharge.

We will want to examine some of the deficiencies of the simple theory in some detail, but first let us discuss a few of the simpler cases. If more than one species of positive ion are present, or if negative ions are present, the only basic difference in a steady state situation has already been suggested in Section 7.1.6. The electron diffusion coefficient will have to be modified. Konyukov (1958), Seeliger (1949) and Wilhelm (1960) have considered such cases.

The role of excited species and two-step ionization is also closely related to the form of the distribution function. Inelastic collisions of electrons with ground-state atoms or molecules will tend to make the distribution function very non-Maxwellian (except at high currents) as was previously shown. This will markedly decrease the one step ionization rate (for the same average electron energy) from that calculated for a Maxwellian distribution. If one step ionization dominated, this would suggest that the average electron energy would have to be higher than that calculated in the previous section. What seems to be the case, however, is that two-step ionization plays an important role. Inelastic collisions produce excited species in the discharge and the lowest energy species are quite often produced in abundance (as will be discussed in the next chapter). These excited species are easily ionized and can, in fact, dominate the ionization process. Vriens (1974) has shown that this two step ionization can dominate in rare gases and it is well known (Waymouth and Bitter, 1956) that two-step ionization dominates in metal vapor discharges. Under such circumstances, the combination of the non-Maxwellian distribution and two step ionization tend to give roughly the same result as was previously obtained for an assumed one step ionization and a Maxwellian distribution.

For two-step ionization, equation (8.9) becomes

$$\nabla^2 n + Cn^2 = 0 \tag{8.42}$$

where C is some constant. The spatial electron distribution corresponding to this equation has been obtained by Spenke (1950) and, in fact, turns out to be somewhat similar to the Bessel function distribution previously obtained, so the Bessel function assumption of the spatial distribution in cylindrical coordinates turns out to be quite good regardless of the ionization process.

As was mentioned in the previous section, an important property of the positive column is that the voltage drop along a positive column is a function of the N R product and is independent of the current. Actually this is not quite true, as might be surmised from the comments on the limitations of the theory. The voltage drop along a positive column actually decreases slightly as the current increases, giving the discharge a negative resistance characteristic which is usually compensated for by a series resistance (ballast). This drop in E/N can be explained in part by the changing distribution function as the current is increased and also by the role of two step ionization. There are at least two more effects that are listed in the deficiencies in the simple theory that also contribute to this effect. The assumption of ambipolar diffusion breaks down at low currents where

$\Lambda/\Lambda_D \simeq 1$, and the assumption of uniform temperature breaks down at high currents when gas heating becomes important. Ingold (1975) considered the role that diffusion plays as the regime goes from free diffusion through the transition regime to ambipolar diffusion and obtained criteria very similar to that of Gerber and Gerardo (1973) for defining where transition diffusion becomes important. At low currents, the diffusion coefficient is greater than the ambipolar value, necessitating a higher ionization rate (or electron energy) resulting in a larger E/N or longitudinal electric field. This effect was shown clearly by Ecker (1954). A comparison of Ecker's theory and experimental results is shown in Fig. 8.2. We see that going to lower currents causes a sharp rise in the longitudinal electric field in accordance with theory.

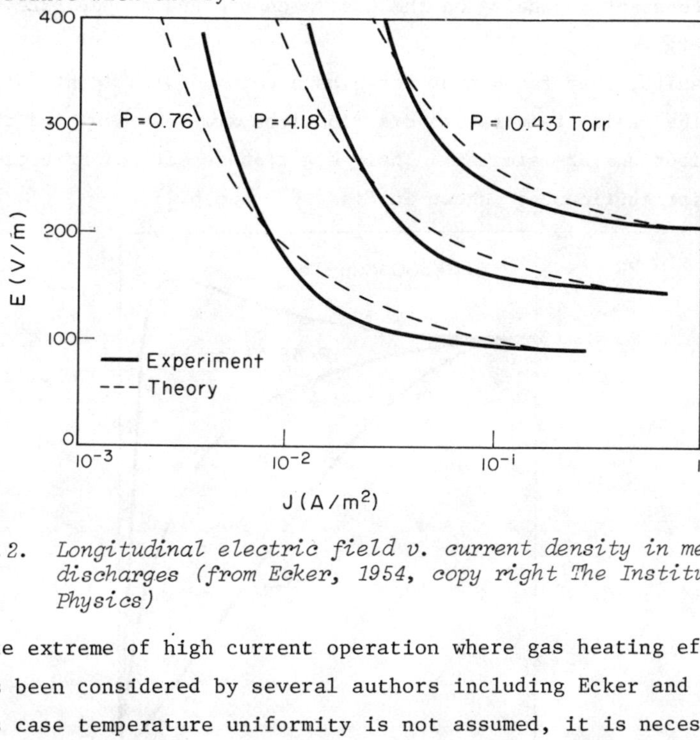


*Fig. 8.2. Longitudinal electric field v. current density in mercury discharges (from Ecker, 1954, copy right The Institute of Physics)*

The opposite extreme of high current operation where gas heating effects become important has been considered by several authors including Ecker and Zoller (1964). Since in this case temperature uniformity is not assumed, it is necessary to simultaneously solve the equations of continuity, momentum conservation and energy conservation for the electrons, ions and neutrals. The continuity equations remain the same as before:

$$\nabla \cdot n_e \bar{v}_e = n_e \nu_I = \nabla \cdot n_+ \bar{v}_+ \,. \tag{8.1}$$

The momentum conservation equations, (8.2) and (8.3) now become

$$\bar{v}_e = -\mu_e \bar{E} - \frac{\mu_e}{en_e} \nabla(n_e kT_e) \,, \tag{8.43}$$

$$\bar{v}_+ = +\mu_+ \bar{E} - \frac{\mu_+}{en_+} \nabla(n_+ kT_+) \,. \tag{8.44}$$

Ecker and Zoller assumed that the ion and neutral temperatures were the same

$$T_+ = T \, . \tag{8.45}$$

For the electron energy balance, they assumed that the electron energy is gained from the electric field and is lost entirely due to elastic collisions. Therefore,

$$\bar{J} \cdot \bar{E} = \frac{3}{2} n_e k(T_e - T)\delta\nu_m \, . \tag{8.46}$$

For the neutral gas, the energy source term is from electron neutral collisions and the loss is due to thermal conduction, therefore

$$K\nabla^2 T + \bar{J} \cdot \bar{E} = 0 \, . \tag{8.47}$$

Ecker and Zoller solved the preceding equations assuming ambipolar diffusion and assuming that the ionization frequency can be expressed as

$$\nu_i = \bar{A} \exp(-V_i/kT_e) \, , \tag{8.48}$$

where $\bar{A}$ is a constant depending on the gas, pressure and temperature and $V_i$ is the ionization energy.

In their results, they found that for plasma columns with identical values of R N and E/N, the radial temperature distributions are identical and the electron density distributions are similar. Their electron density distributions and gas temperature distributions are shown in Figs. 8.3 and 8.4.

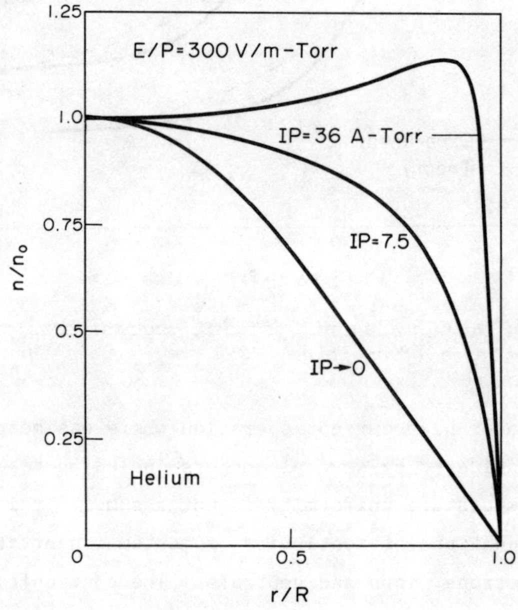


*Fig. 8.3. Normalized electron density distributions as a function of the current pressure product in helium positive columns (from Ecker and Zoller, 1964)*

As can be seen, increasing current can cause significant variations in the electron density distribution. This is, of course, related to the radial variations in electron and neutral temperatures. The exact form of the electron density distribution does depend rather sensitively on the form of the electron temperature variation with E/N. Jaeger, Oster and Phelps (1976) performed calculations similar to that of Ecker and Zoller for a different variation of electron temperature with E/N and ended up with predictions that did not show such a great variation in profile as is shown in Fig. 8.3.

The radial neutral temperature variation predicted by Ecker and Zoller is shown in Fig. 8.4. Increasing current gives increasing neutral temperatures and the temperature distribution is nearly parabolic.

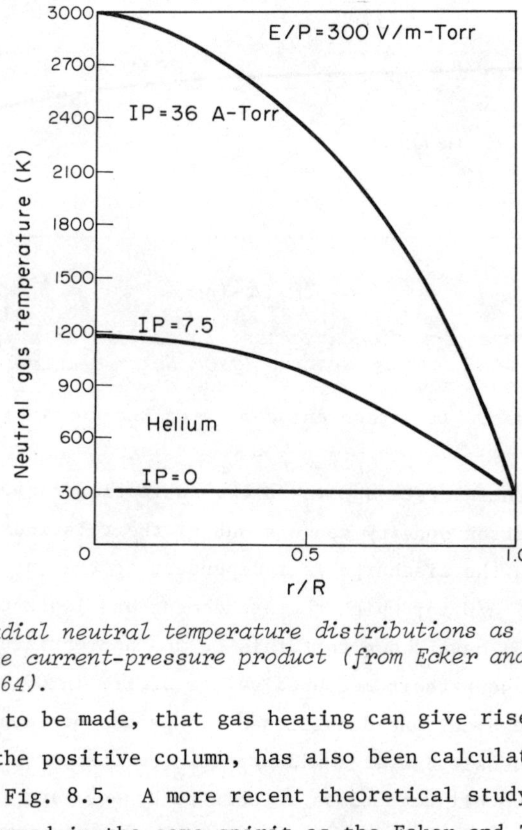


*Fig. 8.4. Radial neutral temperature distributions as a function of the current-pressure product (from Ecker and Zoller, 1964).*

The initial point to be made, that gas heating can give rise to a decreasing voltage drop along the positive column, has also been calculated by Ecker and Zoller and is shown in Fig. 8.5. A more recent theoretical study by Cohen and Whitman (1973) was performed in the same spirit as the Ecker and Zoller paper and showed a similar decrease in the longitudinal electric field with increasing current.

One point to be made in addition is that virtually all theoretical studies have neglected one or more processes that could play an important role in the positive column. To that extent, there is really no comprehensive theory of the positive column available and each case has to be considered individually.

## 8.2 ATTACHMENT- AND RECOMBINATION-DOMINATED DISCHARGES

In high-pressure gas discharges, recombination or attachment are likely to be the

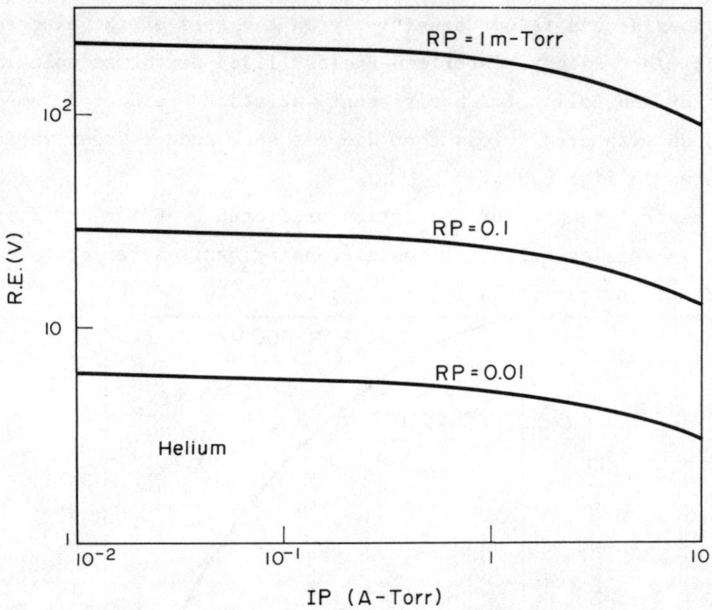


*Fig. 8.5. Voltage drop along the positive column as a function of the current pressure product (from Ecker and Zoller, 1964).*

dominant loss processes. If attachment dominates, the continuity equation becomes

$$n_e \nu_a = n_e \nu_I \tag{8.49}$$

where $\nu_a$ is the attachment rate and $\nu_I$ is the ionization rate.

Obviously, the electron density cancels out of the relationship, so the value of the electric field in the discharge is independent of the current and is determined by the value of E/N for which the attachment and ionization rates are equal. Denes and Lowke (1973) have shown that this is the appropriate regime for $CO_2$ lasers. Figure 8.6 shows their measured values of E/N in $CO_2$:$N_2$:He mixtures over a wide range of reduced current densities J/N. As can be seen, E/N is independent of J/N (or electron density) over a wide range. Furthermore, the measured values of E/N correlate very well with those calculated by equating the attachment rate with the ionization rate. In the Denes and Lowke paper, the ionization rate is written as

$$n_e \nu_I = n_e \alpha W \tag{8.50}$$

where W is the drift velocity, and the attachment rate is written as

$$n_e \nu_a = n_e a W \tag{8.51}$$

Figure 8.7 shows $\alpha$/N and a/N plotted vs. E/N for various gas mixtures, and the intersection points determine the value of E/N measured in Fig. 8.6.

As the electron density, or current, becomes sufficiently high, the value of E/N is no longer independent of current but increases as J/N increases as shown in Fig. 8.6. This is the regime in which recombination becomes important. If

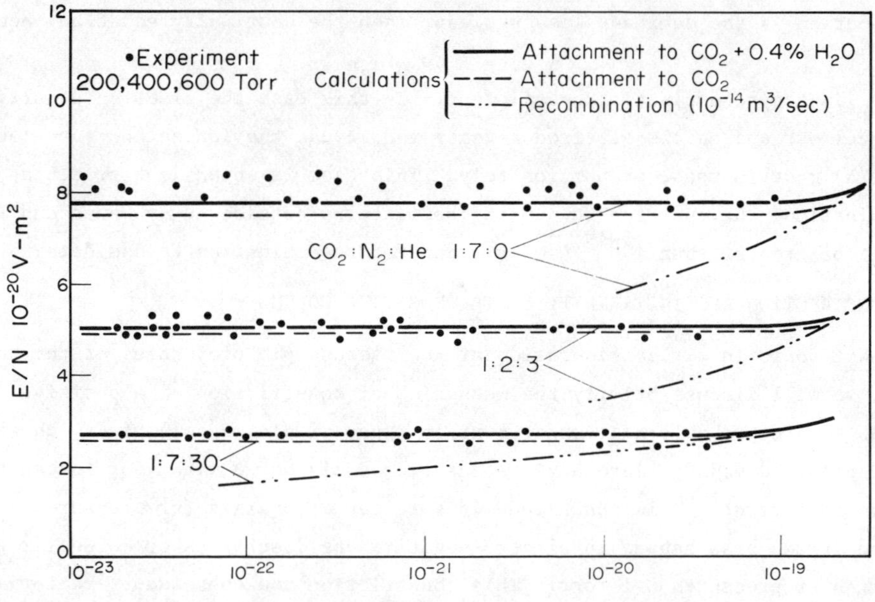


*Fig. 8.6. E/N vs J/N for three $CO_2$:$N_2$:He laser mixtures (from Denes and Lowke, 1973)*

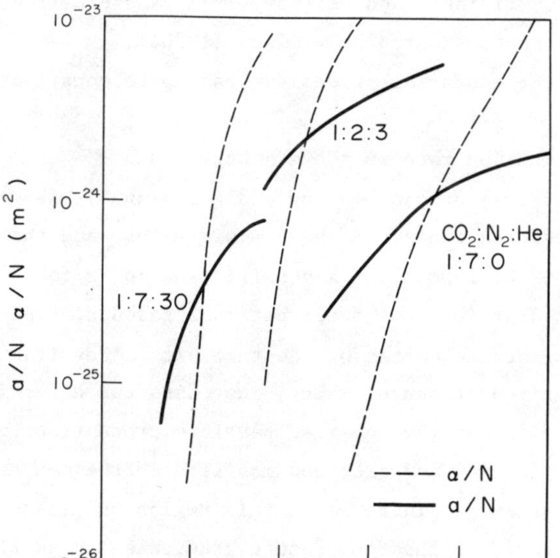


*Fig. 8.7. Calculated values of α/N and a/N as a function of E/N for various gas mixtures (from Denes and Lowke, 1973)*

recombination is the dominant loss process, then the continuity equation becomes

$$n_e \nu_I = n_e \alpha W = \alpha_R n_e^2 \tag{8.52}$$

where $\alpha_R$ is the recombination coefficient. In this case the electron density does not cancel out and as the electron density increases, the ionization rate (and hence E/N) must increase proportionately. This case is shown in Fig. 8.6 as the "recombination" curves. For the $CO_2$:$N_2$:He mixtures considered by Denes and Lowke, J/N must be greater than $10^{-19}$ A m in order for recombination to dominate.

## 8.3 CONSTRICTION AND INSTABILITY OF THE POSITIVE COLUMN

As a last topic in our consideration of the macroscopic properties of the positive column, we will discuss briefly the mechanism of constriction of a positive column. As we are probably all aware, certain types of plasma columns can change from a uniform discharge into a very constricted discharge with, for instance, an increase in current. This can happen in rare gas discharges for pressures $\tilde{>}$ 100 torr and it can also happen in electro-negative gas discharges or molecular gas discharges at pressures < 1 torr. This constriction and subsequent arc formation turns out to be an important limitation on the amount of electrical power that can be put into gas lasers and similar uniform glow media.

Early studies on constriction in rare gas discharges at high pressures were performed by Kenty (1962). Similar studies on constriction in molecular discharges have been performed by Eletskii and Smirnov (1971) and constriction in electronegative gases has been considered by Woolsey (1970).

In these studies, the essential processes leading to constriction were taken to be:

1. a large recombination rate in the discharge, and
2. a significant degree of gas heating and a reasonably large temperature gradient between the center of the discharge tube and the wall.

The basic idea of the development of a constriction is as follows. The usual initial Bessel function distribution of current in a gas discharge will cause preferential gas heating at the center of the tube with a resultant decrease in the neutral density (at fixed pressure). This increases the effective E/N and electron energy resulting in a higher charged particle production rate at the center. The increased charge density and electron mobility in the central region then tends to cause the current to increase in this region and leads to further gas heating near the center. If the temperature gradients are sufficiently high, the recombination rate, which we assume is comparable to or greater than the diffusion rate, increases rapidly away from the center as we enter the cool gas. This rapid increase in recombination with radius then tends to restrict the discharge to the central region of the discharge. To maintain the discharge, the central region must carry the total current and becomes even hotter, making the gradients in temperature and the spatial variation in recombination rate even greater and the effect is cumulative.

We can consider the model in Fig. 8.8 for a constricted discharge to obtain some estimates of the conditions under which constriction will occur. In the region

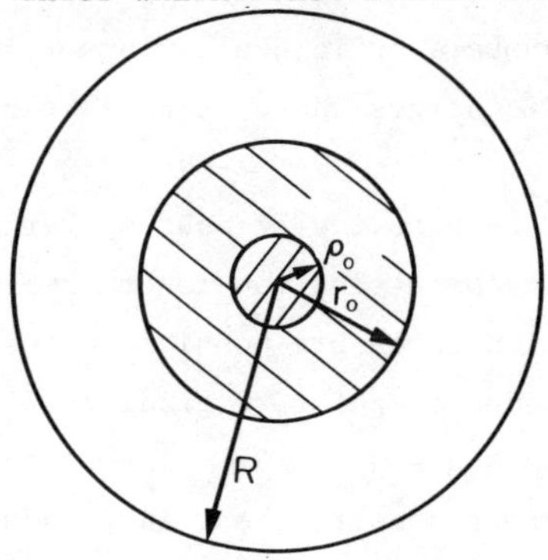


*Fig. 8.8. Model of a constricted discharge*

$r \le \rho_o$, the electrons are produced by ionizing collisions. In this region we need only consider ionization and diffusion, the temperature being too high for recombination. The region $r \le r_o$ is the region occupied by the electrons. This is essentially the extent of the constricted discharge. We can make a crude estimate of the size of the constriction by assuming that $r_o$ is the distance that an electron can diffuse in one recombination time.

$$r_o \simeq (D_a \tau)^{1/2} \simeq \left(\frac{D_a}{\alpha n_e}\right)^{1/2} \tag{8.53}$$

For $\rho_o < r << r_o$, we can consider diffusion only and for $r >> r_o$ we can consider recombination only.

The condition for constriction, then, is that

$$r_o \simeq \left(\frac{D_a}{\alpha n_e}\right)^{1/2} << R \ . \tag{8.54}$$

This also determines Kenty's constriction ratio

$$C = \frac{D_a}{\alpha R^2 n_e} \ .$$

As an example (from Eletskii and Smirnov, 1971) of the conditions under which constriction can occur, consider this case.

Let $R$ = a few cm,

$n_e \simeq 10^{16}-10^{19}/m^3$, typical values for a positive column, and

$p_o \simeq 1$ torr.

Under these conditions, typical values of $D_a$ are

$$D_a \sim 10^{-3}-10^{-2} m^2/sec \ .$$

If $r_o \sim \frac{D_a}{\alpha n_e} \ll R^2$, this requires values of $\alpha$,

$$\alpha \gg 10^{-14} \text{ to } 10^{-18} \text{ m}^3/\text{sec} .$$

Now, values of $\alpha$ in this range can indeed occur for recombination with molecular ions (dissociative) and so constriction of the positive column can readily occur in gases which contain large numbers of molecular ions. This can occur, for instance, in high-pressure rare gas discharges where there is rapid conversion from atomic to molecular ions.

In electronegative gases, even at low pressures, large recombination coefficients are possible by electron attachment and ion-ion recombination. In molecular gases, of course, large recombination rates are possible even at low pressures.

In order to obtain the needed temperature gradients for constriction to occur, it is also necessary to have a gas with a low thermal conductivity. One of the methods, therefore, to suppress constrictions is to add a gas with a large thermal conductivity, such as helium, to a gas discharge in order to smooth out the gas temperature profiles.

The previous picture is, of course, a somewhat simplistic picture of the constriction phenomena. The great importance of this process as a limitation on the power available from $CO_2$ gas lasers has prompted considerably more study. Two interesting papers that consider the time dependent constriction of the glow discharge due to gas heating are by Jaeger, Oster and Phelps (1976) and by Rogoff (1972). Jaeger, Oster and Phelps (1976) considered helium discharges and took into account the spatial and temporal variations in ionization, recombination and ambipolar diffusion to show the constriction as gas heating occurred. Rogoff's calculations provided similar results for hydrogen discharges although he did not take into account thermal dissociation, which he indicates is probably the cause of the final formation of a hot constricted discharge.

A different approach to the constriction problem was taken by Nighan, Wiegand and Haas (1973), Haas (1973), Nighan and Wiegand (1974a,b) and Jacob and Mani (1975). Their basic approach is to linearize the fluid equations describing the positive column in the usual way and study the stability of perturbations on the system. Since they are primarily concerned with instabilities in molecular gas lasers, additional features were found to have an important role. Negative ions formed in the discharge affected the stability through the sensitivity of the attachment and detachment processes to electron temperature. Similarly, the energy stored as vibrational molecular energy was found to have as important a role as the translational kinetic energy of the gas. Refer to the previous references for a complete discussion of the subject.

8.4 REFERENCES

Brown, S. C. (1959), Basic Data of Plasma Physics, MIT Press and Wiley & Sons.

Cohen, I. M., and Whitman, A. M. (1973), "Unified Positive Column Theory of Gas Discharges," Phys. Fluids 16, 307-314.

Denes, L. J., and Lowke, J. J. (1973), "V-I Characteristics of Pulsed $CO_2$ Laser Discharges," Appl. Phys. Letters 23, 130-132.

Ecker, G. (1954), "Theory of the Positive Column," Proc. Phys. Soc. (London) B 67, 485-491.

Ecker, G., and Zoller, O. (1964), "Thermally Inhomogeneous Plasma Column," Phys. Fluids 7, 1996-2000.

Eletskii, A. V., and Smirnov, B. M. (1971), "Contraction of the Positive Column of a Glow Discharge," Sov. Phys. Tech. Phys. 15, 1308-1310.

Gerber, R. A., and Gerardo, J. B. (1973), "Ambipolar to Free Diffusion: The Temporal Behavior of the Electrons and Ions," Phys. Rev. A7, 781-790.

Haas, R. A. (1973), "Plasma Stability of Electric Discharges in Molecular Gases," Phys. Rev. A 8, 1017-1043.

Ingold, J. H. (1975), "Two-Fluid Theory of the Positive Column of a Gas Discharge," Phys. Fluids 15, 75-85.

Jacob, J. H., and Mani, S. A. (1975), "Thermal Instability in High Power Laser Discharges," Appl. Phys. Letters 26, 53-55.

Jaeger, E. F., Oster, L. and Phelps, A. V. (1976), "Growth of Thermal Constrictions in a Weakly Ionized Gas Discharge in Helium," Phys. Fluids 19, 819-830.

Kagan, Y. M. and Lyagushchenko, R. I. (1962), "Electron Velocity Distribution, Excitation and Ionization in the Positive Column of a Neon Discharge," Sov. Phys-Tech. Phys. 7, 134-137

Kenty, C. (1962), "Volume Recombination, Constriction and Volt-Ampere Characteristics of the Positive Column," Phys. Rev. 126, 1235-1238.

Konyukov, M. V. (1958), "Concentration of Negative Ions in the Plasma of a Positive Column," Sov. Phys. JETP 7, 629-631.

Nighan, W. L., and Wiegand, W. J. (1974a), "Influence of Negative-Ion Processes on Steady-state Properties and Striations in Molecular Gas Discharges," Phys. Rev. A10, 922-945.

Nighan, W. L., and Wiegand, W. J. (1974b), "Causes of Arcing in CW $CO_2$ Convection Laser Discharges," Appl. Phys. Letters 25, 633-636.

Nighan, W. L., Wiegand, W. J., and Haas, R. A. (1973), "Ionization Instability in $CO_2$ Laser Discharges," Appl. Phys. Letters 22, 579-582.

Rogoff, G. L. (1972), "Gas Heating Effects in the Constriction of a High Pressure Glow Discharge Column," Phys. Fluids 15, 1931-1940.

Seeliger, R. (1949), "Die Diffusions theorie der positiven Saule in elektronegativen Gasen," Ann. Physik 6, 93-96.

Spenke, E. (1950), "Die Diffusions theorie der positiven Saule mit Berucksichtigung der stufenweisen Ionisierung," Z Phys. 127, 221-242.

Von Engel, A. (1965), Ionized Gases, Oxford.

Vriens, L. (1974), "Two and Three Electron Group Models for Low-pressure Gas Discharges," J. Appl. Phys. 45, 1191-1195.

Waymouth, J. F., and Bitter, F. (1956), "Analysis of the Plasma of Fluorescent Lamps," J. Appl. Phys. 27, 122-131.

Wilhelm, J. (1960), "Zur radialen Tragerdichteverteilung in Saulen mit mehreren arten positiven ionen," Ann. Physik 5, 129-143.

Woolsey, G. A. (1970), "Constriction in the Positive Column," in Discharges in Electronegative Gases, Taylor & Francis, pp. 70-76.

# Chapter 9
# EXCITED SPECIES

Up to now, we have been primarily concerned with the description of the charged species, primarily the electrons, within a discharge. Now it is time to consider the excited atomic and molecular species within the discharge since it is these species which are critical to the determination of light output for lamps or optical gain and available power for lasers. Naturally, the previous description of the electrons is a necessary prerequisite to the study of excited species since electron impact excitation is usually either the primary or secondary source of excitation of the excited species.

To determine the density of the excited species of interest, either on a steady state or time varying basis, it is necessary to evaluate the production and loss mechanisms. In a rather simple fashion, excited species can usually be considered to be produced by electron impact excitation from ground or lower excited species, by collisional energy exchange with excited atoms or molecules, or by radiative cascade from higher excited species. Figure 9.1 gives a simple representation of

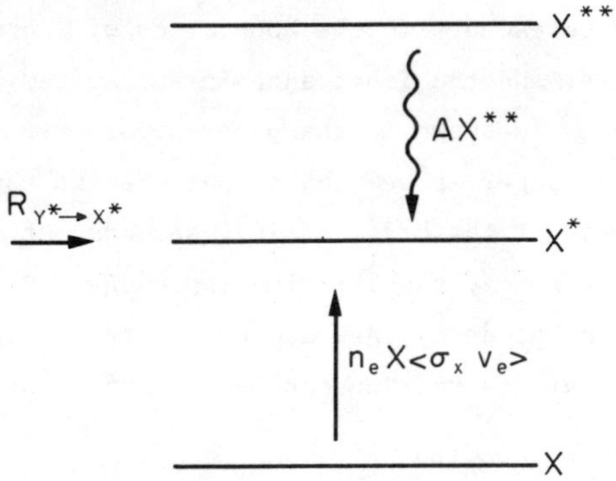


*Fig. 9.1. Representative excitation processes for the excited species* $X^*$

these processes. Once the production processes have been evaluated, then the excited species density can be determined by evaluating the loss processes. There are usually two distinct cases: (a) when radiative decay dominates and (b) when collisional decay dominates.

## 9.1 RADIATIVELY DECAYING SPECIES

In most moderate density atomic systems, excited species often have radiative lifetimes in the order of $10^{-8}$ to $10^{-9}$ seconds. In these cases, radiative decay is usually the dominant loss process and the decay of the excited species is given by

$$\frac{d}{dt} X^* = -AX^* = -X^*/\tau \tag{9.1}$$

where A is the atomic transition probability and $\tau$ is the radiative lifetime of the state. In a steady state, if the excited state has a pumping rate R, then

$$\frac{d}{dt} X^* = 0 = R - AX^* \ . \tag{9.2}$$

Therefore, the density of the excited species

$$X^* = R/A = R\tau \ . \tag{9.3}$$

For atomic systems, many of the A coefficients are tabulated (Wiese, Smith and Glennon, 1966, Wiese, Smith and Miles, 1969) or can be calculated (Bates and Damgaard, 1949), thus the majority of the lifetimes can be found.

If the medium is "optically thick" then the lifetimes for radiative decay are greatly increased and radiative trapping must be taken into account as discussed by Holstein (1947, 1951) and more recently by Van Trigt (1969, 1970, 1971). This effect becomes important when the state to which the excited species radiatively decays is heavily populated (particularly if it is the ground state) and the photon can be rapidly reabsorbed re-creating the original excited state. As an example, the 2537 Å radiation from the $6^3P_1$ state to the ground state in mercury is absorbed within a distance of 0.001 cm at 1 torr of mercury. The actual decay of the $6^3P_1$ state then must take into account multiple absorption and re-emission processes as the radiation escapes to the boundaries of the medium. The importance of this effect in mercury, and the importance in any system where a large number of absorbers is available, is shown in the paper by Alpert, McCoubrey and Holstein (1949). In this paper it was shown that whereas the natural radiative lifetime of a $6^3P_1$ atom is of the order of $10^{-7}$ seconds, at densities of around $10^{22}$ ground state atoms per $m^3$, the lifetime approaches $10^{-4}$ seconds. This is a rather profound change in the decay rate and can decrease the radiative decay to the point where collisional destruction can become important.

## 9.2 COLLISIONALLY DECAYING SPECIES

In atomic systems, excited species whose radiative decay is heavily trapped (primarily transitions to ground) or excited species that do not radiatively decay (metastables) can build up large densities and play an important role in gas discharges. They can act as stores of energy which can de-activate on the walls or by collisions with other species in the discharge. They are, of course, very important as pumping mechanisms in gas lasers. We shall discuss this in detail later.

Since radiative decay no longer dominates, a significant number of previously negligible processes need to be taken into account. Rather than discuss the decay processes for such species in a general context, we will instead examine two representative cases. The two cases are: (a) the neon $1s_5$ metastable species and (b) the helium $2^3S$ metastable species.

### 9.2.1 The Neon ($1s_5$) Metastable Species

This particular case has been studied in detail and provides an excellent example of collisionally dominated decay. Figure 9.2 shows a partial energy level diagram for neon (not to scale) showing some relevant levels and some of the processes involved in the excitation and de-excitation of the important species. In Fig. 9.2 the spectroscopic notation for the energy levels is used ($^3P_2$) as well as the fairly common Paschen notation ($1s_5$, 2p, etc.).

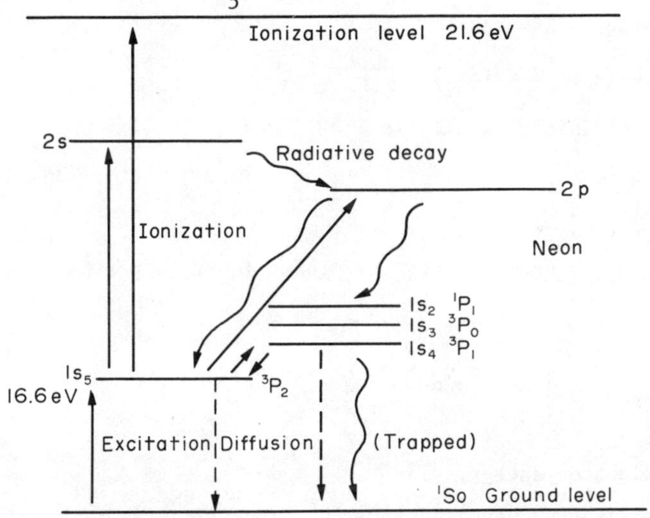


*Fig. 9.2. Partial energy-level diagram for neon and some rate processes*

Of the important excitation and de-excitation processes shown, the solid lines indicate electron collisional processes, the dotted lines indicate diffusion processes and the wavy lines indicate radiative processes. Now consider the important production and destruction processes for the $1s_5$ metastables and their consequences. Fairly complete discussions have been presented by Salinger and Rowe (1968) and by Schlie, Verdeyen, Cherrington and Wisniewski (1970). The important metastable production terms are:

1. Direct electron impact excitation from the ground state:

   $e + Ne \rightarrow M + e - (16.6\ eV)$

   (we will use M to indicate the metastable).

2. Indirect excitation by radiative decay from higher lying levels:

   $Ne(3s_2, 2s_2) \rightarrow M + h\nu$ .

The important metastable destruction terms are:

3. Diffusion to the container walls $D_m \nabla^2 M$.
4. De-excitation by three-body collisions $\gamma N^2 M$.
5. Metastable-metastable collisional de-excitation:

   $< \sigma_{m-m} v_{av} > M^2$ .

6. Collisional conversion to a radiative state and radiative decay

   $M + N(\text{or } e) \rightarrow Ne(1s_4,\ 1s_2) \rightarrow Ne + h\nu(744, 736\ \text{Å})$.

7. Superelastic collisions:

   $M + e(\text{slow}) \rightarrow Ne + e(\text{fast})$

8. Electron collisional ionization:

   $M + e \rightarrow Ne^+ + 2e$.

Consider each of these terms in order:

1. Electron impact excitation. The production rate per particle $< \sigma_p v_e >$ is given by

$$< \sigma_p v_e > = \int_{\varepsilon_x}^{\infty} v(\varepsilon)\sigma_p(\varepsilon)f(\varepsilon)d\varepsilon \qquad (9.4)$$

where $\varepsilon$ = electron energy,

$\sigma_p(\varepsilon)$ = excitation cross section,

$\varepsilon_x$ = excitation energy of the metastable state,

f = electron energy distribution.

A straight line approximation to the cross section for electron impact excitation to the $1s_5$ and $1s_3$ (metastable) states is given in Figure 9.3. For a more exact figure see the paper by Salinger and Rowe (1968). If we assume a Maxwellian electron energy distribution, the production rate per particle as a function of the average electron energy is shown in Figure 9.4.

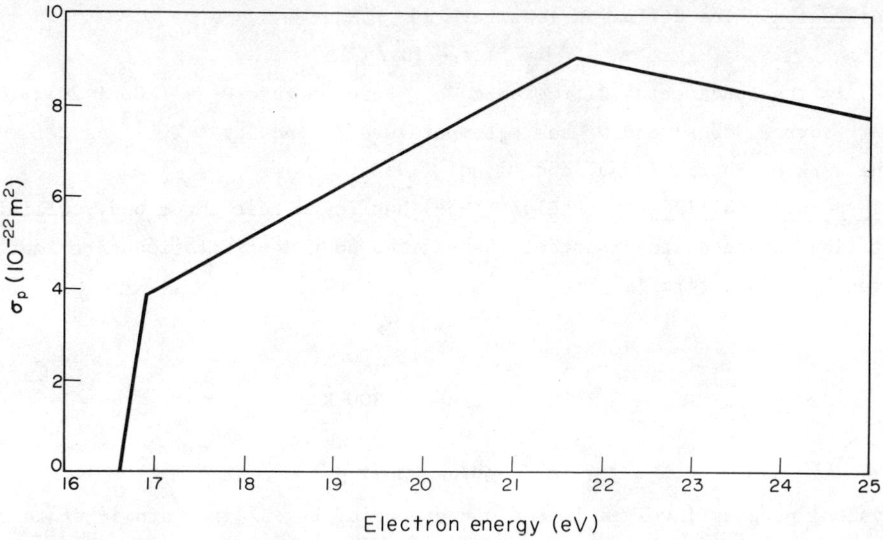


*Fig. 9.3. Cross section for excitation of metastable states of neon*

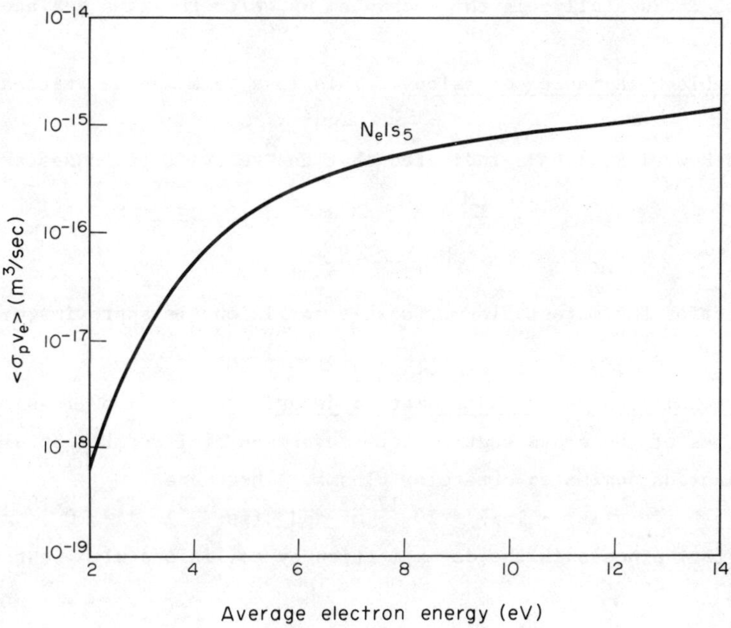


*Fig. 9.4. Excitation rate as a function of electron energy for the metastable states of neon*

2. Indirect excitation by radiative decay from higher lying levels. Most of the higher lying levels are populated by step-wise excitation from the 1s levels with direct impact being negligible. These states also quickly decay (radiatively) to the 1s states so these two processes cancel out and can be neglected.

3. Diffusion. The diffusion loss term is

$$D_m \nabla^2 M \simeq - (D_m/\Lambda_1^2)M \tag{9.5}$$

where $\Lambda_1$ is the fundamental diffusion mode. Here we take $D_m = (200 \pm 20) \times 10^{-4}$ $m^2$/sec (1 torr @ 300 K) and $D_m$ has a temperature dependence ~ $T^{0.73}$ as determined from the work of Phelps (1959) and Biondi (1953).

4. Three-body collisions. Phelps (1959) has found that three body collisional deexcitation can be quite important whereas two body de-excitations are improbable. The three body loss term is

$$\frac{dM}{dt} = -\gamma N^2 M \tag{9.6}$$

where

$$\gamma = 5 \times 10^{-46} \ m^6/\text{sec @ 300 K}$$

and

$$\gamma = 5 \times 10^{-47} \ m^6/\text{sec @ 77 K} .$$

The physical process involved is the conversion of the excited atomic state to an excited molecular state which subsequently radiatively decays to ground

$$X^* + 2X \rightarrow X_2^* + X. \tag{9.7}$$

This process, incidentally, is the mechanism by which the rare gas excimer lasers are pumped.

5. Metastable-metastable collisions. This loss term can be written as

$$\frac{dM}{dt} = -\beta M^2 . \tag{9.8}$$

Salinger and Rowe (1968) have indicated that the relevant processes are

$$X^M + X^M \rightarrow X + X^+ + e \tag{9.9}$$

or, more probably,

$$X^M + X^M \rightarrow X_2^+ + e . \tag{9.10}$$

The cross section for metastable-metastable collisions is approximately

$$\sigma_{M-M} \simeq 1.4 \times 10^{-18} \ m^2 .$$

6. Collisional conversion and radiative decay. Phelps (1959) has made some rough estimates of the cross section for conversion of 1s states by slow electrons in a recombination dominated afterglow plasma. These are

$$\sigma(1s_4 - 1s_5) \simeq 10^{-17} \ m^2 \text{ and } \sigma(1s_3 - 1s_4) \simeq 10^{-18} \ m^2 .$$

A more important process is the de-excitation by collisions with neutral atoms. Phelps gives

$$\sigma(1s_4 - 1s_5) \doteq 5 \times 10^{-23} \ m^2,$$
$$\sigma(1s_3 - 1s_4) = 6 \times 10^{-24} \ m^2,$$

and

$$\sigma(1s_3 - 1s_5) = 6 \times 10^{-24} \ m^2 \text{ at 300 K.}$$

If the temperature increases from 200 K to 500 K, $\sigma(1s_4 - 1s_5)$ increases by a factor of 10. Incidentally, the cross section for conversion by helium atoms

$$\sigma(1s_4 - 1s_5) = 1.5 \times 10^{-22} \ m^2 \text{ @ 300 K.}$$

The converse process, exciting the $1s_5$ atoms to the $1s_4$ or $1s_2$ states, could be important if the radiative decay time for these states is shorter than the diffusion decay time for the $1s_5$'s. In most cases this is not a significant loss term since the radiation at 744 Å and 736 Å is usually strongly trapped and the time constant for the decay of the imprisoned resonance radiation is usually of the order of the diffusion time of the $1s_5$'s. This correction can then be incorporated into a modified diffusion time constant. In general the decay rate of the resonant state atoms by the escape of resonance radiation is expressed as

$$-\frac{n(x)}{\tau} + \frac{1}{\tau}\int_{x'} n(x')G(x',x)dx \simeq -\nu n(x) \tag{9.11}$$

where $\tau$ = spontaneous emission lifetime of the resonance state,

$n$ = density of the resonance state,

$G(x',x)$ = probability that a quantum emitted at x is absorbed at x',

$\nu$ = decay constant for the diffusion of resonance radiation.

For an authoritative discussion see the articles by Holstein (1947, 1951) and by Van Trigt (1969, 1970, 1971).

7. Superelastic collisions. This is the reverse of the electron collisional excitation term so the rate coefficient for electron-collisional de-excitation can be calculated from the excitation process using detailed balance.

At thermodynamic equilibrium electron collisional excitation and de-excitation balance so

$$< \sigma_d v_e > n_e M = <\sigma_p v_e > n_e N \tag{9.12}$$

and

$$M = N \frac{g_1}{g_o} e^{-\varepsilon_x/kT_e} \tag{9.13}$$

where $\varepsilon_x$ = excitation energy for the $1s_5$ state.

Therefore,

$$< \sigma_p v_e > = \frac{g_1}{g_o} < \sigma_d v_e > e^{-\varepsilon_x/kT_e} \tag{9.14}$$

and the destruction rate turns out to be

$$< \sigma_d v_e > \simeq 7 \times 10^{-16} m^3/sec . \tag{9.15}$$

8. Electron collisional ionization. The theoretical calculations of Vriens (1964) indicate an ionization cross section peak value of $6 \times 10^{-20}$ $m^2$ as shown in Figure 9.5. If a Maxwellian distribution exists, then the ionization rate $< \sigma_I v_e >$ is as shown in Figure 9.6. We notice, for instance, that this process would normally dominate superelastic collisions as a loss process for neon metastables. This is largely a consequence of the fact that the neon metastable is at 16.6 eV above the ground state and requires only about 5 eV electron energy to

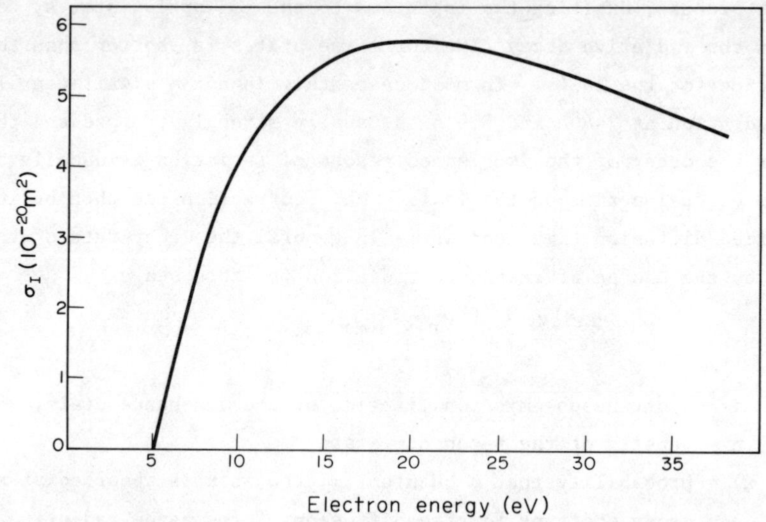


*Fig. 9.5. Electron impact ionization cross section for Ne metastables (from Vriens, 1964).*

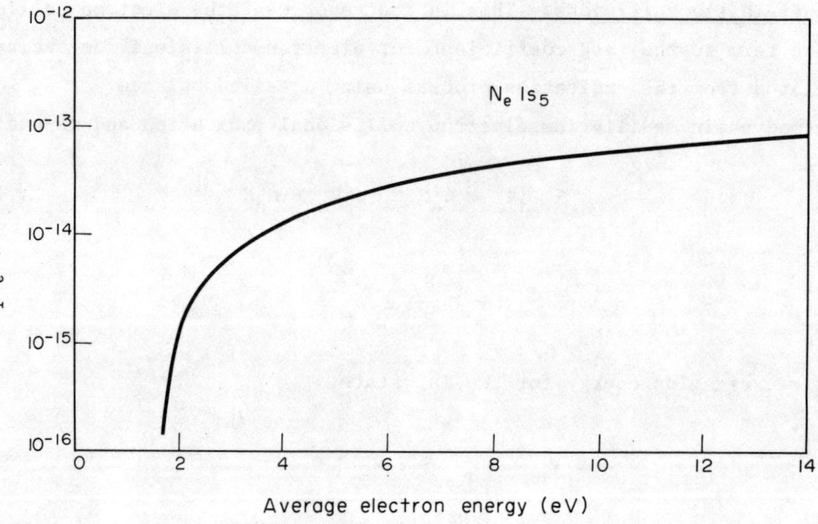


*Fig. 9.6. Ionization rate for neon metastables as a function of electron energy*

ionize it. In other media, notably mercury and cesium, superelastic collisions are a very important loss process for excited states and may well dominate all other loss processes.

9.2.1.1 Neon metastables in the DC discharge. Now that we have established the various possible processes, which are dominant for the particular regime that is of interest? For DC discharges in the few torr range, we can eliminate many of these processes.

Obviously, electron collisional excitation is the important production process. In this range the dominant loss processes are diffusion and electron collisional ionization. Three body collisions, metastable collisions, collisional conversion and superelastic collisions are all negligible in comparison. The steady state rate equation describing the behavior of the neon $1s_5$ atoms is therefore

$$- D_m \nabla^2 M(r) + \langle\sigma_I v_e\rangle n_e(r) M(r) = \langle\sigma_p v_e\rangle n_e(r) N_{Ne} \quad . \tag{9.16}$$

Now, obviously, at low electron densities (i.e. low currents) the diffusion loss term will dominate the collisional ionization term, the metastables will be distributed in the fundamental diffusion mode and the metastable density will be linearly dependent on the electron density (or current). At high densities, the electron collisional destruction term will dominate. Thus the collisional production and destruction terms will balance, the metastable density will be independent of current and the metastable density distribution will be flat. The experimental measurements verify these predictions. The $1s_5$ radial profiles for a DC discharge in a cylindrical tube 5.5 mm in ID are shown in Figure 9.7. As we

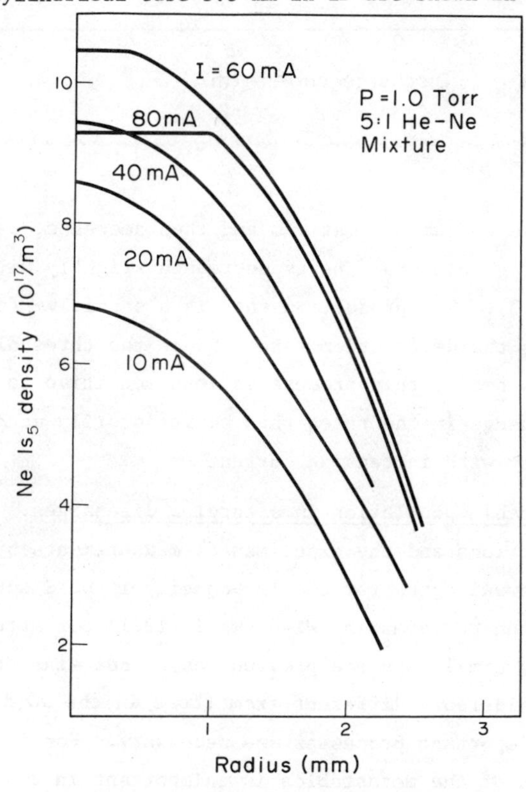


*Fig. 9.7. Neon metastable density distribution as a function of discharge current*

can see, at low currents we have the fundamental diffusion mode and at higher currents we begin to saturate. (The presence of He does not change the basic physics here so it can be ignored.)

The axial metastable density as a function of discharge current is shown in Figure 9.8. As we can see, the predicted saturation does occur. The behavior at

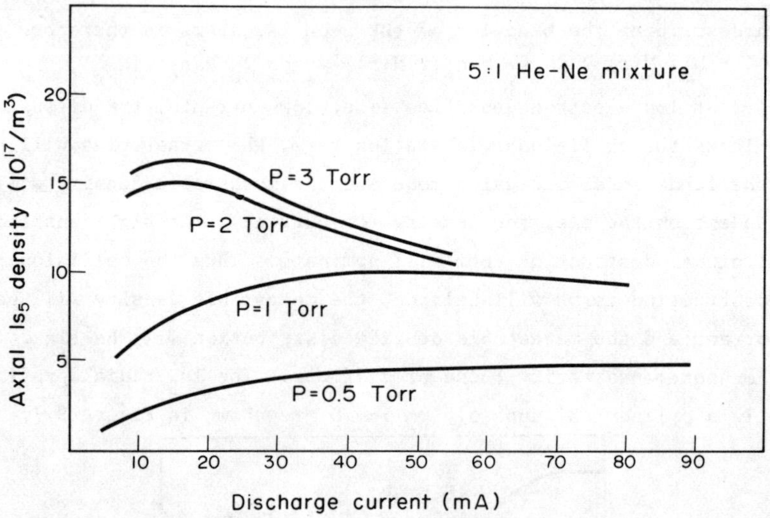


*Fig. 9.8. Axial neon metastable density variation with discharge current*

higher pressures, where we reach a maximum and then decrease, can be explained by the fact that the average electron energy decreases slightly with increasing current (see Section 8.1.3). The production rate is a sensitive function of electron temperature as is the destruction rate. Since the threshold energy for the destruction process is small, this process is less sensitive to electron temperature changes. This change in the rates then satisfactorily accounts for the decrease in $1s_5$ density with increasing current.

9.2.1.2 Neon metastable population in afterglow discharges. We have seen how the theoretical predictions and the experimental measurements by Schlie, Verdeyen, Cherrington and Wisniewski agree for the $1s_5$ density in DC discharges. Similar investigations have been performed by Wisniewski (1971) for afterglow discharges. The arguments are very similar to the previous case, but since the conditions in the afterglow are considerably different from those in the DC discharge, some re-evaluation of the important processes are necessary. For instance, electron collisional ionization of the metastables is unimportant in cool afterglow plasmas. The decay of the metastables can be given by:

$$\frac{dM}{dt} = P + \left\{ D_m \nabla^2 - \gamma N_e^2 - N_e \sum_M^n < \sigma v > \right\} M$$
$$- n_e M \sum_M^P < \sigma v > - n_e M < \sigma v >_M^o - \beta M^2 \quad . \tag{9.17}$$

P is the metastable production term in the afterglow. We will discuss this in detail shortly. The bracketed terms indicate the dominant loss terms in the late afterglow (diffusion, three-body collisions, neutral collisions to the $1s_4$ and other adjacent states). The remaining loss terms are more relevant to the early afterglow where electron densities and metastable atom densities are high. These are, in order, electron collisional excitation to higher-lying excited states ($1s_4$, etc.), superelastic electron collisional de-excitation, and metastable-metastable collisions.

Typical data for the neon $1s_5$ density variation as a function of time in the afterglow of a discharge produced in 10 torr neon in a cylindrical discharge tube 8 mm in diameter are shown in Fig. 9.9. We also plot the corresponding intensity of the visible recombination light as observed at the side of the discharge tube since this will enter our explanation.

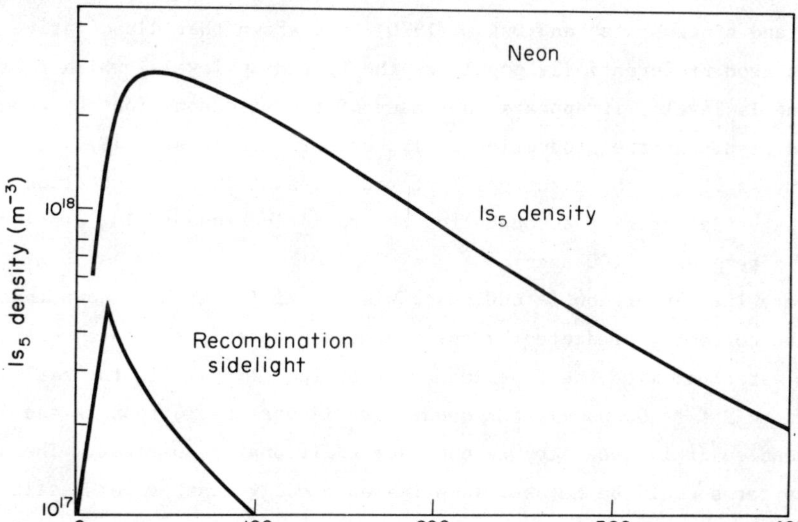


*Fig. 9.9. Neon $1s_5$ density variation as a function of time in an afterglow*

First of all we notice that the metastable density shows a sharp increase with time in the early afterglow and reaches a peak which is much greater than that in the DC discharge. The $1s_5$ density then decays as we would expect. This indicates that there is a $1s_5$ production term which is not present in the DC discharge and which is important in the early afterglow. The logical source for these $1s_5$ atoms is through electron-ion recombination. The electron temperature in the DC discharge is too high to allow any appreciable recombination to occur, but in the afterglow the electron temperature drops rapidly to room temperature (typically within about 50 μsec) and the electrons can be captured by the ions to produce excited atoms (the first step in the recombination process) and these excited atoms can radiatively decay to the $1s_5$ state. The plot of the recombination sidelight intensity indicates the rate at which these processes occur in the afterglow. You will also note that this is a consistent explanation since the maximum sidelight intensity (maximum number of recombination events) corresponds to the maximum rate of increase of $1s_5$ density with time (maximum production).

The specific production process is assumed to be dissociative recombination

$$Ne_2^+ + e \rightarrow N_e + N_e^* \tag{9.18}$$

since this process has a large recombination coefficient and, as shown by Gaur and Chanin (1969), the densities of $Ne_2^+$ and $Ne^+$ are approximately equal in neon at 10 torr.

Since recombination is the dominant electron loss process in neon at these pressures, and since Veatch and Oskam (1970) have shown that dissociative recombination in neon preferentially populates the 2p and 3p levels, which radiatively decay to the 1s levels, it appears that many of the electrons lost by recombination could result in the production of $1s_5$ atoms. Wisniewski (1971) estimates that as many as 78% of the electrons lost would result in the production of $1s_5$ atoms. This satisfactorily accounts for the rapid increase in $1s_5$ density in the very early afterglow (< 100 μsec).

In the very late afterglow (> 600 μsec) Wisniewski (1971) has shown that the decay can be correctly predicted by the bracketed terms in the rate equation. This is in agreement with the experiments of Phelps (1959). In the early afterglow (100 μsec < t < 400 μsec), the decay rate is more rapid than in the late afterglow and so it is necessary to consider additional loss terms. The additional loss terms could be metastable-metastable collisions, superelastic collisions or electron collisional excitation to adjacent states. If a metastable-metastable collision cross section of $10^{-18}$ $m^2$ is used, then it would appear that this process could not account for the observed loss rate. It appears that superelastic collisions or electron collisional excitation would then be important.

### 9.2.2 The Helium ($2^3S$) Metastable Species

The previous discussion for neon outlined the basic principles that must be used in order to determine population densities in systems of interest. The studies of the helium ($2^3S$) metastables are similar but there are some interesting additional considerations that make it worthwhile to examine this as well. This section summarizes the paper by Miller, Verdeyen and Cherrington (1971).

The partial energy level diagram for helium is shown in Figure 9.10.

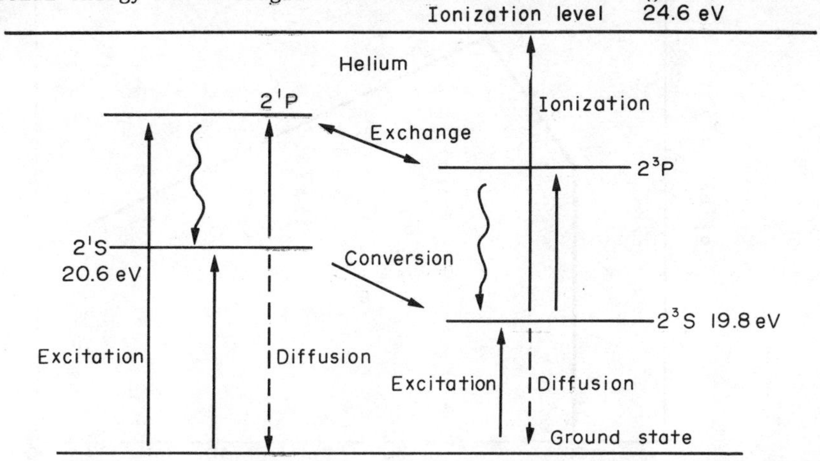


*Fig. 9.10. Partial energy-level diagram for helium and some rate processes*

9.2.2.1 <u>Helium metastables in the DC discharge</u>. The continuity equation for the triplet metastable species in DC discharges is given as

$$\frac{dM}{dt} = D\nabla^2 M + \langle \sigma_p v_e \rangle n_e N - \langle \sigma_I v_e \rangle n_e M - \beta M^2 \qquad (9.19)$$

where $M$ = He($2^3S$) density,

$n_e$ = electron density,

$D$ = metastable diffusion coefficient,

$\langle \sigma_I v_e \rangle$ = rate coefficient for loss of metastables by ionizing collisions with electrons,

$\beta$ = rate coefficient for the loss of metastables by metastable-metastable collisions.

An approximate cross section for excitation of the triplet metastables by direct and indirect processes is shown in Figure 9.11. The rate coefficients were determined by averaging over the electron energy distribution functions that were computed by Smit (1936) and which were discussed earlier. The appropriate variable, as used by Smit, is E/P. As you will remember from our discussion of the positive column, this defines an electron temperature. The rate coefficient $\langle \sigma_p v_e \rangle$ is then shown in Figure 9.12. Similarly, the rate coefficient for ionization, $\langle \sigma_I v_e \rangle$, is shown in Fig. 9.13.

As in the case of the neon metastables, we expect the He($2^3S$) population to have a diffusion distribution at low currents and to saturate at high currents. The experimental measurements confirm this as shown in Figure 9.14. We can also

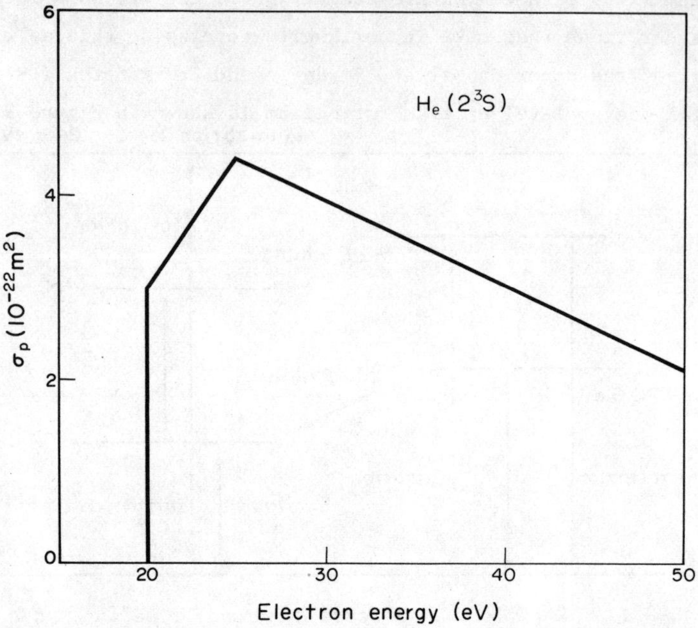


*Fig. 9.11. Electron impact excitation cross section for the He($2^3S$) metastable species*

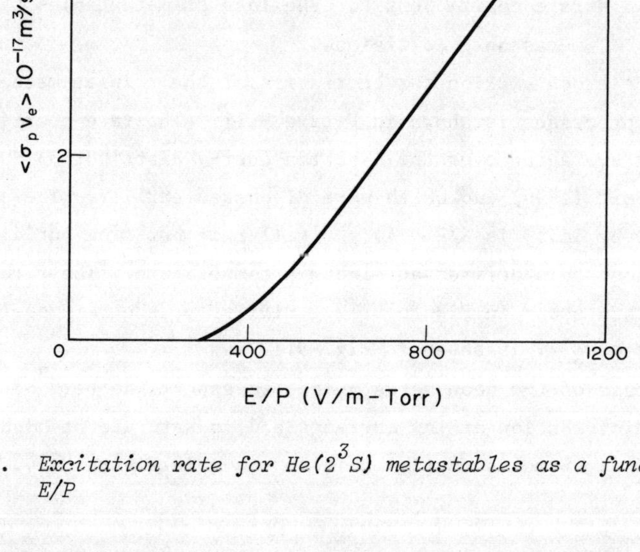


*Fig. 9.12. Excitation rate for He($2^3S$) metastables as a function of E/P*

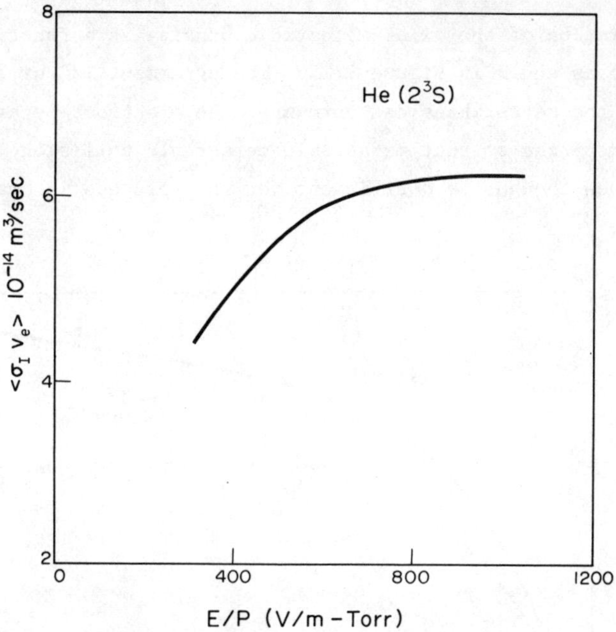


*Fig. 9.13. Ionization rate for He($2^3S$) metastables as a function of E/P*

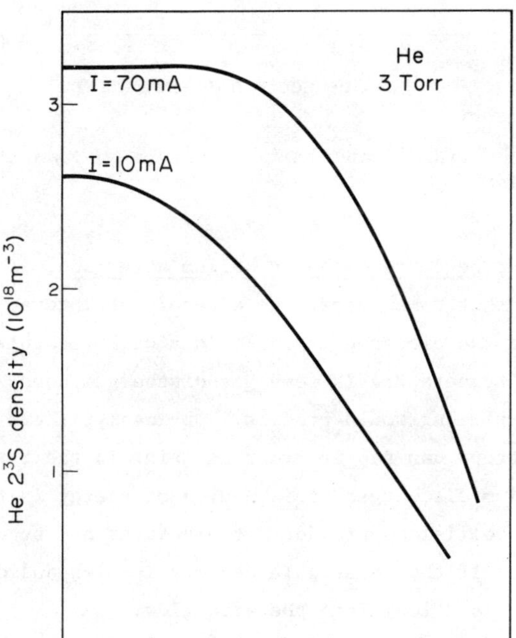


*Fig. 9.14. He($2^3S$) metastable distributions as a function of discharge current*

compare the saturation of the axial metastable density as a function of electron density (current) as shown in Figure 9.15. The agreement is quite good. You will also notice that the metastables can outnumber the electrons by as much as 50:1 under some circumstances so that metastable-metastable collisions can be an important loss term even though the coefficient for this process is small ($\beta \simeq 1.8 \times 10^{-15}$ $m^3$/sec).

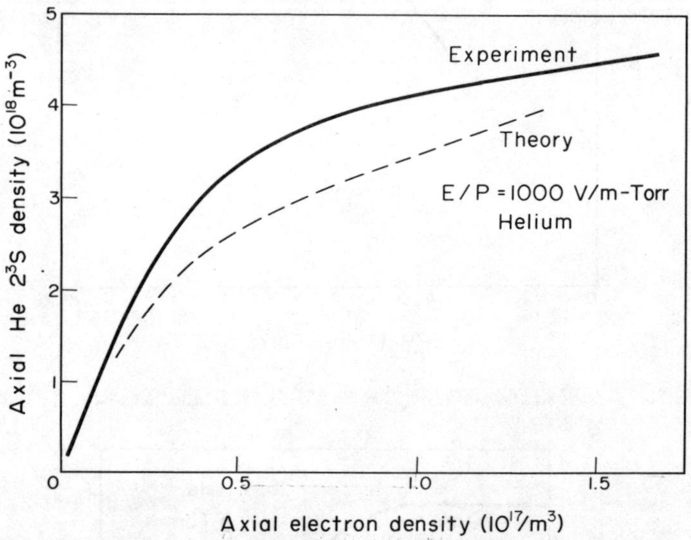


*Fig. 9.15. Axial He($2^3S$) metastable density as a function of electron density*

9.2.2.2. Electron temperature decay in helium afterglow. Now consider another facet of this problem that comes under the category of interactions between the metastable species and the electron gas. It is a well recognized fact that the decay of the electron temperature in some gas discharges does not occur as rapidly as simple collisional cooling would predict. Obviously, there is some source of energy which the electrons can tap in order to maintain their average energy at a higher value. In helium discharges, this source of energy is the metastables. Metastable-metastable collisions provide hot electrons and superelastic collisions provide hot electrons. If the metastable density is high and decays slowly, then this energy source can last long into the afterglow.

The electron energy balance equation (neglecting conduction losses) can be written as follows for a volume of plasma in the helium afterglow

$$\frac{d}{dt}\int \frac{3}{2} n_e k T_e d^3 r = \int \left( \frac{\beta}{2} M^2 + < \sigma_d v_e > n_e M \right) \Delta\varepsilon d^3 r$$
$$- \int < \nu_m > \delta \frac{3}{2} n_e k (T_e - T_g) d^3 r \ , \qquad (9.20)$$

where $\beta$ is the rate coefficient for metastable-metastable collisions,

$< \sigma_d v_e >$ is the rate coefficient for superelastic electron collisional de-excitation of the metastables,

$< \nu_m >$ is the average electron neutral collision frequency,

$\delta$ is the fractional energy loss per elastic collision,

$\Delta\varepsilon$ is the average energy produced in either of the metastable processes which is transferred to the electron gas.

$$\Delta\varepsilon(\text{eV}) = 4[n_e(\text{m}^{-3})/10^{17}]^{1/2} \ . \qquad (9.21)$$

Actually, the electrons produced in these processes have 15-20 eV of energy, $\Delta\varepsilon$ is the part that is transferred to the electron gas (see Montgomery and Tidman, 1964). The rate coefficient $< \sigma_d v_e >$ can be computed from the rate coefficient for excitation using microscopic reversibility and gives

$$< \sigma_d v_e > (\text{m}^3/\text{sec}) = (2.7 \times 10^{-17})\ Te^{1/2} \quad (Te \text{ in K}) \ . \qquad (9.22)$$

In solving the energy balance equation, the time derivative can be dropped since the electron temperature decay is much slower than would occur due to elastic collisions only. The electron temperature can then be computed as a function of time by balancing the two terms on the right hand side of the equation. Miller (1971) measured the electron density and metastable density as a function of time in the afterglow and, using the previous equation, computed the theoretical electron temperature as a function of time in the afterglow. The comparison between measured and calculated values is shown in Figure 9.16. Once again, good agreement is found.

These studies show that good agreement exists between the actual properties of excited species in gas discharges and the predictions based on the rate equation approach.

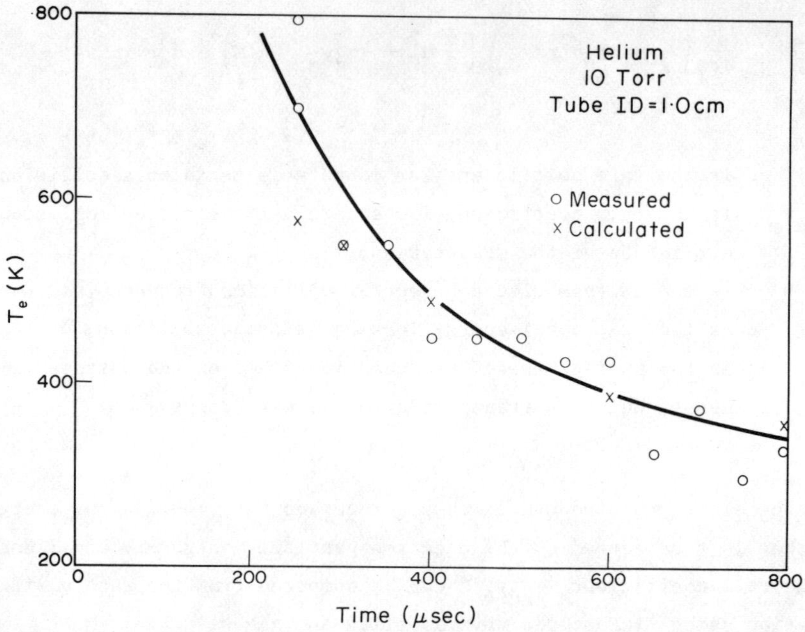


*Fig. 9.16. Electron temperature as a function of time in a helium afterglow*

## 9.3 REFERENCES

Alpert, D. A., McCoubrey, A. O. and Holstein, T. (1949), "Imprisonment of Resonance Radiation in Mercury Vapor," Phys. Rev. 76, 1257-1259.

Bates, D. R., and Damgaard, A. (1949), "The Calculation of the Absolute Strengths of Spectral Lines," Phil. Trans. A242, 101-122.

Biondi, M. A. (1953), "Diffusion, De-excitation and Ionization Cross Sections for Metastable Atoms," Phys. Rev. 88, 660-665.

Gaur, J. P., and Chanin, L. M. (1969), "Rate Coefficient for $X^+ + 2X \rightarrow X_2^+ + X$ in Neon and Argon," Phys. Rev. 182, 167-175.

Holstein, T. (1947/1951), "Imprisonment of Resonance Radiation in Gases, I," Phys. Rev. 72, 1212-1233 (1947); II, ibid. 83, 1159-1168 (1951).

Miller, P. A., Verdeyen, J. T., and Cherrington, B. E. (1971), "Behavior of He ($2^3S$) Metastable Atoms in Weakly Ionized Helium Plasmas," Phys. Rev. A4, 692-700.

Montgomery, D. C., and Tidman, D. A. (1964), Plasma Kinetic Theory, McGraw-Hill.

Phelps, A. V. (1959), "Diffusion, De-excitation and Three-body Collision Coefficients for Excited Neon Atoms," Phys. Rev. 114, 1011-1025.

Salinger, S. N., and Rowe, J. E. (1968), "Determination of the Predominant Ionization and Loss Mechanisms for the Low Voltage Arc Mode in a Neon Plasma Diode," J. Appl. Phys. 39, 4299-4307.

Schlie, L. A., Verdeyen, J. T., Cherrington, B. E., and Wisniewski, E. (1970), "Behavior of Neon $1s_5$ Atoms in He-Ne Active DC Discharges," United States Air Force, Aerospace Research Laboratories research report ARL 70-0123, July 1970 (unpublished).

Smit, J. A. (1936), "Berechnung der Geschwindigkeitsverteilung der elektronen bei Gasentladungen in Helium," Physica 3, 543-560.

VanTrigt, C. (1969, 1970, 1971), "Analytically Solvable Problems in Radiative Transfer" I, Phys. Rev. 181 97-144, II, Phys. Rev. A1 1298-1314, III, Phys. Rev. A4 1303-1316.

Veatch, G. E., and Oskam, H. J. (1970), "Recombination and Ion Conversion Processes in Helium-Neon Mixtures," Phys. Rev. A2, 1422-1428.

Vriens, L. (1964), "Calculations of Absolute Ionization Cross Sections of He, $He^*$, $He^+$, Ne, $Ne^*$, $Ne^+$, Ar, $Ar^*$, Hg, and $Hg^*$," Phys. Lett. 8, 260-261.

Wiese, W. L., Smith, M. W., and Glennon, B. M. (1966), "Atomic Transition Probabilities Vol. I Hydrogen through Neon," National Standard Reference Data Series, NSRDS-NBS 4, U.S. Dept. of Commerce.

Wiese, W. L., Smith, M. W., and Miles, B. M., (1969), "Atomic Transition Probabilities, Vol. II, Sodium through Calcium," National Standard Reference Data Series, NSRDS-NBS 22, U. S. Dept. of Commerce.

Wisniewski, E. (1971), "Investigation of the $1s_5$ Metastable Atom Density in Afterglow plasmas in Neon," Ph.D. thesis, University of Illinois (unpublished).

# Chapter 10

# ATOMIC NEUTRAL GAS LASERS

In this and succeeding chapters, the fundamental concepts developed so far will be applied to the analysis of a number of gas laser systems. Actually, this is easier said than done since few gas lasers have been analyzed in detail on the basis of fundamental gas discharge processes. For this reason, only a few lasers will be discussed as representative models. For an excellent and thorough treatment of a wide variety of gas lasers, see the book by Willett (1974).

## 10.1 THE LASER CONCEPT

In rather simple terms, a laser is nothing more than an optical oscillator and as such it must consist of:

(1) a medium with optical gain and

(2) a positive feedback system so that oscillation can be produced. The optical cavity system that produces feedback will not be considered here. We will simply assume that the medium has sufficient optical gain to overcome the mirror transmission losses, optical surface scattering losses and all other loss processes within the cavity. We will concentrate solely on the necessary conditions within the medium to achieve optical gain.

### 10.1.1 <u>Population Inversion and Gain</u>

Consider propagation of light through a medium where excited species can make transitions between two levels so that optical absorption or emission at the wavelength of the light can occur.

We can have:

(1) Spontaneous emission (noise)

$$\frac{dN_2}{dt} = - A_{21}N_2 = - \frac{dN_1}{dt} \quad . \tag{10.1}$$

(2) Absorption (optical loss)

$$\frac{dN_2}{dt} = B_{12}N_1\rho_\nu = - \frac{dN_1}{dt} \quad . \tag{10.2}$$

(3) Stimulated emission (optical gain)

$$\frac{dN_2}{dt} = - B_{21}N_2\rho_\nu = - \frac{dN_1}{dt} \quad . \tag{10.3}$$

The A and B's are the Einstein coefficients for spontaneous emission, absorption and stimulated emission. $\rho_\nu$ is the optical energy density (joules/$m^3$) in a unit frequency interval $d\nu$ and is assumed to be uniform over the atomic (or molecular) transition.

If we wish to consider the rate of change of the optical energy density at a particular frequency, $\nu$, within such a medium, then remember that each photon emitted (or absorbed) has an energy $h\nu$ and

$$\frac{d\rho_\nu}{dt} = \left[(B_{21}N_2 - B_{12}N_1)\,\rho_\nu + A_{21}N_2\right] g(\nu)\ h\nu \quad . \tag{10.4}$$

$g(\nu)$ is a lineshape factor that expresses the probability that the transition will occur at a frequency between $\nu$ and $\nu + d\nu$ within the linewidth of the transition and $g(\nu)$ is defined such that

$$\int_0^\infty g(\nu)\ d\nu = 1 \quad . \tag{10.5}$$

The term involving $A_{21}$ simply represents noise that is isotropically radiated within the system, and it will be ignored for purposes of this discussion. The remaining terms in equation (10.4) can now be used to determine the change in optical intensity as light propagates through the medium. The light intensity $I_\nu$ (watts/$m^2$) = c $\rho_\nu$ and the distance of propagation dZ = c dt. Therefore,

$$\frac{dI_\nu}{dZ} = (B_{21}N_2 - B_{12}N_1)\ I_\nu\ g(\nu)\ h\nu/c \quad . \tag{10.6}$$

Now, as Yariv (1967) has shown, the Einstein A and B coefficients are related by:

$$g_1B_{12} = g_2B_{21} \tag{10.7}$$

and

$$\frac{A_{21}}{B_{21}} = \frac{8\pi\ h\nu^3}{c^3} \tag{10.8}$$

where the g's are the statistical weights of the upper and lower levels. Inserting these in equation (10.6) gives

$$\frac{dI_\nu}{dZ} = \{A_{21}\ (\frac{c^2}{8\pi\ \nu^2})\ [N_2 - \frac{g_2}{g_1}\ N_1]\ g(\nu)\}\ I_\nu \quad . \tag{10.9}$$

Obviously, for gain we must have

$$N_2 > \frac{g_2}{g_1}\ N_1 \quad , \tag{10.10}$$

the inversion condition. Equation (10.9) can also be written as

$$\frac{dI_\nu}{dZ} = \gamma\ (\nu)\ I_\nu \quad . \tag{10.11}$$

Therefore

$$I_\nu\ (Z) = I_\nu\ (o)\ e^{\gamma(\nu)\ Z} \quad . \tag{10.12}$$

10.1.2 Small Signal Gain, Saturation Intensity and Amplitude of Oscillation

If we assume that the optical cavity into which the laser medium is placed can be described by a distributed loss coefficient $\alpha$, then

$$\frac{dI_\nu}{dZ} = \gamma(\nu)\ I_\nu - \alpha\ I_\nu \quad . \tag{10.13}$$

Obviously, for laser action to occur $\gamma$ must exceed $\alpha$, and not just be positive as indicated by equation (10.10). Consider the intensity as a function of time.

$$\frac{dI_\nu}{dt} = c\ \frac{dI_\nu}{dZ} = c\ [\gamma(\nu) - \alpha]\ I_\nu \quad . \tag{10.14}$$

At steady state, $dI_\nu/dt = 0$ so the gain must equal the loss at the operating intensity. This suggests that the gain will be a function of intensity and that saturation will occur.

Let us look at the rate processes involved. Assume that the upper level (2) is pumped at a rate $R_2$ and spontaneously decays with a time constant $\tau_{20}$ to levels other than level 1. Correspondingly, level 2 decays spontaneously to level 1 with a time constant $\tau_{21} = 1/A_{21}$. Similarly, level 1 is pumped at a rate $R_1$ and decays with a time constant $\tau_1$. Therefore,

$$\frac{dN_2}{dt} = R_2 - N_2/\tau_2 - (N_2 - N_1)\ B_{21}\ g(\nu)\ \rho_\nu \tag{10.15}$$

and

$$\frac{dN_1}{dt} = R_1 - N_1/\tau_1 + N_2/\tau_{21} + (N_2 - N_1)\ B_{21}\ g(\nu)\ \rho_\nu \tag{10.16}$$

where

$$\frac{1}{\tau_2} = \frac{1}{\tau_{20}} + \frac{1}{\tau_{21}} \quad .$$

Define

$$W_s = B_{21}\ g(\nu)\ \rho_\nu = B_{21}\ g(\nu)\ I_\nu/c \tag{10.17}$$

as the stimulated emission rate.

Consider first the case when there is no stimulated emission and steady state exists (small signal condition).

$$\frac{dN_{20}}{dt} = 0 = R_2 - \frac{N_{20}}{\tau_2} \quad . \tag{10.18}$$

Therefore,

$$N_{20} = R_2 \tau_2 \quad . \tag{10.19}$$

$$\frac{dN_1}{dt} = 0 = R_1 - \frac{N_{10}}{\tau_1} + \frac{N_{20}}{\tau_{21}} \quad . \tag{10.20}$$

Therefore,

$$N_{10} = \tau_1\ (R_1 + \frac{\tau_2}{\tau_{21}}\ R_2) \quad . \tag{10.21}$$

The small signal gain, $\gamma_o$, is proportional to $N_{20} - N_{10}$, or

$$\gamma_o(\nu) \sim N_{20} - N_{10} = R_2\tau_2(1 - \tau_1/\tau_{21}) - R_1\tau_1 \quad . \tag{10.22}$$

Neglecting $\tau_1/\tau_{21}$ for now, we have

$$N_{20} - N_{10} = R_2\tau_2 - R_1\tau_1 \quad . \tag{10.23}$$

and we see what is necessary for laser action.

The upper state should have a long lifetime (quasi-forbidden transition) and should be strongly pumped. The lower state should have a very short lifetime (strongly allowed transition) and should be weakly pumped.

Consider equations (10.15) and (10.16) when the laser intensity is large and we again have a steady state. To simplify slightly, assume $\tau_{21} >> \tau_1$. (In other words, we assume the previous criteria hold).

We now have

$$\frac{dN_2}{dt} = R_2 - N_2/\tau_2 - W_s(N_2 - N_1) = 0 \quad , \tag{10.24}$$

$$\frac{dN_1}{dt} = R_1 - N_1/\tau_1 + W_s(N_2 - N_1) = 0 \tag{10.25}$$

or

$$R_2\tau_2 - R_1\tau_1 = N_2 - N_1 + W_s(N_2 - N_1)\ (\tau_2 + \tau_1) \quad . \tag{10.26}$$

Therefore,

$$N_2 - N_1 = \frac{R_2\tau_2 - R_1\tau_1}{1 + W_s(\tau_2+\tau_1)} = \frac{N_{20} - N_{10}}{1 + W_s(\tau_2+\tau_1)} \quad . \tag{10.27}$$

Providing the pumping rates are not changed by the optical intensity, we can see that the gain is, indeed, a function of optical intensity and

$$\gamma(\nu) = \frac{\gamma_o(\nu)}{1 + W_s(\tau_2+\tau_1)} \quad . \tag{10.28}$$

Since $W_s = B_{21}\, g(\nu)\, I_\nu/c$ and $A_{21}/B_{21} = 8\pi\, h\nu^3/c^3$ then,

$$W_s = \frac{c^2}{8\pi\, h\nu^3}\, g(\nu)\, A_{21}\, I_\nu \tag{10.29}$$

and

$$\gamma(\nu) = \frac{\gamma_o(\nu)}{1 + I_\nu/I_s(\nu)} \tag{10.30}$$

where

$$I_s^{-1} = A_{21}\frac{c^2}{8\pi\, h\nu^3}\, g(\nu)\, (\tau_2+\tau_1) \quad . \tag{10.31}$$

$I_s$ is the saturation intensity, or value of the input flux for which the gain reduces to one half of the small signal gain.

[For what is known as an "inhomogeneously" broadened line, the appropriate equation is

$$\gamma(\nu) = \frac{\gamma_o(\nu)}{\sqrt{1 + I_\nu/I_s(\nu)}} .$$

We won't go into this aspect here. See Yariv (1967) for details].

Returning to equation (10.14) we can now write

$$\frac{c\,dI_\nu}{dZ} = \frac{dI_\nu}{dt} = c\left[\frac{\gamma_o(\nu)}{1 + I_\nu/I_s} - \alpha\right] I_\nu \tag{10.32}$$

and at steady state

$$I_\nu = \left[\frac{\gamma_o(\nu)}{\alpha} - 1\right] I_s . \tag{10.33}$$

This represents the oscillation amplitude of a "homogeneously" broadened laser.

### 10.1.3 Power Available from a Laser Amplifier

The concept of saturation plays an important role in laser amplifiers as well as oscillators. If we ignore the loss term, from equation (10.32) we can write

$$\int_{I_{in}}^{I_{out}} \frac{dI_\nu}{I_\nu}\left(1 + \frac{I_\nu}{I_s}\right) = \int_o^d \gamma_o\, dZ \tag{10.34}$$

or,

$$\ell n \frac{I_{out}}{I_{in}} + \frac{I_{out} - I_{in}}{I_s} = \gamma_o d . \tag{10.35}$$

For the small signal condition, $I_{out}/I_s << 1$,

$$\frac{I_{out}}{I_{in}} = e^{\gamma_o d} .$$

For the large signal condition, $I_{out} \simeq I_s$,

$$I_{out} = I_{in} + (\gamma_o I_s)\, d . \tag{10.36}$$

The product of the small signal gain, $\gamma_o$, and the saturation intensity, $I_s$, then represents the power that can be extracted (watts/$m^3$) per unit length of the amplifier.

From equations (10.9) and (10.31)

$$\gamma_o I_s = h\nu\left[\frac{N_{20} - (g_2/g_1) N_{10}}{\tau_2 + \tau_1}\right] . \tag{10.37}$$

For a high gain medium, $N_{20} >> g_2 N_{10}/g_1$ and $\tau_2 > \tau_1$. Therefore, using equation (10.19), we obtain

$$\gamma_o I_s = h\nu R_2 \tag{10.38}$$

where $R_2$ is the pumping rate into the upper laser or amplifier level.

Obviously, this is the maximum power density extractable from an amplifier and the actual power available will be some fraction of $h\nu R_2$.

## 10.2 THE HELIUM-NEON LASER

The previous discussions of the neon and helium systems can now be applied to the analysis of the He-Ne laser with the important addition that we now have collisions between the helium metastables and neon ground state atoms to produce neon atoms in excited states corresponding to the upper levels of conventional laser transitions. A partial energy level diagram showing some of the important processes is given in Fig. 10.1.

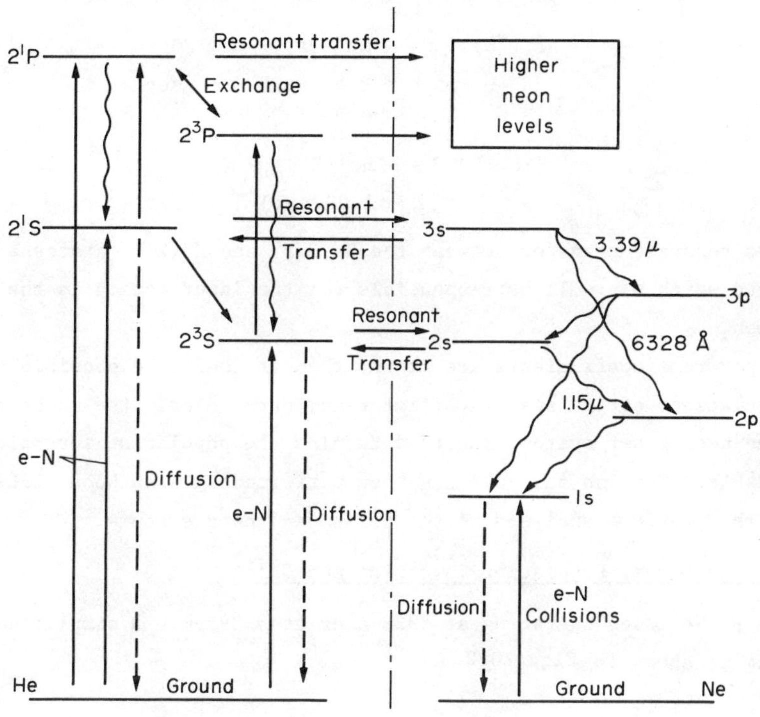


*Fig. 10.1. Partial energy level diagram of the helium-neon system (Not to scale)*

The important processes for pumping the He-Ne laser involve the resonant transfer of energy by inelastic atom-atom collisions between excited helium atoms and ground state neon atoms. As you can see from Fig. 10.1, a helium atom excited to the $2^1S$ metastable state can collide with a ground state neon atom to produce a neon atom excited to the 3s state and a ground state helium atom. This is a very selective excitation mechanism and is largely responsible for the resulting population inversion in neon at 6328 Å and at 3.39 μm. There is a similar resonant transfer between the $He(2^3S)$ state and the Ne(2s) state which gives rise to the population inversion in neon at 1.15 μm. The important rates of excitation exchange (and their inverses) have been measured by Schmeltekopf and Fehsenfeld (1970) to be

$$He\,(2^1S) + Ne \rightarrow Ne\,(3s_2) + He \quad \text{at } k = 6.46 \times 10^{-17}\,m^3/sec \quad , \tag{10.39}$$

$$Ne(3s_2) + He \rightarrow He(2^1S) + Ne \quad \text{at } k = 2.53 \times 10^{-16}\,m^3/sec \quad , \tag{10.40}$$

$$He(2^3S) + Ne \rightarrow Ne\,(2s) + He \quad \text{at } k = 3.85 \times 10^{-18}\,m^3/sec \tag{10.41}$$

and

$$Ne(2s) + He \rightarrow He\,(2^3S) + Ne \quad \text{at } k = 1.20 \times 10^{-18}\,m^3/sec \quad . \tag{10.42}$$

There is also resonant transfer between the $He(2^1P)$ and $He(2^3P)$ states and upper levels of neon which may well be responsible for the laser action in the 6p-6d system of neon.

If the proper rate coefficients are known, then it should be possible to determine the populations of the species of interest, particularly the upper and lower levels of the neon laser system, and to determine the population inversions and powers available. Such an analysis has been performed by Arrathoon (1969) and can be based on the previous analysis of the helium and neon systems.

### 10.2.1 Rate Equations, Population Inversion and Gain

Consider the He-Ne laser operating at 6328 Å or at 3.39 μm. A simplified energy level diagram is shown in Fig. 10.2.

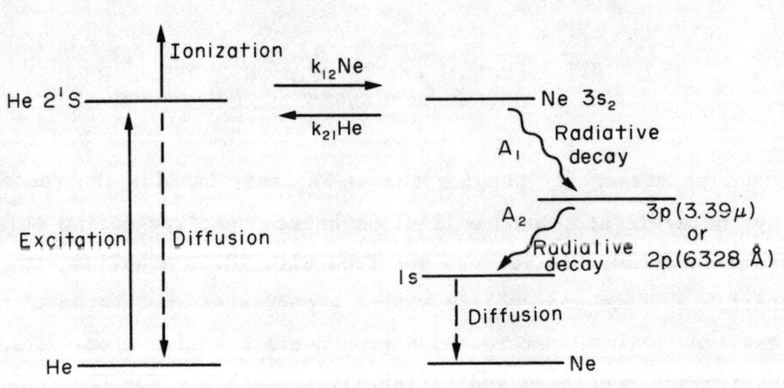


*Fig. 10.2. Partial energy level diagram and some rate processes for the He-Ne laser at 3.39 μm or 6328 Å*

Figure 10.2 basically shows a classical four level laser system in which we assume that because of the selective pumping mechanism, the upper laser level (3s) is pumped much more strongly than the lower laser level (3p or 2p). We are also aided in obtaining a population inversion by the fact that the A coefficient for the 3s level is approximately $1.3 \times 10^7$ $sec^{-1}$ (see Arrathoon, 1969) which is smaller than the A coefficient for the $2p_4$ level which is $5.24 \times 10^7$ $sec^{-1}$ (see Bennett and Kindlmann, 1966) so that the lower laser level radiatively decays at a faster rate than the upper level. This would produce a population inversion even in the absence of a selective excitation process.

To determine the density of the Ne(3s) species, the rate equation for the $He(2^1S)$ and Ne(3s) species must be solved. In the steady state

$$\frac{d}{dt} Ne(3s_2) = 0 = k_{12}\ Ne\ He(2^1S) - k_{21}\ He\ Ne(3s_2) - A_1\ Ne(3s_2) \quad , \qquad (10.43)$$

where Ne and He represent the neutral gas densities of neon and helium. From this equation, we find

$$Ne(3s_2) = \frac{k_{12}\ Ne}{[k_{21}\ He + A_1]}\ He(2^1S) \quad . \qquad (10.44)$$

(Note that we assume no pumping of the $Ne(3s_2)$ level through the neon system). To find the $He(2^1S)$ density, we solve the rate equation

$$\frac{d}{dt} He(2^1S) = 0 = n_e <\sigma_p v_e> He - n_e <\sigma_d v_e> He(2^1S) - k_{12}\ Ne\ He(2^1S)$$
$$+ k_{21} He\ Ne(3s_2) + D_m \nabla^2 He(2^1S) \quad . \qquad (10.45)$$

In this equation

$<\sigma_p v_e>$ is the rate coefficient for electron impact excitation from the ground state,

$<\sigma_d v_e>$ is the rate coefficient for electron impact destruction (excitation, ionization, de-excitation),

$k_{12}$ = the resonant transfer rate $He(2^1S) \rightarrow Ne(3s_2)$,

$k_{21}$ = the resonant transfer rate $Ne(3s_2) \rightarrow He(2^1S)$,

$D_m$ = the diffusion coefficient for the helium$(2^1S)$ metastable species.

If it is assumed that the $He(2^1S)$ metastables are distributed in the fundamental diffusion mode, then

$$\nabla^2 He(2^1S) = -\frac{He(2^1S)}{\Lambda^2} = -\left(\frac{2.405}{R}\right)^2 He(2^1S) \qquad (10.46)$$

and the solution to equation (10.45) is

$$He(2^1S) = \frac{n_e <\sigma_p v_e> He + k_{21}\ He\ Ne(3s_2)}{n_e <\sigma_d v_e> + k_{12}\ N_e + D_m (\frac{2.405}{R})^2} \quad . \qquad (10.47)$$

Using equation (10.44) to express the Ne($3s_2$) density in terms of the He($2^1S$) density, equation (10.47) can be simplified to

$$He(2^1S) = \frac{n_e \langle \sigma_p v_e \rangle He}{n_e \langle \sigma_d v_e \rangle + \frac{k_{12}}{k_{21}} \frac{Ne\, A_1}{He + A_1} + D_m \left(\frac{2.405}{R}\right)^2} \quad . \tag{10.48}$$

This equation is valid in the vicinity of the tube center.

The He($2^1S$) density can be obtained from a knowledge of the rate coefficients and, of course, the Ne($3s_2$) population will follow the He($2^1S$) population. The excitation cross sections for the He($2^1S$) and He($2^3S$) states are known (see Massey and Burhop, 1969) and the electron collisional metastable ionization cross sections can be estimated from the work of Vriens (1964) or Long and Geballe (1970). The electron density is an independent parameter and in some cases has been measured (Labuda and Gordon, 1964) as a function of discharge current in specific mixtures and tube radii. The distribution functions, however, are generally unknown and even though electron temperature measurements are available (Labuda and Gordon, 1964), calculations based on an assumed Maxwellian will result in a considerable overestimate of the excitation rate. The $A_1$ coefficient has already been mentioned, as well as the values for $k_{12}$ and $k_{21}$. The metastable diffusion coefficients have been obtained by Phelps (1955) and are

for He($2^1S$) in He, $D_m = [440 \times 10^{-4}]/P\ m^2/sec$,

and for He($2^3S$) in He, $D_m = [470 \times 10^{-4}]/P\ m^2/sec$

where P is the helium pressure in torr.

Since the distribution functions are not known, we will not attempt a complete calculation at this point, however, as was shown in Section 9.2.2, the helium metastable density can be expected to vary linearly with discharge current (electron density) at low electron densities, but it will saturate as the current increases to the point where the electron collisional destruction processes become important. This was verified by White and Gordon (1963). Their results for the He($2^1S$) density as a function of current in a 0.5 torr He - 0.1 torr Ne gas mixture are shown in Fig. 10.3. The predicted direct relationship between the Ne($3s_2$) density and the He($2^1S$) density was also verified by White and Gordon by observing the spontaneous light emission (proportional to excited state density) from the $3s_2$ state. This is shown in Fig. 10.4. The first important point is that the spontaneous light emitted at 6328 Å ($3s_2$-$2p_4$), which is an indicator of the population of the $3s_2$ state, follows exactly the same current dependence as the He($2^1S$) density. This is in agreement with the previous formulation. The second point to notice is that the $5s_2$-$2p_4$ radiation (proportional to the $5s_2$ population) increases

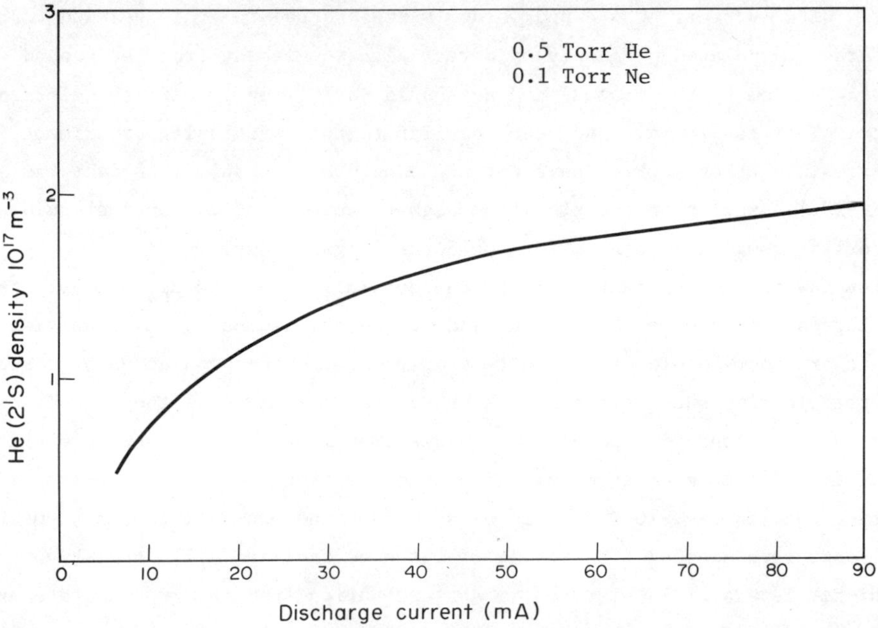


*Fig. 10.3. He ($2^1S$) density as a function of discharge current in a He-Ne mixture. The tube diameter is 6 mm. (from White and Gordon, 1963)*

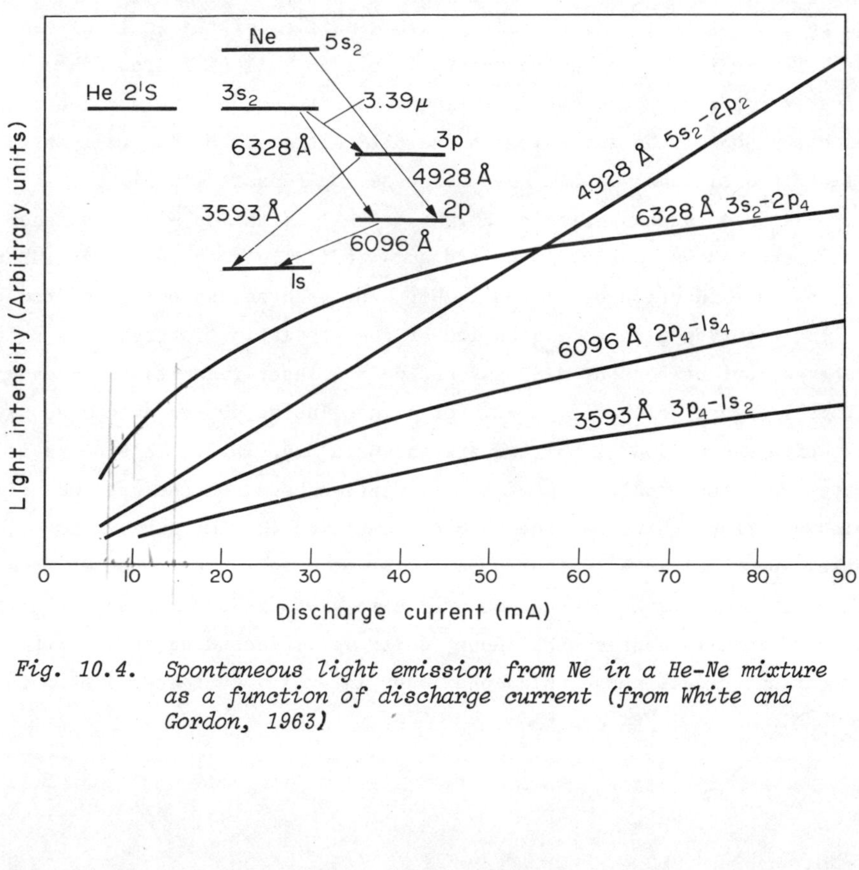


*Fig. 10.4. Spontaneous light emission from Ne in a He-Ne mixture as a function of discharge current (from White and Gordon, 1963)*

linearly with current. White and Gordon's explanation for this assumed that the $5s_2$ state is populated primarily by direct electron impact from the ground state (or cascade from higher states) and so should vary linearly with the electron density. This result may have been peculiar to this particular experiment, however, since other workers have not reproduced this result. In fact, as shown in Fig. 10.1, excitation transfer from higher excited helium species could well be an important pumping mechanism for the 5s and higher levels.

Notice now the curves that indicate the populations of the $2p_4$ and $3p_4$ states. These curves also show a saturation, and the determination of the densities of these lower laser levels can be quite complex due to the population by cascading from upper levels, such as the $3s_2$ which saturates because of the saturation of the helium pumping mechanism, population by electron impact from lower levels and the depopulation due to atom-atom collisions and electron collisional de-excitation. Referring back to Fig. 10.2, it can be seen that the populating mechanisms for the lower laser levels consist basically of three mechanisms: (1) radiative cascade from higher levels, (2) electron impact excitation from the ground state and (3) stepwise excitation from the 1s levels. Radiative cascade from the 3s levels is, of course, no problem since the favorable ratio of A coefficients ensures that the 2p level depletes more rapidly than it is pumped. The 1s levels are only about 2 eV in energy below the 2p levels and therefore can be relatively easily excited to the 2p level. Since fairly high densities of the 1s level can be built up, both electronic excitation to the 2p levels as well as the possible role of radiative trapping reducing the radiative decay of the 2p levels can strongly influence the population inversion on the 6328 Å transition (as well as the other He-Ne laser transitions) as has been discussed by Bennett (1965).

The $2p_4$-$1s_4$ spontaneous emission curve, as well as the $3p_4$-$1s_2$ spontaneous emission curve, shown in Fig. 10.4, are good indications of the lower laser level populations as a function of current. With the spontaneous emission from the $3s_2$ levels, these curves give an indication of the population inversion as a function of discharge current for the 3.39 μm. and 6328 Å laser transitions. Measurements of the small signal gain at 3.39 μm. for a (non-lasing) He-Ne discharge have been made by White and Gordon (1963) and are shown in Fig. 10.5. This curve is consistent with the predictions of the population based on the previous figure. The gain vs current curve for the 6328 Å transition is similar in nature, but is, of course, much smaller - typically on the order of a few percent per meter. Obviously, then, laser action can be predicted at both 3.39 μm. and at 6328 Å and experimental results confirm the theory quite well, including the fact that an optimum current exists above which the laser intensity decreases due to the combination of the saturation of the upper laser level and the increased population of the lower laser level.

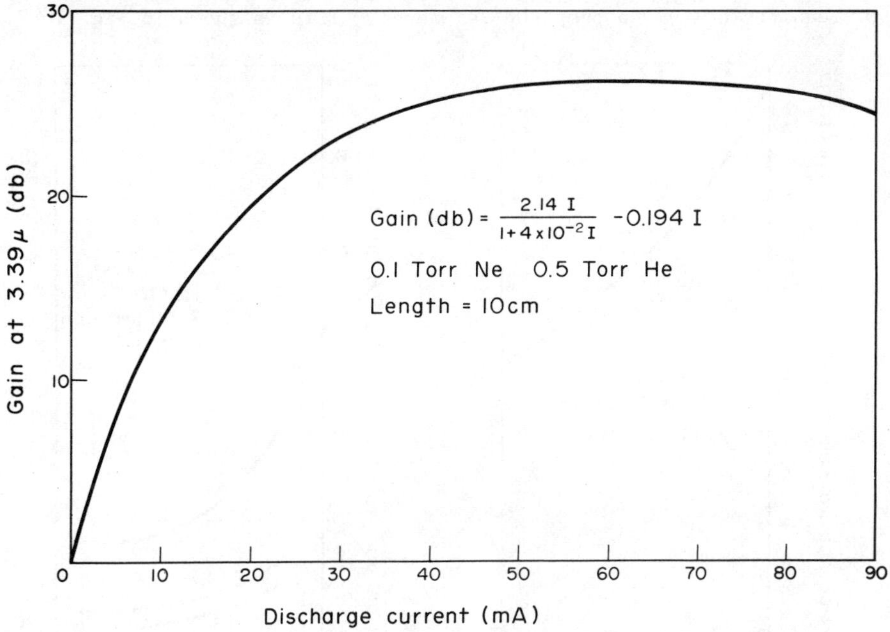


*Fig. 10.5. Small signal gain at 3.39 μm vs discharge current. I is in mA. The tube diameter is 6 mm. (from White and Gordon, 1963)*

Now consider the third well known He-Ne laser transition, the 1.15 μm. transition. Oscillation around 1.15 μm.occurs due to the population inversion between the 2s and 2p states in neon. We would expect, based on the previous discussions, that the primary excitation mechanism for the 2s levels is by neutral Ne collisions with He($2^3S$) species. As it turns out, however, this is not quite true. In this case, electron collisions with ground state Ne atoms can be roughly as important as the collisional transfer between Ne and He($2^3S$). Young, Willett and Maupin (1970) have calculated the excitation rates of the 3s and 2s neon states due to electron impact and due to helium. The relative excitation ratios are given in Fig. 10.6. We can see two important facts here. First of all, the $3s_2$ state is most strongly pumped by helium and so in helium-neon lasers, the laser lines seen originating from the 3s levels come only from the $3s_2$ levels (6328 Å, 3.39 μm.). Indeed, it is somewhat surprising that the $3s_2$ level is the only one of the 3s levels that is selectively excited by the helium $2^1S$ species. The more uniform population mechanisms for the 2s states, however, result in laser action from <u>all</u> 2s levels. There is an additional important point. Since electron impact excitation of the 2s levels is important, it is possible to obtain laser action from the 2s levels (1.15 μm.) in pure Ne, whereas laser action at 6328 Å and 3.39 μm. requires helium to pump the $3s_2$ level. This is shown in Fig. 10.7.

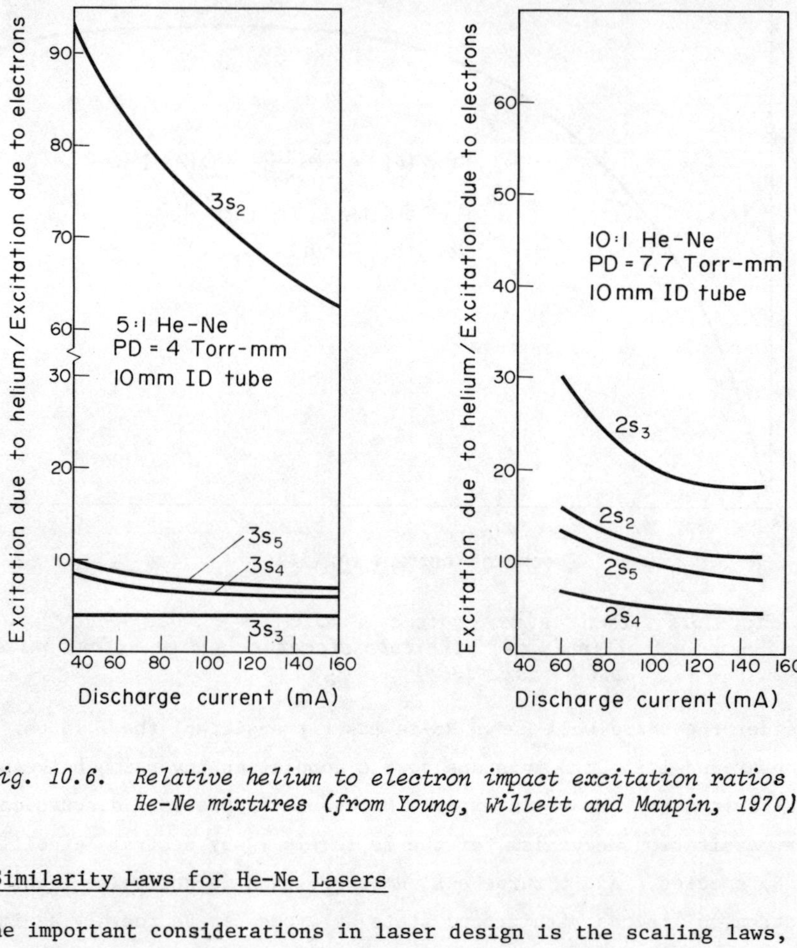


*Fig. 10.6. Relative helium to electron impact excitation ratios in He-Ne mixtures (from Young, Willett and Maupin, 1970)*

## 10.2.2 Similarity Laws for He-Ne Lasers

One of the important considerations in laser design is the scaling laws, or similarity laws, that allow the correct prediction of the optimum operating conditions as the design parameters are changed. In the case of the He-Ne lasers operating at 6328 Å and 3.39 μm, such similarity laws are known and can be explained on the basis of the previous analyses. Gordon and White (1963) stated that the optimum laser gain for a range of tube diameters extending from 1-15 mm is found for

$$PD \simeq 2.9\text{-}3.6 \text{ torr-mm} \quad ,$$

$$P_{He} \simeq 5P_{Ne} \quad .$$

A qualitative analysis of the scaling relationships is given in Gordon and White's paper and experimental verification is provided in the paper by Field (1967). Field obtained the results shown in Figs. 10.8 through 10.10 for the 6328 Å laser.

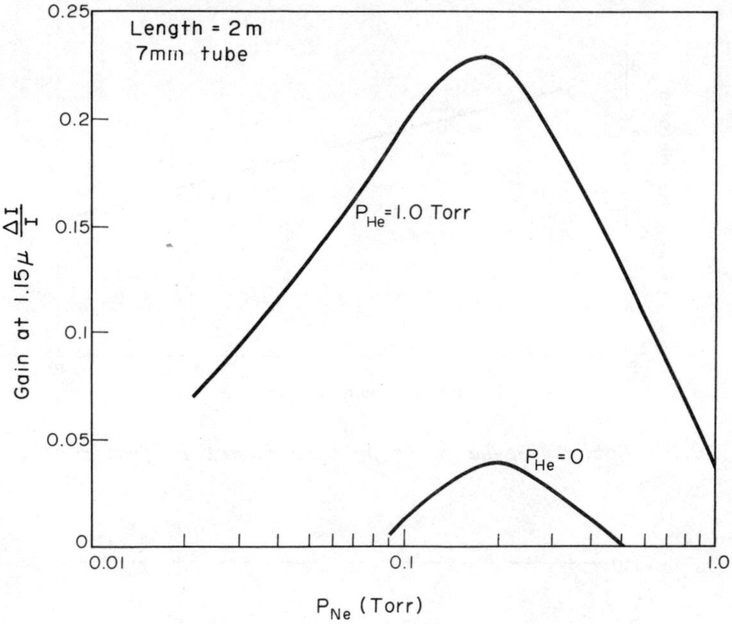


*Fig. 10.7. Gain at 1.15 μm in He-Ne mixtures as well as pure Ne (from Bennett, 1962)*

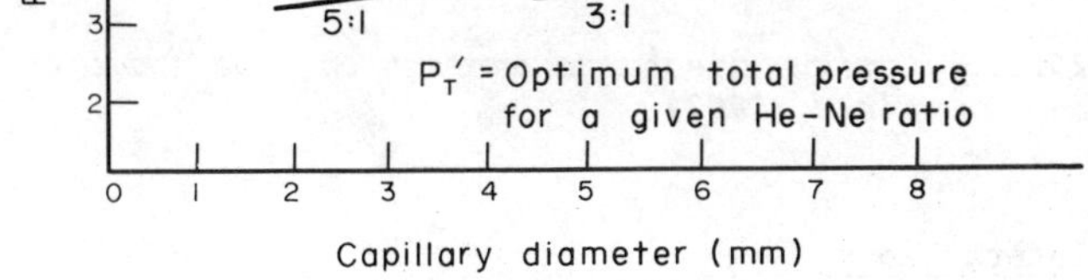


*Fig. 10.8. Optimum PD ratio vs tube diameter for various He-Ne ratios (from Field, 1967)*

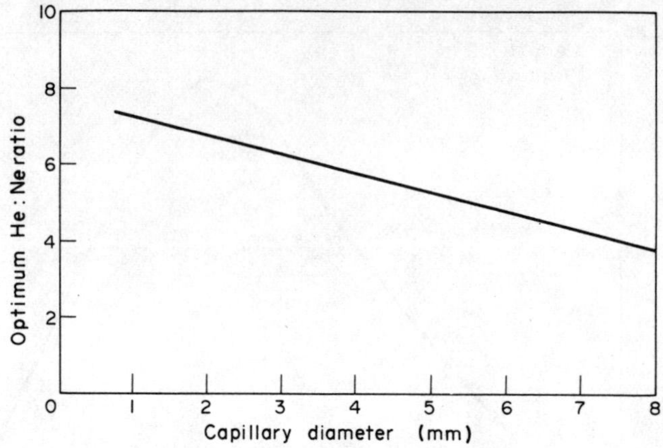


*Fig. 10.9. Optimum He-Ne ratio vs tube diameter (from Field, 1967)*

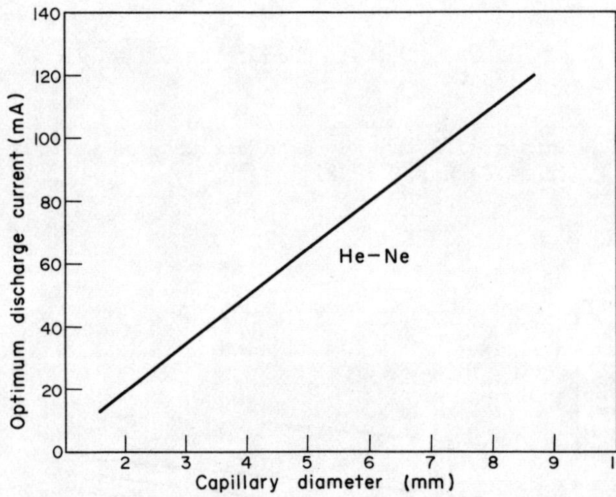


*Fig. 10.10. Optimum discharge current vs tube diameter (from Field, 1967)*

10.2.3 Cascade Interactions

All of the calculations that we have discussed so far have dealt with determining the small signal gain of a potential laser medium, which involves calculating population inversions under non-lasing conditions. When the medium is placed within a resonant cavity so that stimulated emission occurs, some very profound changes can occur in the excited state populations and in the discharge parameters.

Some idea of the effects can be seen in the papers by Weaver and Freiberg (1966) and Freiberg and Weaver (1967). One of the most outstanding examples is the effect of laser behavior upon the discharge properties of $CO_2$ lasers. We shall examine this later.

The changes in excited state populations can also result in new laser transitions that can only be maintained with this type of stimulated emission pumping. As an example, take the paper by Gerritsen and Goedertier (1964) on the cascade effect in He-Ne lasers. The appropriate energy level diagram is shown in Fig. 10.11.

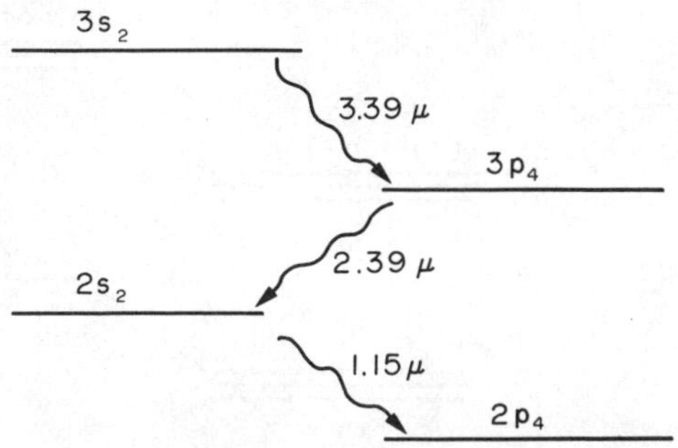


*Fig. 10.11. Energy level diagram for the cascade effect in He-Ne lasers*

Without a laser cavity, the transition at 2.396 μm is absorptive. When a laser cavity allows oscillation at 3.39 μm, the strong stimulated emission at 3.39 μm pumps the $3p_4$ level and inverts the $3p_4$-$2s_2$ transition allowing laser oscillation at 2.396 μm and 1.1526 μm. The oscillation at 1.1526 μm aids, of course, in the de-population of the lower 2.396 μm laser level. The gain at 1.15 μm using the cascade transitions is considerably greater than that which could be obtained without oscillation at 3.39 μm. Bennett (1965) also discusses this case, only he considers the 2.035 μm laser oscillation ($3p_4$-$2s_4$) and the 1.525 μm laser oscillation ($2s_4$-$2p_7$).

## 10.3 ELECTRON COLLISION PUMPED LASERS

Laser oscillation in the pure noble gases, Ne, Ar, Kr and Xe has been obtained using electron collisional pumping alone. In this case, the ground states consist of closed shell $(np)^6$ configurations and there are strongly allowed optical transitions (and hence electron impact pumping) between ground state and higher $(np)^5$ ms and $(np)^5$ md levels. An energy level diagram as given by Bennett (1962) is shown in Fig. 10.12. There is effectively no electron pumping of the p states from ground (forbidden transition) and so inversion can occur and lasing is possible on the s-p and d-p transitions provided that the gas pressure is sufficiently high so that spontaneous radiation from the s and d states to ground is trapped, and the gas pressure is sufficiently low so that excitation transfer collisions are unimportant and that the superelastic electron collision rate is much less than the spontaneous decay rate. As it turns out, optimum pressures are of the order of a few tens of microns, and the gain varies inversely with tube diameter. The radiative lifetimes of the p levels are short ($10^{-8}$ sec) and hence the lower levels rapidly decay down to the 1s levels and population inversion can be achieved on a CW basis. It is also possible to invert the 2p-1s transition on a pulsed basis but not continuously because of the long lifetime of the 1s levels.

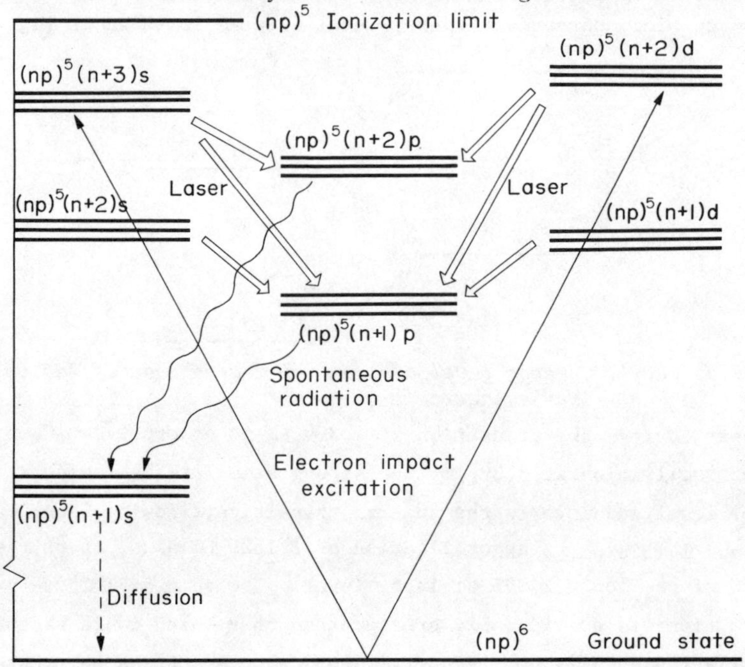


*Fig. 10.12. Energy level diagram typical of pure noble gases (from Bennett, 1962)*

Oscillation in the heavier noble gases usually occurs in the d-p configurations. As Bennett (1962) has pointed out, there are several reasons for this. First, in the heavier gases, the d shells become more and more depressed with respect to the s shells and hence require less excitation energy and the excitation rate for d levels is about an order of magnitude greater than for s levels. Secondly, the transition probabilities for d→p transitions are larger than for s→p transitions. The article by Faust, McFarlane, Patel and Garrett (1964) discussed observations on Ne, Ar, Kr and Xe. They point out the Xe 5d-6p system, with twenty lines between 2 and 13 μm as being particularly notable. An energy level diagram for Xenon with laser transitions indicated is shown in Fig. 10.13.

A similar diagram for neon is shown in Fig. 10.14. In this case the 5s-4p transitions are probably due to helium metastable pumping and not electron collisional pumping.

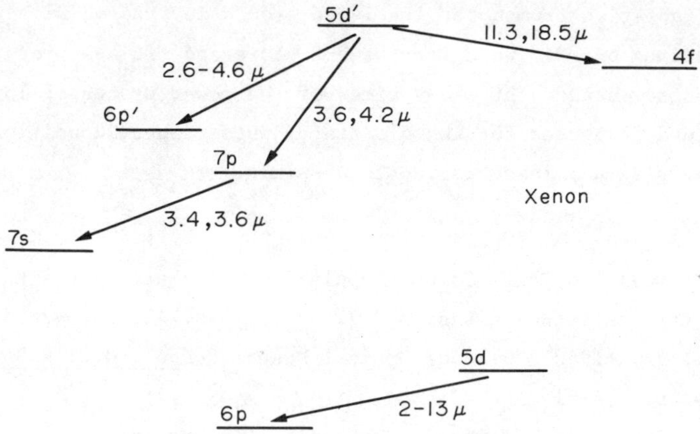


*Fig. 10.13. Energy level diagram and laser lines in pure xenon (from Faust, McFarlane, Patel and Garrett, 1964)*

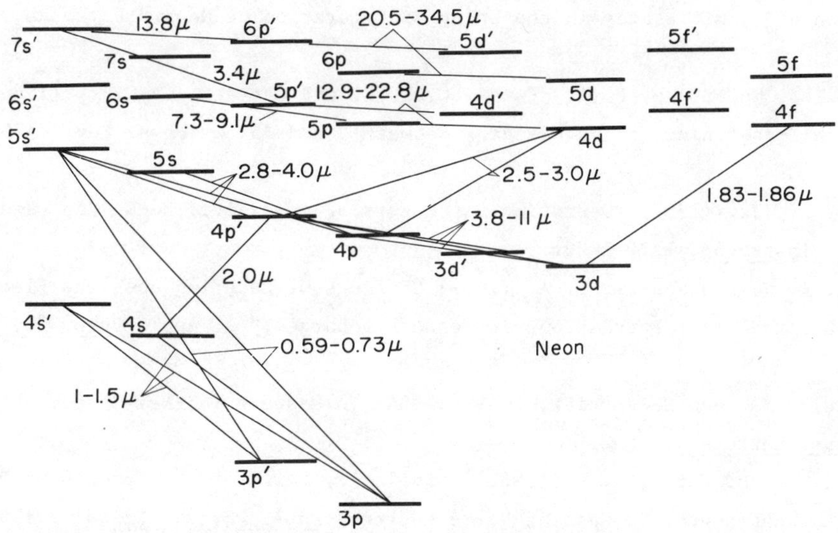


*Fig. 10.14. Energy level diagram and laser lines in pure neon (from Faust, McFarlane, Patel and Garrett, 1964)*

The electron configuration in helium is quite different from that of the other rare gases - however, laser action is still possible in He as has been shown by Levine and Javan (1969). In this case laser action was obtained at 95.8 μm on the $3^1P-3^1D$ transition and at 216.3 μm on the $4^1P-4^1D$ transition. The excitation mechanism is straightforward. Transitions from the upper $3^1D$ and $4^1P$ levels to the ground state are highly allowed (lifetimes of 1.7 and 3.9 nsec) and hence they are strongly pumped by electron impact. The lower states are not optically connected to ground and hence are not pumped by electrons. At the operating pressure, the radiation from upper levels to ground is trapped with resulting radiative lifetimes of 73.3 nsec for the $3^1P$ state and 116 nsec for the $4^1P$ state. These lifetimes now exceed the decay times of the lower states of 15.4 nsec for the $3^1D$ state and 37.8 nsec for the $4^1D$ state. Under these conditions continuous population inversion and laser action is possible.

## 10.4 REFERENCES


Arrathoon, R. (1969) "Positive Column Population Calculations for the Evaluation of Dispersive Effects in He-Ne Lasers," J. Appl. Phys. 40, 2875-2883.

Bennett, W. R., Jr. (1962) "Gaseous Optical Masers," Appl. Optics Supplement on Optical Masers, pp. 24-61.

Bennett, W. R., Jr. (1965) "Inversion Mechanisms in Gas Lasers," Appl. Optics Supplement 2 of Chemical Lasers. pp. 3-33.

Bennett, W. R., Jr. and Kindlmann, P. J. (1966) "Radiative and Collision-Induced Relaxation of Atomic States in the $2p^5 3p$ Configuration of Neon," Phys. Rev. 149, 38-51.

Faust, W. L., McFarlane, R. A., Patel, C. K. N. and Garrett, C. G. B. (1964) "Noble Gas Optical Maser Lines at Wavelengths between 2 and 35 μ," Phys. Rev. 6A 1476-1486.

Field, R. L., Jr. (1967) "Operating Parameters of dc-excited He-Ne Gas Lasers," Rev. Sci. Instr. 38, 1720-1722.

Freiberg, R. J. and Weaver, L. A. (1967) "Effects of Lasering upon the Electron Gas and Excited State Populations in Xenon Discharges," J. Appl. Phys. 38, 250-262.

Gerritsen, H. J. and Goedertier, P. V. (1964) "A Gaseous (He-Ne) Cascade Laser," Appl. Phys. Letters 4, 20-21.

Gordon, E. I. and White, A. D. (1963) "Similarity Laws for the Effects of Pressure and Discharge Diameter on Gain of He-Ne Lasers," Appl. Phys. Letters 3, 199-201.

Labuda, E. F. and Gordon, E. I. (1964) "Microwave Determination of Average Electron Energy and Density in He-Ne Discharges," J. Appl. Phys. 35, 1647-1648.

Levine, J. S. and Javan, A (1969) "Far Infrared Continuous Wave Laser Oscillation in Pure Helium" Appl. Phys. Letters 14, 348-350.

Long, D. R. and Geballe, R. (1970) "Electron Impact Ionization of He($2s^3S$)," Phys. Rev. A1, 260-265.

Massey, H. S. W. and Burhop, E. H. S. (1969) Electronic and Ionic Impact Phenomena, Vol. 1, Oxford.

Phelps, A. V. (1955) "Absorption Studies of Helium Metastable Atoms and Molecules," Phys. Rev. 99, 1307-1312.

Schmeltekopf, A. L. and Fehsenfeld, F. C. (1970) "De-excitation Rate Constants for Helium Metastable Atoms with Several Atoms and Molecules," J. Chem. Phys. 53, 3173-3177.

Vriens, L. (1964) "Calculations of Absolute Ionization Cross Sections of He, $He^*$, $He^+$, Ne, $Ne^*$, $Ne^+$, Ar, $Ar^*$, Hg and $Hg^*$," Phys. Lett. 8, 260-261.

Weaver, L. A. and Freiberg, R. J. (1966) "Laser Induced Perturbation of Excited State Populations in a He-Ne Discharge," J. Appl. Phys. 37, 1528-1535.

White, A. D. and Gordon, E. I. (1963) "Excitation Mechanisms and Current Dependence of the Population Inversion in He-Ne Lasers," Appl. Phys. Letters 3, 197-199.

Willett, Colin S. (1974) Introduction to Gas Lasers:Population Inversion Mechanisms, Pergamon Press.

Yariv, Amnon (1967) Quantum Electronics, John Wiley & Sons.

Young, R. T., Jr., Willett, C. S. and Maupin, R. T. (1970) "Effect of Helium on Population Inversion in the Helium-Neon Laser," J. Appl. Phys. 41, 2936-2942.

# Chapter 11
# ION LASERS

There are a number of gaseous media in which laser action has been obtained in the ionized species. These can be broken down basically into the rare gas (argon and krypton) and the metal vapor (lead, calcium, germanium, indium, tin, mercury, cadmium and zinc) systems. For a good review, see the article by Davis and King (1975).

## 11.1 METAL VAPOR LASERS

A very good review of these species has been presented by Bennett, (1969). The most interesting case here is the helium-cadmium laser, since it is simple to construct, operates at reasonably high efficiency and gives approximately 200 mW CW at 4416 Å with a gain of 20%/m and it gives approximately 20 mW CW at 3250 Å (see Silfvast, 1968, and Goldsborough, 1969). This laser uses inelastic atom-atom collisions as an excitation method, just as in the He-Ne laser, and the proposed pumping mechanism is

$$He^{M} + Cd \longrightarrow (Cd^{+})^{*} + e + He \tag{11.1}$$

where $H_e^{M}$ is the atomic helium metastable species. Kamin and Chanin (1976) have also presented evidence that pumping occurs through the molecular species as well.

$$He_2^{M} + Cd \longrightarrow (Cd^{+})^{*} + 2He + e \quad , \tag{11.2}$$

$$He_2^{+} + Cd \longrightarrow (Cd^{+})^{*} + 2He \quad . \tag{11.3}$$

A partial energy level diagram for $Cd^{+}$ is shown in Fig. 11.1.

The operation of the He-Cd laser obviously is very similar to the He-Ne laser, and, in fact, a rate equation analysis similar to Arrathoon's, which was presented earlier, can be performed to satisfactorily predict the behavior of the He-Cd laser. See the paper by Giallorenzi and Ahmed (1971).

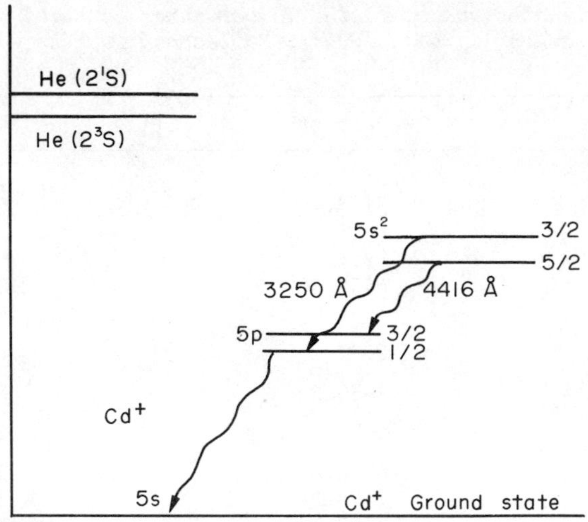


*Fig. 11.1. Energy level diagram for $Cd^+$ and the appropriate laser lines*

There are some additional aspects of metal vapor lasers which should be mentioned. One is the fact that at normal operating temperatures most of these metals are solids and so the discharge regions must be heated in order to produce the metal vapor for the laser action. The effect is particularly striking in He-Cd lasers where an increase in temperature and hence in Cd vapor pressure changes the color of the discharge from the helium straw color to a cadmium blue color and, due to the lower ionization potential of the cadmium, the impedance (or electron temperature) of the discharge drops sharply.

As an example of the temperatures needed for metal vapor lasers, let us examine part of Table 11.1 (presented by Silfvast, Fowles and Hopkins, 1966). As you can see, Cd lasers are the simplest to handle but fairly advanced technology would be required to handle Sn.

Another aspect of metal vapor lasers that is again very obvious with He-Cd lasers is the cataphoretic pumping of the Cd from the anode to the cathode. Unless you can reverse the current in a He-Cd laser, the amount of operating time is severely limited. Schuebel (1970a) has presented a nice solution to this problem by using a hollow cathode discharge. This configuration eliminates the cataphoresis problem and, because of the electron distribution function characteristic of hollow cathode discharges with their non-Maxwellian property, additional CW laser lines have been found (see Schuebel, 1970 b,c).

## 11.2 RARE-GAS ION LASERS

This class of lasers includes the well known argon and krypton ion lasers and is of considerable interest since high powers (~ 100 W) are available on a CW basis in

*Table 11-1. Wavelengths and Operating Temperatures of Metal Vapor Ion Lasers (from Silfvast, Fowles and Hopkins, 1966)*

| Element | Wavelength | Temperature (°C) |
|---|---|---|
| Ge II | 5131.5 | 900 - 1000 |
| | 5178.4 | 900 - 1000 |
| Sn II | 5798.7 | 900 - 1000 |
| | 6453.0 | 1000 |
| | 6844.0 | 1000 |
| Pb II | 5372.1 | 650 - 800 |
| In II | 4680.5 | 875 - 1000 |
| Cd II | 4415.6 | 200 - 300 |
| Zn II | 6102.8 | 300 - 400 |
| | 7478.3 | 300 - 500 |
| | 7587.5 | 300 - 500 |
| | 7611.8 | 300 - 400 |

the visible through ultraviolet spectrum. These lasers are of quite low efficiency (~ $10^{-3}$) so the power supplied must be very high in order to obtain appreciable power out. The high powers required have resulted in a considerable amount of technological development in argon ion and krypton ion laser construction. Two very good review papers that discuss the technology as well as fundamental processes are Kitaeva, Odintsov, and Sobolev (1970) and Bridges, Chester, Halsted and Parker (1971).

11.2.1 Argon Ion Lasers

This discussion will be confined to argon ion lasers since these lasers have been thoroughly studied. The energy level diagram for the common argon ion laser lines has been presented by Bridges (1964) and is shown in Fig. 11.2. All of these lines can be readily obtained in argon ion lasers on a CW basis and even the early experiments showed gains of approximately 200%/meter for the 4880 Å, 5145 Å and 4765 Å lines on a pulsed basis. It is possible to obtain CW outputs of up to 100 W and pulsed outputs in the tens of kW.

There seem to be a number of different hypotheses concerning the excitation of the upper laser levels, but it appears that the excitation mechanism for pulsed lasers is by direct electron impact from the argon ground state and the excitation mechanism for CW lasers is by multi-step excitation. The mechanism for pulsed lasers has been presented by Bennett, Knutson, Mercer and Detch (1964).

The energy level diagram is shown in Fig. 11.3. In this case, direct electron impact from the argon ground state excites the upper level

$$e + Ar \longrightarrow (Ar^{+})^{*} + e + e \qquad (11.4)$$

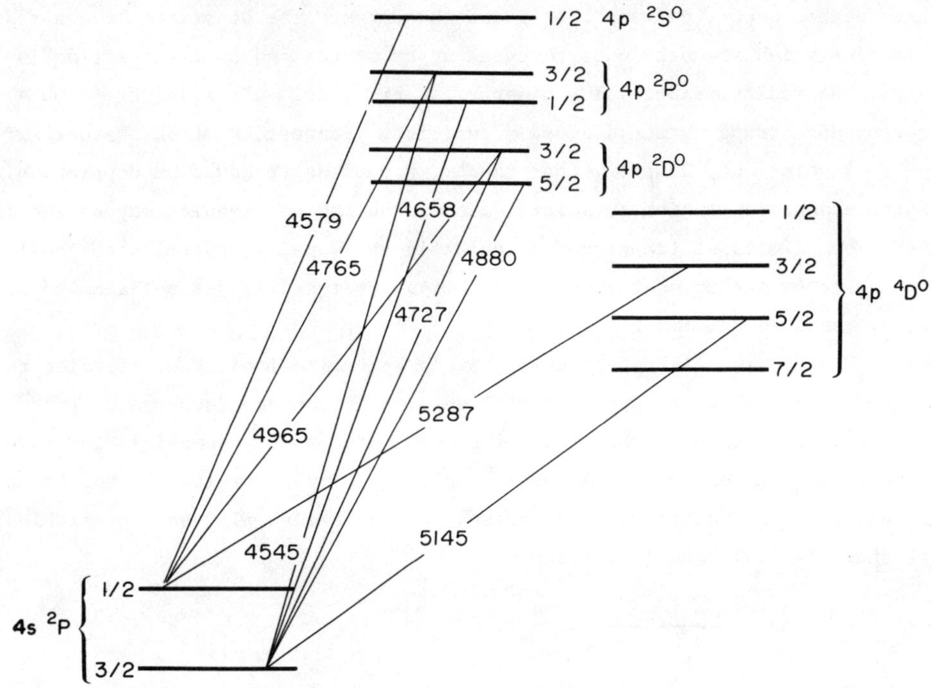


*Fig. 11.2. Energy level diagram and laser lines for the argon ion laser (from Bridges, 1964, with a modified assignment provided by W. B. Bridges)*

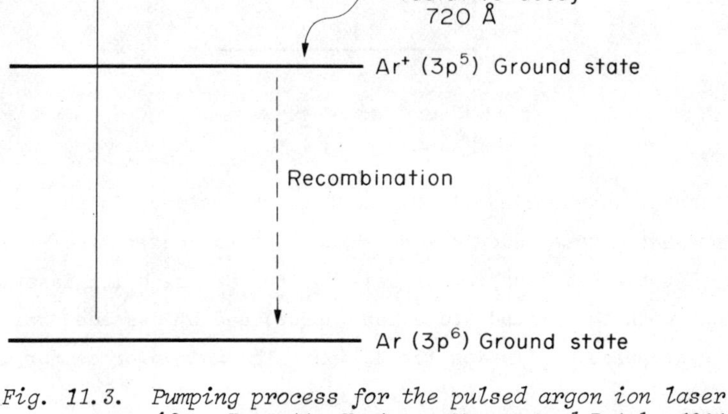


*Fig. 11.3. Pumping process for the pulsed argon ion laser (from Bennett, Knutson, Mercer and Detch, 1964)*

The laser action occurs between the 4p and 4s states. The 4s states have a radiative lifetime which is much lower than the upper states and so laser action is possible. The radiative decay is, however, in the vacuum ultraviolet (~ 700 Å) and so resonance trapping could cause saturation. Bennett, Knutson, Mercer and Detch (1964) point out, however, that the $Ar^+$ ground state could be Doppler shifted out of the absorption region by several linewidths and so trapping may not be important. The effects of trapping and inelastic and superelastic electron collisions with the 4s states on the decay of the 4s level still does not seem to be completely resolved, however.

For the CW argon ion laser, there is a two step excitation process leading to a quadratic dependence of the upper laser level population on discharge current. Figure 11.4 indicates the combination of pumping processes responsible for laser action in the argon ion laser. As well as electron impact excitation to the 4p level, radiative cascade from higher argon ion levels also play an appreciable role as shown by Rudko and Tang (1966).

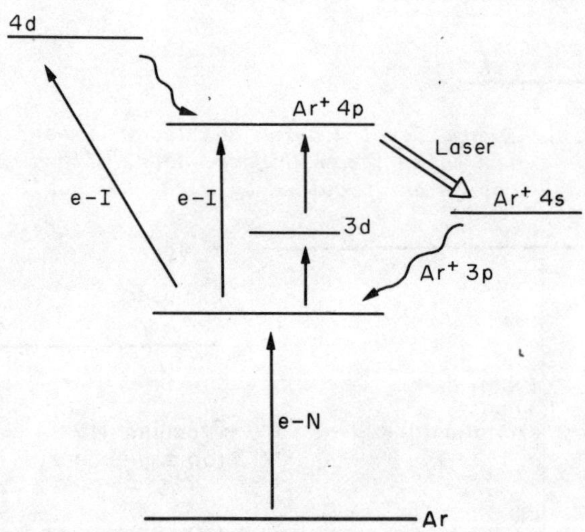


*Fig. 11.4. Multistep excitation processes for the CW argon ion laser*

Kitaeva, Odintsov and Sobolev (1970) and Kitaeva, Osipov, Rubin and Sobolev (1969) have presented some calculations which tend to confirm the previous theories and indicate that the main pumping process for the CW argon ion laser is by electron collisions with the ground state ion (3p-4p) and by cascade excitation (3p-4d-4p) and stepwise excitation (3p-3d-4p). The main process for depopulation of the lower level is radiative decay (4s-3p).

The radiative decay rates as presented by Kitaeva, Odintsov and Sobolev (1970) are shown in Fig. 11.5. Since the 4s state radiatively decays at a rate that is approximately 25 times greater than the 4p state, this is obviously the main source of the population inversion.

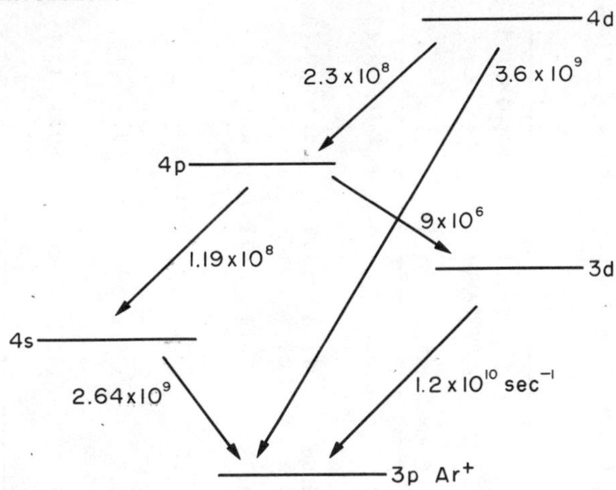


*Fig. 11.5. Radiative decay rates for the argon ion configuration (from Kitaeva, Odintsov and Sobolev, 1970)*

The calculated electronic excitation rates $< \sigma v_e >$, have also been calculated by Kitaeva, Odintsov and Sobolev (1970) and are shown in Table 11-2. Combining the

*Table 11-2. Values of $< \sigma v_e >$ ($10^{-16}$ $m^3$/sec) for $Ar^+$ (from Kitaeva Odintsov and Sobolev, 1970)*

| | T (10,000 K) | | | |
|---|---|---|---|---|
| Transition | 3 | 5 | 8 | 10 |
| 3p-4p | 0.18 | 3.0 | 14 | 23 |
| 3p-3d | 3.0 | 44 | 190 | 300 |
| 3p-4d | 0.078 | 2.4 | 16 | 28 |
| 3p-5d | 0.0047 | 0.22 | 1.8 | 3.4 |
| 3p-4s | 0.23 | 2.7 | 11 | 17 |
| 3p-5s | 0.003 | 0.10 | 0.62 | 11 |
| 3d-4p | 1900 | 2000 | 2000 | 1900 |
| 4s-4p | 4800 | 6600 | 7700 | 8000 |

information given in Fig. 11.5 and Table 11-2, Kitaeva, Odintsov and Sobolev (1970) have calculated pumping rates and inversions per unit statistical weight for various discharge conditions as shown in Table 11-3. In this table, the

*Table 11-3. Pumping Rates, Populations and Inversions for the CW Argon Ion Laser (from Kitaeva, Odintsov and Sobolev, 1970)*

| $P_o$ (torr) | J (A/cm$^2$) | Te($10^4$ K) Ne($10^{19}$ m$^{-3}$) | Pumping rate 3p-4s | $10^{23}$ m$^{-3}$ sec$^{-1}$ 3p-4p | 3p-4d-4p | 3p-3d-4p | Total pumping rate 4p | 4s | Population $10^{15}$ m$^{-3}$ 4p | 4s | Inversion per unit stat wt $10^{14}$ m$^{-3}$ |
|---|---|---|---|---|---|---|---|---|---|---|---|
| 0.21 | 160 | Te = 5.6, Ne = 2.9 | 1.7 | 2.4 | 0.14 | 0.02 | 2.6 | 4.3 | 2.0 | 0.12 | 0.9 |
| | 330 | Te = 9.2, Ne = 2.9 | 7 | 10.8 | 0.91 | 0.06 | 11.8 | 18.8 | 8.3 | 0.66 | 3.5 |
| .37 | 160 | Te = 5.2, Ne = 5.4 | 5.2 | 5.8 | 0.36 | 0.29 | 6.5 | 11.7 | 4.7 | 0.42 | 1.9 |
| | 330 | Te = 9.0, Ne = 5.2 | 23 | 31.6 | 2.7 | 0.90 | 35.2 | 58.2 | 25.6 | 2.0 | 10.9 |
| .62 | 160 | Te = 3.2, Ne = 14.4 | 4.6 | 4.0 | 0.10 | 3.06 | 7.1 | 11.8 | 2.9 | 1.6 | - 1.0 |
| | 330 | Te = 6.5, Ne = 13 | 61 | 99.6 | 6.2 | 35 | 141 | 202 | 80.3 | 15.4 | 19 |

3p-3d-4p pumping rate allows for the radiative trapping of the 3d-3p transition, which at an ion density of $10^{20}$ $m^{-3}$ increases the 3d lifetime by a factor of 14-18.

These calculations turn out to be quite good when compared to the results of experiments. Kitaeva, Odintsov and Sobolev (1970) have shown this agreement in Table 11-4.

*Table 11-4. Comparison of Theoretical and Experimental Values of 4p and 4s Populations in a CW Argon Ion Laser (from Kitaeva, Odintsov and Sobolev, 1970)*

| Discharge parameters | | Population per unit statistical weight | |
|---|---|---|---|
| $P_o$ (torr) | J ($A/cm^2$) | Experiment | Theory |
| 0.2 | 220 | 2.5 - 2.7 x $10^{14}$ | 2.5 x $10^{14}$ |
| 0.67 | 160 | 2-3 x $10^{14}$ | 1.3 x $10^{14}$ |
| 0.3 | 160 | 4-8 x $10^{14}$ | 3 x $10^{14}$ |
| | | 5 x $10^{14}$ (4s) | 0.7 x $10^{14}$ (4s) |

Detailed calculations of the population processes for pulsed argon ion lasers have been presented by Goto, Kawahara and Hattori (1971) and Hattori and Goto (1969). Hattori and Goto (1968) have also shown the saturation effect of radiation trapping of the lower levels when pulsed discharge currents are high. Their curve is shown in Fig. 11.6. The saturation is obvious for the 4965 Å and 4880 Å

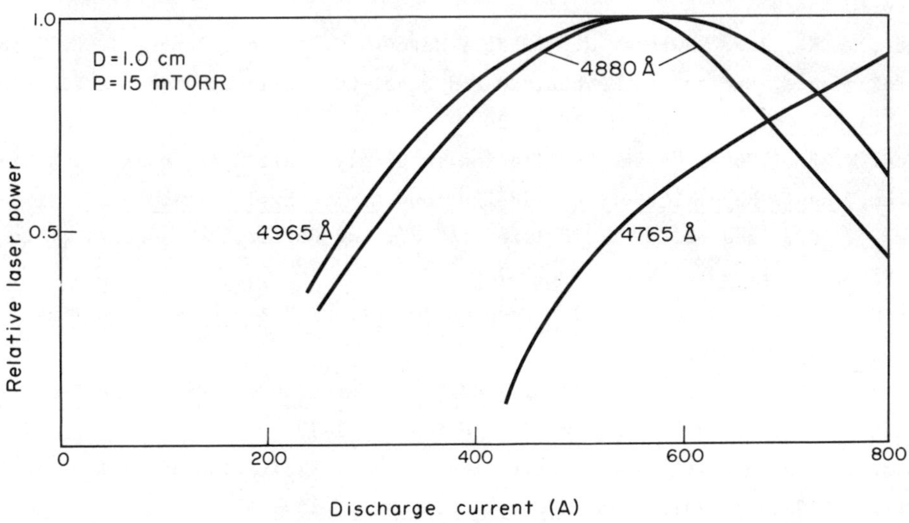


*Fig. 11.6. Saturation effect in pulsed argon ion lasers (from Hattori and Goto, 1968)*

transitions. Obviously, some mechanism must be present to help increase the population of the upper level of the 4765 Å transition to overcome the increased lower state trapping. Of course, similar considerations hold for other rare gas lasers and, in fact, laser operation has been obtained in the doubly ionized species of Kr and Ar. Latimer (1968) presents the curves as shown in Fig. 11.7.

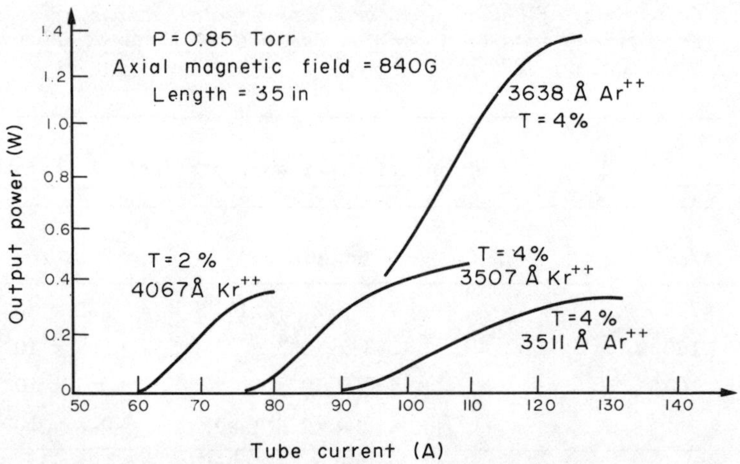


*Fig. 11.7. Output power vs discharge current for doubly ionized argon and krypton (from Latimer, 1968)*

## 11.3 REFERENCES

Bennett, W. R., Jr. (1969), "Metal Vapor Lasers," Comments on Atomic and Molecular Physics 1, 15-26.

Bennett, W. R., Jr., Knutson, J. W., Jr., Mercer, G. N., and Detch, J. L. (1964), "Super-Radiance, Excitation Mechanisms and Quasi-CW Oscillation in the Visible $Ar^+$ Laser," Appl. Phys. Letters 4, 180-182.

Bridges, W. B. (1964), "Laser Oscillation in Singly Ionized Argon in the Visible Spectrum," Appl. Phys. Letters 4, 128-130, and Appl. Phys. Letters 5, 39 (1964)

Bridges, W. B., Chester, A. N., Halsted, A. S., and Parker, J. V. (1971), "Ion Laser Plasmas," Proc. IEEE 59, 724-737.

Davis, C. C., King, T. A. (1975), "Gaseous Ion Lasers," Advances in Quantum Electronics 3, 169-454.

Giallorenzi, T. G., and Ahmed, S. A. (1971), "Saturation and Discharge Studies in the He-Cd Laser," IEEE J. Quantum Electronics 7, 11-17.

Goldsborough, John P. (1969), "Stable, Long Life CW Excitation of Helium-Cadmium Lasers by DC Cataphoresis," Appl. Phys. Letters 15, 159-161.

Goto, T., Kawahara, A., and Hattori, S. (1971), "Quantitative Level Population

Mechanism in Gas Discharges for Pulsed Argon-Ion Lasers," IEEE J. Quantum Electronics 7, 555-560.

Hattori, S., and Goto, T. (1968), "Saturation of Power Output of a Pulsed Ion Laser in Very High Current Regions," Appl. Phys. Letters 12, 131-133.

Hattori, S., and Goto, T. (1969), "Excitation Mechanism and Plasma Parameters in Pulsed Argon-Ion Lasers," IEEE J. Quantum Electronics 5, 531-538.

Kamin, M., Chanin, L. (1976), "Production Processes for Excited States of $Cd^+$ in Helium-Cadmium Afterglows," Appl. Phys. Letters 29, 756-758.

Kitaeva, V. F., Odintsov, A. N., and Sobolev, N. N. (1970), "Continuously Operating Argon Ion Lasers," Sov. Phys. Uspekhi 12, 699-730.

Kitaeva, V. F., Osipov, Yu I., Rubin, P. L., and Sobolev, N. N. (1969), "On the Inversion Mechanism in the CW Argon-Ion Laser," IEEE J. Quantum Electronics 5, 72-77.

Latimer, I. D. (1968), "High Power Quasi-CW Ultraviolet Ion Laser," Appl. Phys. Letters 13, 333-335.

Rudko, R. I., and Tang, C. L. (1966), "Effects of Cascade in the Excitation of the Ar II Laser," Appl. Phys. Letters 9, 41-44.

Schuebel, W. K. (1970a), "Transverse Discharge Slotted Hollow-cathode Laser," IEEE J. Quantum Electronics 6, 574-575.

Schuebel, W. K. (1970b), "New CW Cd-Vapor Laser Transitions in a Hollow Cathode Structure," Appl. Phys. Letters 16, 470-472.

Schuebel, W. K. (1970c), "New CW Laser Transitions in Singly Ionized Cadmium and Zinc," IEEE J. Quantum Electronics 6, 654-655.

Silfvast, W. T. (1968), "Efficient CW Laser Oscillations at 4416 Å in Cd II," Appl. Phys. Letters 13, 169-171.

Silfvast, W. T., Fowles, G. R., and Hopkins, B. D. (1966), "Laser Action in Singly Ionized Ge, Sn, Pb, In, Cd and Zn," Appl. Phys. Letters 8, 318-319.

# Chapter 12

# MOLECULAR GAS LASERS

Molecular gas lasers, particularly $CO_2$ lasers, have been in the limelight because of their very high efficiencies (of the order of 30%) and their extremely high powers (megawatts pulsed and tens to probably hundreds of kilowatts CW). Before we begin to discuss molecular gas lasers, it seems worthwhile to briefly review some of the nomenclature and fundamental ideas of molecular structure and spectra.

## 12.1 MOLECULAR STRUCTURE AND NOMENCLATURE

Probably the best available references are the superb trilogy by Herzberg on "Molecular Spectra and Molecular Structure," "I Spectra of Diatomic Molecules," "II Infrared and Raman Spectra of Polyatomic Molecules," and "III Electronic Spectra and Electronic Structure of Polyatomic Molecules." A good introduction is volume I. We will review diatomic molecules since these seem to be reasonably well understood and the concepts are fairly general.

Diatomic molecules can be excited into various electronic states in the same manner that atoms are excited, in that the electrons are in a more energetic orbit about the nucleus. Transitions between these electronic energy levels require or result in substantial amounts of energy corresponding to photons in the visible to ultraviolet region of the spectrum. The molecule, however, has two additional degrees of freedom, and therefore energy-level structures, corresponding to the vibration of the molecule along its internuclear axis and to the rotation of the molecule as a whole. The differences in vibrational energy levels roughly correspond to photon energies in the infrared portion of the spectrum and the differences in rotational energy levels roughly correspond to photon energies in the microwave portion of the spectrum. Let us examine these cases.

### 12.1.1 Rotation

The rotational energy levels are given by the term value F(J) expressed in units of $cm^{-1}$. For the rigid rotator

$$F(J) = \frac{\varepsilon}{hc} = BJ(J + 1) \quad (12.1)$$

where B is the rotational constant and J is the rotational quantum number

$$J = 0, 1, 2 \ldots .$$

The energy levels are shown in Fig. 12.1.

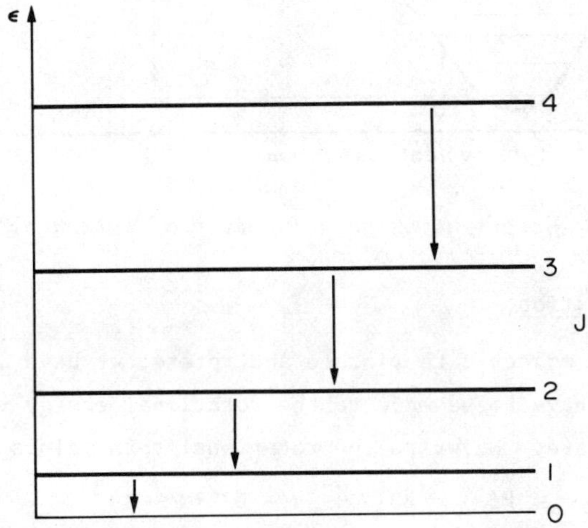


*Fig. 12.1. Energy level structure for rotational transitions*

For rotational transitions, the selection rule is

$$\Delta J = \pm 1 . \quad (12.2)$$

### 12.1.2 Vibration

If the vibrator is described as a simple harmonic oscillator, the term values can be written as

$$G(v) = \omega(v + 1/2) \quad (12.3)$$

where v is the vibrational quantum number v = 0, 1, 2, ... and ω is the vibrational constant. The vibrator is better described by an anharmonic oscillator where the energy varies as a function of nuclear separation as shown in Fig. 12.2. The vibrational term values are now given as

$$G(v) = \omega_e(v + 1/2) - \omega_e x_e(v + 1/2)^2 + \omega_e y_e(v + 1/2)^3 + \text{---} . \quad (12.4)$$

For the simple harmonic oscillator the selection rule was $\Delta v = \pm 1$. For the anharmonic oscillator we can have $\Delta v = \pm 2, \pm 3$, etc. but with decreasing intensity. The energy $D_o$ shown on Fig. 12.2 is the dissociation energy.

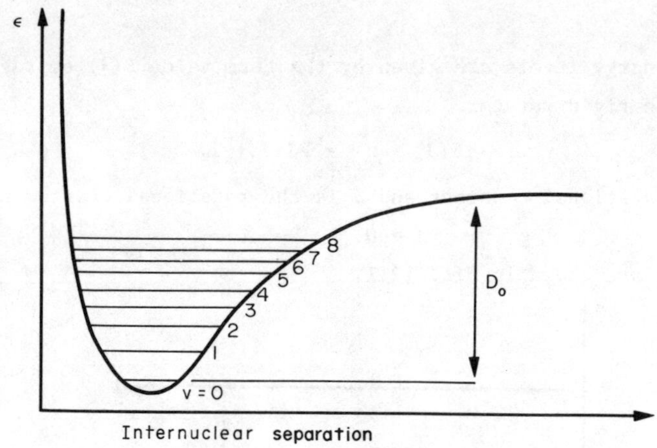


*Fig. 12.2. Potential energy as a function of internuclear separation for a diatomic molecule*

### 12.1.3 Vibration-Rotation

Now that we allow the molecule to vibrate and rotate, we have a non-rigid rotor and some corrections have to be made to the rotational energy values depending on the vibrational state. We write the rotational term values now as

$$F(J) = BJ(J + 1) - DJ^2(J + 1)^2 \qquad (12.5)$$

where B and D depend on the vibrational state.

Consider the particularly interesting case of the rotational structure on vibrational transitions. For a vibrational transition v' → v'', there are a large number of upper and lower levels corresponding to the rotational levels. We have the selection rule $\Delta J = \pm 1$ so we can define two branches of the spectrum.

$$R_{branch} \quad J_{upper} - J_{lower} = +1 , \qquad (12.6)$$

$$P_{branch} \quad J_{upper} - J_{lower} = -1 . \qquad (12.7)$$

We associate a particular P or R branch transition with the J value of the lower state as shown in Fig. 12.3. We can also have a Q branch ($\Delta J = 0$) if v' and v'' belong to different electronic states (say, $^1\Pi - {}^1\Sigma$ transition).

### 12.1.4 Electronic Structure

The electronic structure for diatomic molecules is similar to the energy level structure of atoms only the energies depend on the internuclear distance as shown in Fig. 12.4. The vibrational structure then corresponds to the fine structure of these electronic levels.

Let us deal briefly with the notation used for the electronic levels. It is almost exactly analogous to the spectroscopic notation used in atomic energy levels, so the Σ, Π, Δ, Φ, notation corresponds to the atomic S, P, D, F notation.

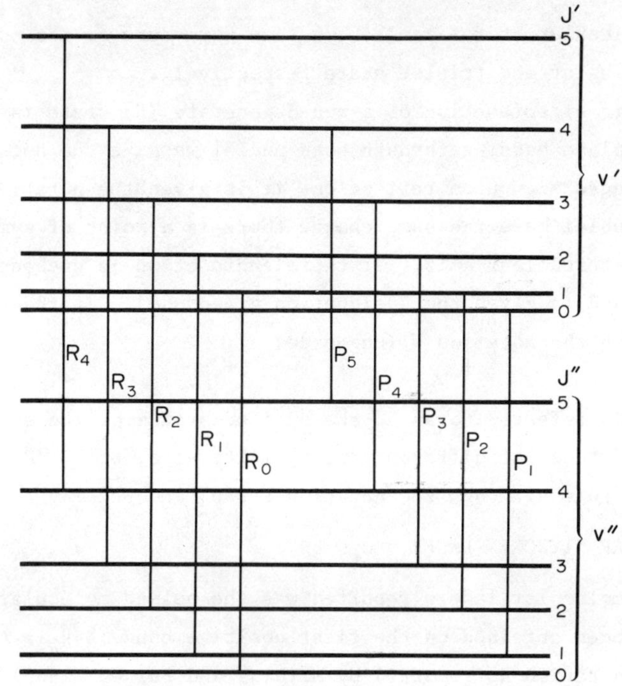


*Fig. 12.3. P and R branch vibrational-rotational transitions*

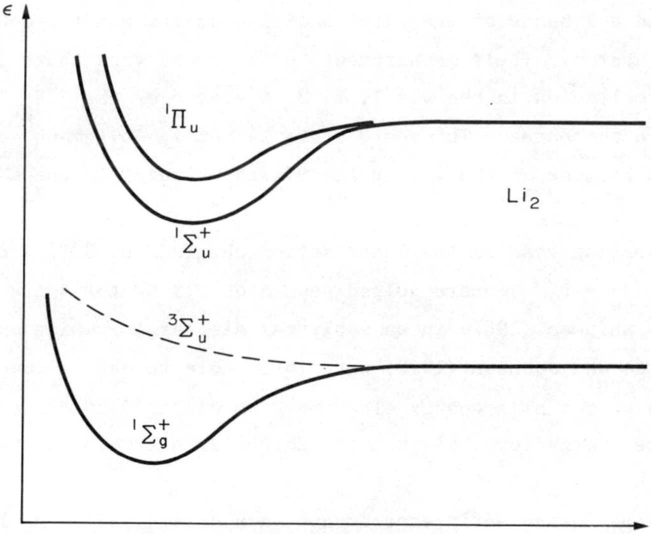


*Fig. 12.4. Potential energy versus internuclear separation for various electronic configurations*

Similarly, the molecular states can have a fine structure so the notation $^1\Pi$, $^3\Sigma$ etc. indicate a singlet and triplet state respectively.

If the electronic eigenfunction of a non-degenerate ($\Sigma$) state is unchanged when reflected at any plane passing through both nuclei we give the notation +. If the eigenfunction changes sign upon reflection it is given the notation - i.e., $\Sigma^+$, $\Sigma^-$. If the two nuclei have the same charge there is a point of symmetry at the midpoint of the internuclear axis. If the eigenfunction is unchanged on reflection, it is even and is given the designation g (gerade). If the sign changes, it is odd and is given the notation u (ungerade)

$$\text{i.e.}\quad \Sigma_g^+ , \qquad \Sigma_u^+ .$$

The ground state is referred to as X, the excited states of the same multiplicity as A, B, and C and those of different multiplicity as a, b, c. $N_2$ is the only exception to this rule although the notation for $H_2$ and $He_2$ may be different.

## 12.2 THE MOLECULAR NITROGEN LASER

One of the first molecular lasers reported was the pulsed molecular nitrogen laser. Laser action has been obtained on the first positive band ($B^3\Pi_g - A^3\Sigma_u^+$) in the 8600 Å to 12,400 Å region as reported by Mathias and Parker (1963). Since the lower state is metastable, this oscillation and population inversion can only be obtained on a pulsed basis.

Kasuya and Lide (1967) have observed stimulated emission on the 0-0, 0-1, 1-0, 2-0, 2-1, 3-1 and 4-2 bands of the first positive system and the 0-0 bands of the second positive system. Their measurements also showed that there is a Boltzmann distribution of molecules in the v = 1, 2, 3, 4 states of the $B^3\Pi_g$ level before laser oscillation commences. The v = 0 state is not in Boltzmann equilibrium with the other states because of the strong stimulated emission in the $C^3\Pi_u - B^3\Pi_g$ (0-0) band.

The more interesting case is the laser action obtained at 3371 Å on the second positive band ($C^3\Pi_u - B^3\Pi_g$) where pulsed powers of 2.5 MW for 4 nsec durations were achieved by Shipman (1967) in an early gas discharge pumping experiment. Patterson, Gerardo and Johnson (1972) were later able to obtain powers of 24 MW in 6 nsec pulses when a high energy electron beam was used to excite the gas.

The appropriate energy level diagram for molecular nitrogen is shown in Fig. 12.5.

Leonard (1965) performed early experiments on a pulsed 3371 Å $N_2$ laser where he obtained powers of 200 kW for 20 nsec. He also showed that the power output is approximately proportional to the electric field intensity, so high field strengths are desirable. Of particular interest are Leonard's curves of laser power as a function of the active length of the discharge as shown in Fig. 12.6. For powers

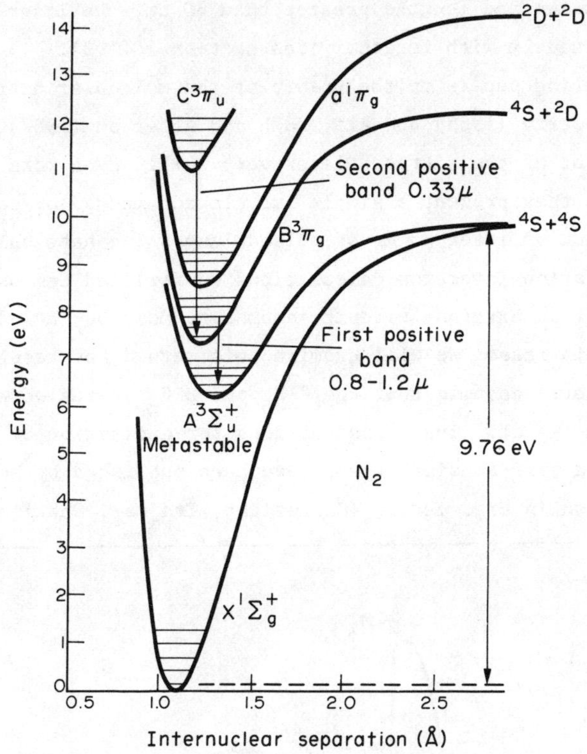


*Fig. 12.5. Energy level diagram for the molecular nitrogen laser*

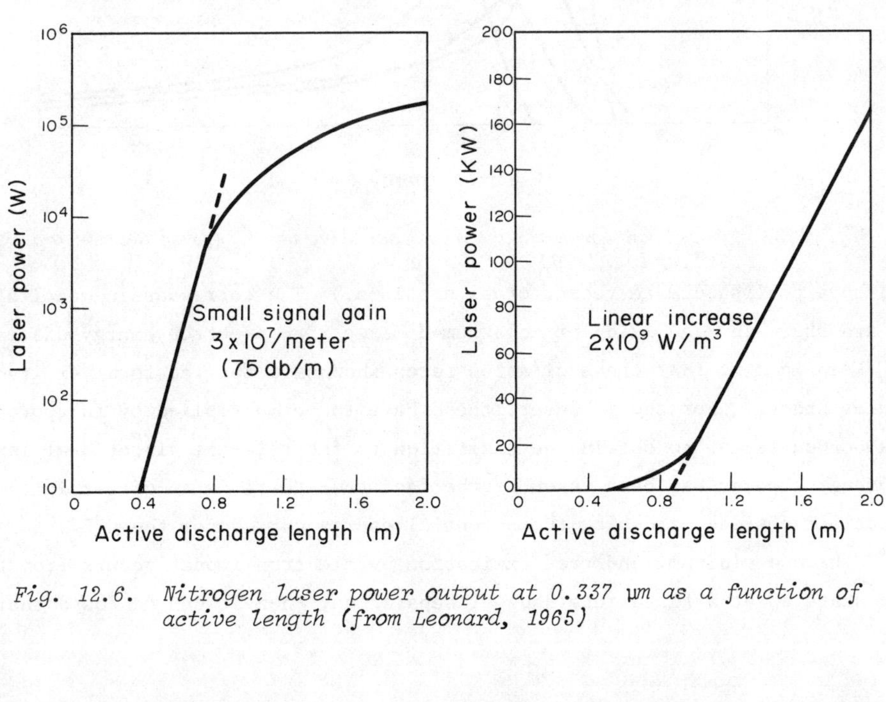


*Fig. 12.6. Nitrogen laser power output at 0.337 μm as a function of active length (from Leonard, 1965)*

greater than $10^4$ watts and lengths greater than 80 cm, the laser saturates and the power increases linearly with length. (See section 10.1.3).

Two very interesting papers on the theory of the molecular nitrogen laser have been published by Gerry (1965) and Ali, Kolb and Anderson (1967). These theories predict the behavior of the nitrogen laser very well. Even more important from our point of view, they present a simple but elegant method of calculating the population inversion or laser power of pulsed lasers (we have only considered steady state population inversion calculations so far) and demonstrate the method in which the important external circuit parameters must be included for pulsed operation. For this reason we will examine this method and results.

To begin with, Gerry assumed that the $C^3\Pi_u$ and $B^3\Pi_g$ states were both populated by direct electron impact. The cross sections for excitation by this process have been measured and a self-consistent set have been published by Newman and DeTemple (1976) and are shown in Fig. 12.7. (Cartwright, Trajmar, Chutjian and Williams

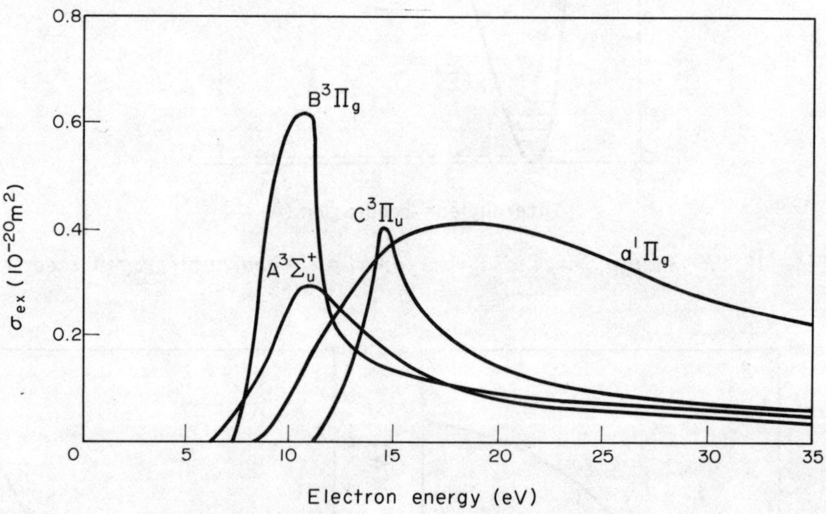


*Fig. 12.7. Electron excitation cross section in* $N_2$ *(from Newman and DeTemple, 1976)*

(1977) have published more recent cross sections.) The corresponding excitation rates are shown in Fig. 12.8 for an assumed Maxwellian electron energy distribution. Keep in mind that the excitation rates shown are for the total excited molecular states. For the $N_2$ laser, these have to be multiplied by an appropriate Franck-Condon factor to obtain the excitation to the relevant vibrational levels. When proper account has been taken of the fact that there is an offset in the internuclear distance at which the potential energy curves for the $C^3\Pi_u$, $B^3\Pi_g$, and $X^1\Sigma_g$ have a minimum, and that excitation by electron impact occurs from the ground state up at a fixed internuclear separation, then it can be found that a

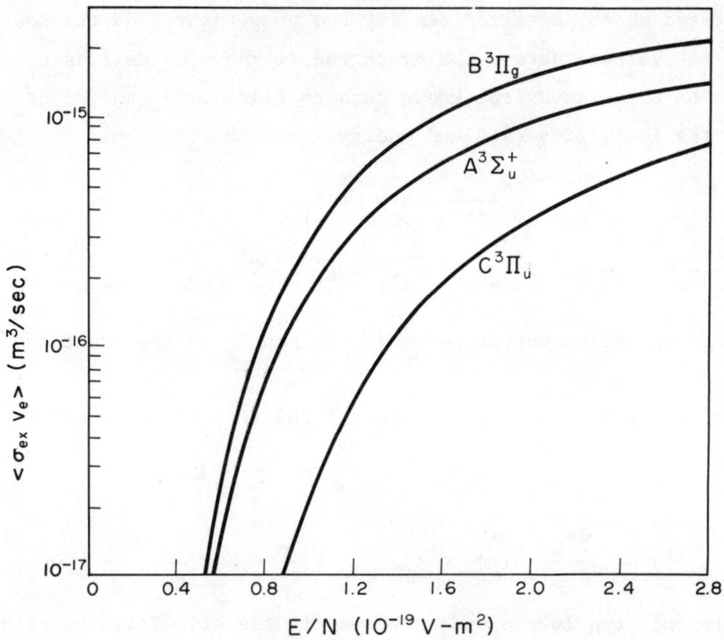


*Fig. 12.8. Calculated excitation rates for $N_2$ (from Newman and DeTemple, 1976)*

population inversion does indeed occur between the $C^3\Pi_u$ and $B^3\Pi_g$ states for the low vibrational levels involved in the laser. (A population inversion also occurs for the B-A, first positive laser system.)

Gerry estimated from the Franck-Condon factor that the cross section for excitation to the $B^3\Pi_g$ lower laser level is approximately one tenth of the cross section for excitation to the $C^3\Pi_u$ upper laser level. In this case, it is possible to obtain population inversions for times on the order of 40 nsec, the radiative lifetime of the $C^3\Pi_u$ state. Since the $B^3\Pi_g$ state has a 10 μsec radiative decay time, CW inversion is not possible. Gerry (1965) and Ali, Kolb and Anderson (1967) then solved the combined circuit and discharge equations as follows.

Assume that the circuit consists of a capacitor C charged to a voltage $V_o$ and discharged through an inductance L and the discharge tube. We can then write

$$V_c = LA \frac{d}{dt} J + \rho \ell J \tag{12.8}$$

and

$$\frac{dV_c}{dt} = -\frac{JA}{C} \tag{12.9}$$

where J is the discharge current density,

ρ is the plasma resistivity,

ℓ is the discharge length,

A is the cross-sectional area of the discharge.

Now J and $\rho$ depend on the electron density and temperature and the neutral gas pressure. If the voltage across the discharge is known as well as $n_e$ and $T_e$, the mobilities can be calculated from known data to find J. $n_e$ and $T_e$ have to be obtained from the ionization rate and energy equations.

$$\frac{dn_e}{dt} = n_e N_o < \sigma_I v_e > \tag{12.10}$$

and

$$\frac{d}{dt}\left(\frac{3}{2} n_e kTe\right) = \rho J^2 - n_e N_o < \sigma_I v_e > \varepsilon_i - \sum_k n_e N_o < \sigma_{xk} v_e > \varepsilon_k \tag{12.11}$$

where $\varepsilon_i$ is the ionization energy per molecule and $\varepsilon_k$ is the excitation energy for the excited molecules.

The densities of excited species are then found from the excitation rate equations

$$\frac{dN_c}{dt} = n_e N_o < \sigma_{oc} v_e > - \frac{N_c}{\tau_c} - P \quad , \tag{12.12}$$

$$\frac{dN_B}{dt} = n_e N_o < \sigma_{oB} v_e > + \frac{N_c}{\tau_c} + P - \frac{N_B}{\tau_B} \ . \tag{12.13}$$

This deals with only two levels and P represents the stimulated emission term. At this point there are two possible simple approaches. We can calculate the non-lasing population inversion by letting P = 0 and calculating $N_C$ and $N_B$ <u>or</u> we can calculate the <u>saturated laser power</u> (which is pretty much the case for the $N_2$ laser) by assuming the P is strong enough to equalize $N_C$ and $N_B$. We can then let $N_C - N_B << N_C$ and set $N_C = N_B = N$ and calculate P. This latter case is what Gerry was particularly interested in. If we let $N_C = N_B = N$ and substitute in (12.12) and (12.13) by adding we obtain:

$$\frac{dN}{dt} + \frac{N}{2\tau_B} = \frac{1}{2}\left[n_e N_o < \sigma_{oB} v_e > + n_e N_o < \sigma_{oc} v_e > \right] \ . \tag{12.14}$$

By subtracting we obtain

$$P = \frac{1}{2}\left[n_e N_o < \sigma_{oc} v_e > - n_e N_o < \sigma_{oB} v_e > - \frac{2 - (\tau_c/\tau_B)}{\tau_c} N\right] \ . \tag{12.15}$$

Gerry solved these equations in the saturation approximation for the given circuit and laser configuration he was using. The results of his calculation are shown in Fig. 12.9. The calculated peak power of $2.1 \times 10^6$ kw/m$^3$ agreed quite well with the experimental value of $2 \times 10^6$ kw/m$^3$. The calculated efficiency of 2.1% compared to the experimental value of 0.6% to 0.9% indicates, however, that the analysis is not precise in all facets although it is certainly very good.

Ali, Kolb and Anderson (1967) followed Gerry's analysis but allowed for collisional mixing of the laser levels, collisional ionization of the $C^3\Pi_u$ state and vibrational excitation of ground state nitrogen. Their calculations were slightly different in that they calculated the saturated laser power output assuming a fast

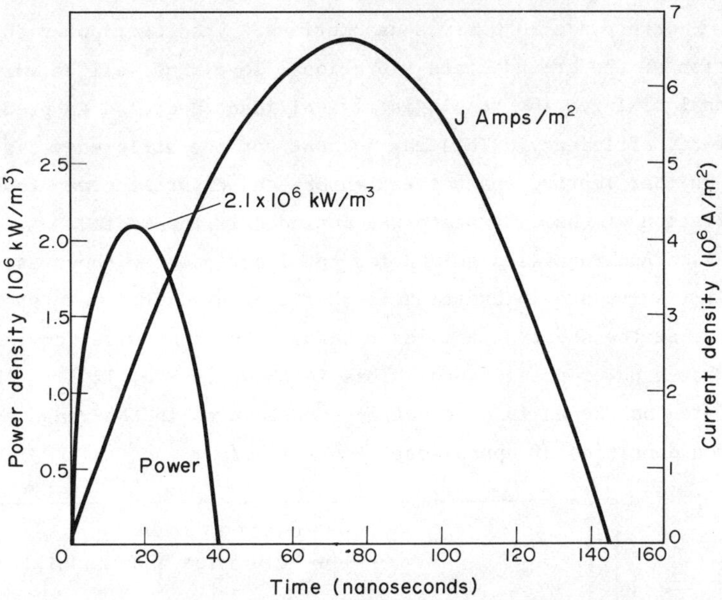


*Fig. 12.9. $N_2$ laser power density and current density vs time (from Gerry, 1965)*

rise (3 nsec) excitation pulse with a slow decay. This corresponds to the Shipman excitation method that Ali, Kolb and Anderson (1967) were using. Their calculated laser power waveforms are as shown in Fig. 12.10. The highest power assumes no vibrational excitation of the ground state and no collisional mixing of the laser

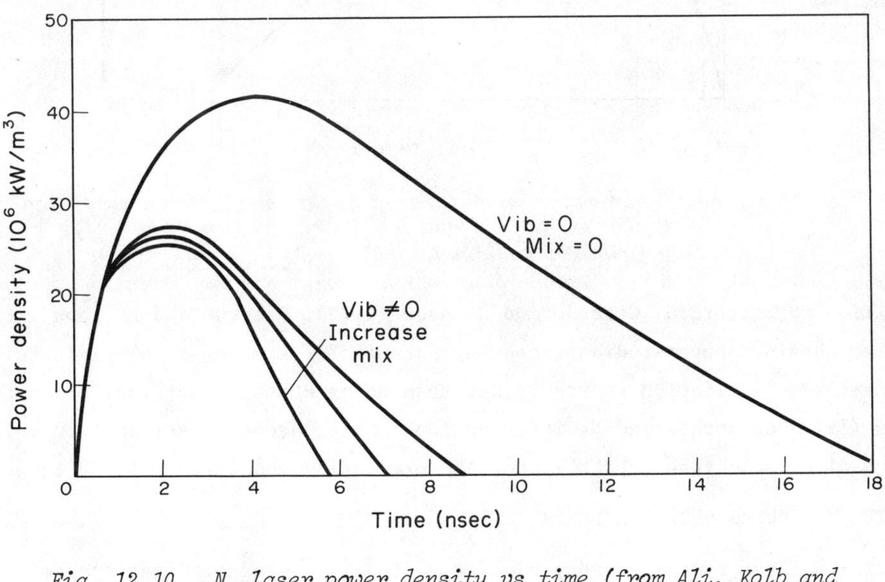


*Fig. 12.10. $N_2$ laser power density vs time (from Ali, Kolb and Anderson, 1967)*

levels. This corresponds to Gerry's assumptions. The lower power curves account for excitation of the ground state vibrational levels as well as various degrees of collisional mixing. The result is a significant decrease in predicted power and, therefore, efficiency. This may account for the efficiency discrepancy found by Gerry. Further improvement between theory and experiment was found when collisional ionization of the $C^3\Pi_u$ state was included in the model.

Ali, Kolb and Anderson also solved for the laser power without assuming saturation. Their results are interesting in that they predict a very sharp (picosecond) pulse at the start of the laser output, but otherwise very similar results to the saturated power predictions. This is shown in Fig. 12.11. These results were calculated on the basis of electron temperatures in the range of 3 to 5 eV with electron densities of approximately $6 \times 10^{20}/m^3$.

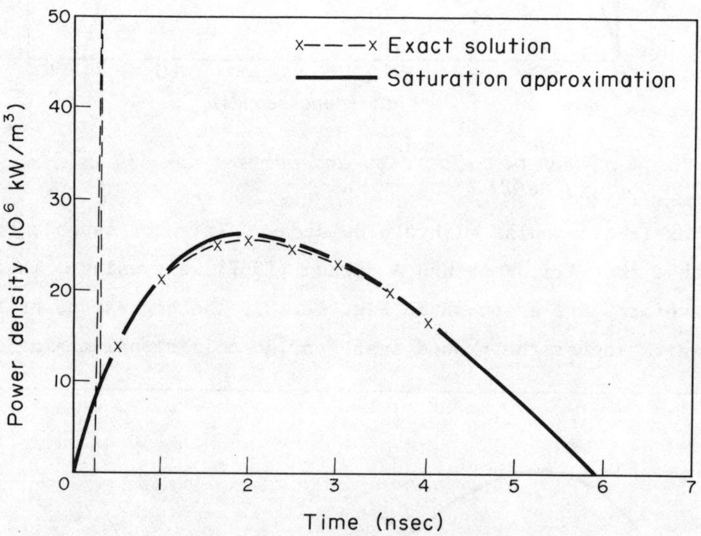


*Fig. 12.11. $N_2$ laser power density as a function of time calculated by the exact method and by the saturation method (from Ali, Kolb and Anderson, 1967)*

In addition to Patterson, Gerardo and Johnson (1972), Dreyfus and Hodgson (1972) have also obtained super-radiant laser output at 3371 Å using a relativistic electron beam as an excitation source rather than using a fast electrical discharge. Searles (1974) also obtained $N_2$ laser action in high pressure $Ar$-$N_2$ mixtures, although direct electron impact is not the excitation mechanism in this case.

## 12.3 THE MOLECULAR HYDROGEN LASER

The hydrogen laser system is very similar to the molecular nitrogen system. Ali and Kolb (1968) and Ali and Kepple (1972) performed an analysis of the hydrogen system which is similar to the previously described analyses of the nitrogen system. The energy level diagram is shown in Fig. 12.12. The analysis in the 1968 paper was based upon the Werner band of $H_2$ (C-X) in the region of 1025 - 1239 Å. The analysis in the 1972 paper is based on both the Werner and Lyman (B-X) band at around 1600 Å. At room temperature, most of the $H_2$ molecules are in the v = 0 level of the ground state and so a very fast rise excitation pulse (a few

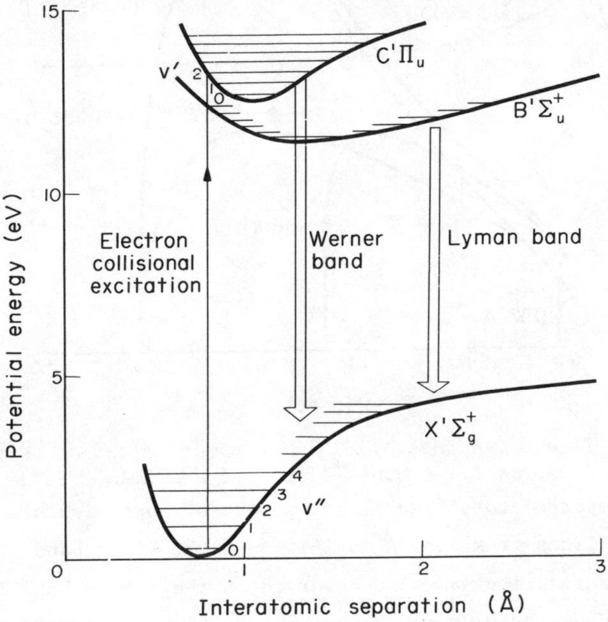


*Fig. 12.12. Energy level diagram for the* $H_2$ *laser*

nsec) could conceivably produce a population inversion between the electron pumped vibrational levels of the $C^1\Pi_u$ and/or $B^1\Sigma_u^+$ states and the empty vibrational states of the $X^1\Sigma_g^+$ state. Ali and Kolb expected inversion between the C(v' = 0-4) states and the X(v'' = 1-8) states. Ali and Kolb solved the coupled equations (about 31 in all) for an initial capacitor voltage of 150 kV and a fill pressure of 10 torr. Some of their predictions are summed up in Fig. 12.13. In the paper by Ali and Kepple (1972), similar results were presented for both Lyman and Werner bands based on the solution of 202 equations and 150 laser lines.

Laser action was first observed on the Lyman band by Hodgson (1970) and by Waynant, Shipman, Elton and Ali (1970, 1971). In these experiments, hundreds of kilowatts in approximately 1 nsec have been obtained in the 1600 Å wavelength region. The v' = 2 to 11 levels of the $B^1\Sigma_u^+$ state should be preferentially populated by electron impact, so laser action is expected between B(v' = 2-11, J' = 0,2) and X(v'' = 9-14, J'' = 1,3). The v' = 5 to v'' = 12 and the v' = 6 to v'' = 13 transitions are expected to be the strongest. Waynant, Ali and Julienne

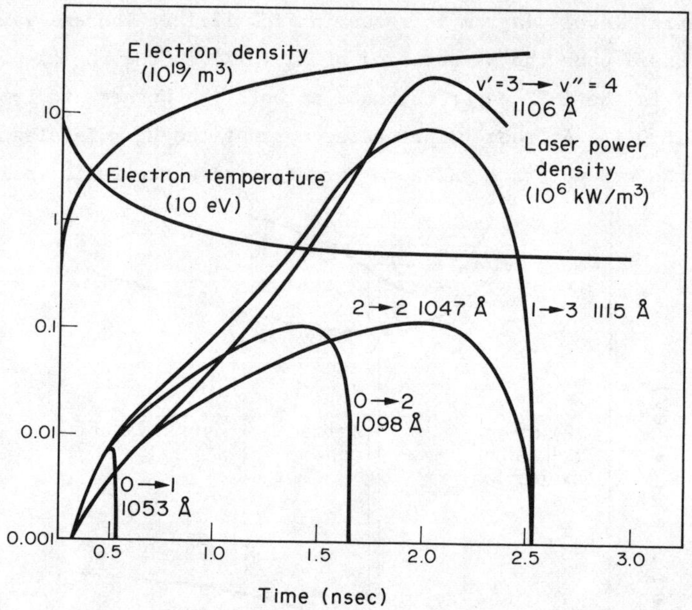


*Fig. 12.13. Time dependence of $H_2$ laser power, electron density and electron temperature (from Ali and Kolb, 1968)*

(1971) have also reported total peak laser powers of approximately 1 MW in 1 nsec pulses in fourteen lines from 1568 Å to 1614 Å on the Lyman band of $D_2$.

Subsequently, laser action was also observed on the Werner band in the 1161-1240 Å region by Hodgson and Dreyfus (1972) as well as by Waynant (1972).

## 12.4 $CO_2$ LASERS

$CO_2$ lasers are of particular interest because of their high efficiencies, high powers (~ 100 kW CW, megawatts pulsed) and the atmospheric window that exists for $CO_2$ laser signal propagation. There is a fantastic amount of literature available on $CO_2$ lasers as well as a great deal of current research so it is virtually impossible to keep up to date. Nevertheless, we can present some of the basic ideas. There are several interesting review papers by Sobolev and Sokovikov (1967) (which is rapidly becoming out of date), by Tyte (1970) (a little more recent) and by DeMaria (1973).

The primary $CO_2$ laser output wavelengths are around 9.4 μm and 10.4 μm and these are vibrational rotational transitions. Before we give the appropriate energy level diagrams, let us describe the notation for the vibrational levels of $CO_2$. Since $CO_2$ is a linear triatomic molecule, three different modes of vibration are possible.

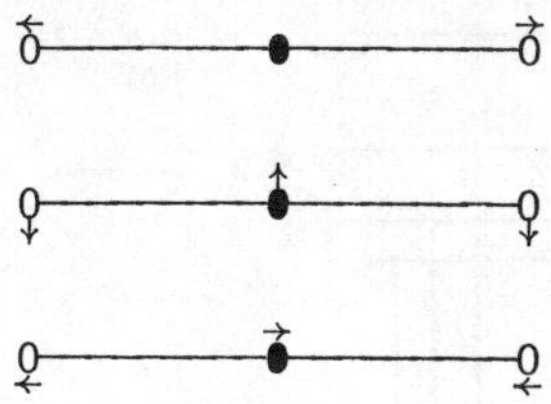

symmetric stretch ($\nu_1$)

deformational stretch ($\nu_2$)
(bending)

asymmetric stretch ($\nu_3$)

The number of quanta in each mode are given by the numbers $\nu_1$, $\nu_2$ and $\nu_3$ and the notation for a particular vibrational level is

$$\nu_1 \; \nu_2^{\Lambda} \; \nu_3 \, .$$

The $\Lambda$ superscript indicates the electronic orbital angular momentum of the molecular state. The appropriate energy level diagram is shown in Fig. 12.14. The traditional assignment of the 10.4 μm laser transition is $00^{0}1 - 10^{0}0$ and the traditional assignment of the 9.4 μm laser transition is $00^{0}1 - 02^{0}0$. The $10^{0}0$ and

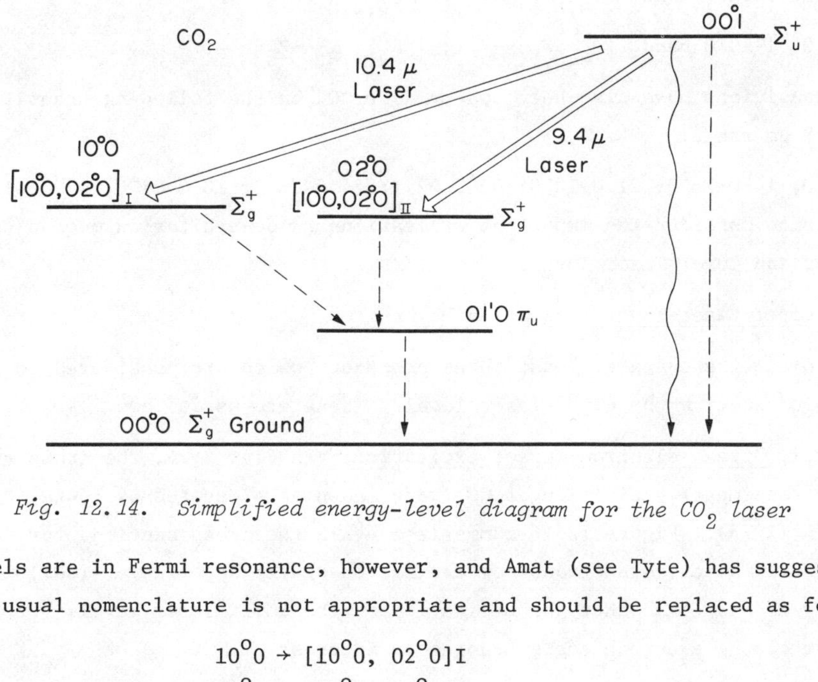


*Fig. 12.14. Simplified energy-level diagram for the $CO_2$ laser*

$02^{0}0$ levels are in Fermi resonance, however, and Amat (see Tyte) has suggested that the usual nomenclature is not appropriate and should be replaced as follows:

$$10^{0}0 \rightarrow [10^{0}0,\ 02^{0}0]\text{I}$$
$$02^{0}0 \rightarrow [10^{0}0,\ 02^{0}0]\text{II}$$

This has also been shown on Fig. 12.14.

Let us now consider some of the details of the laser transitions. We have in $CO_2$ where the spins of the oxygen nuclei are zero, zero statistical weight for all antisymmetric levels. Therefore, for example, we have only even levels of $\Sigma_g^{+}$ and only odd levels of $\Sigma_u^{+}$. The rotational structure of the $00^{0}1$-$10^{0}0$ transition will then be as shown in Fig. 12.15, and we have even values only of the R and P

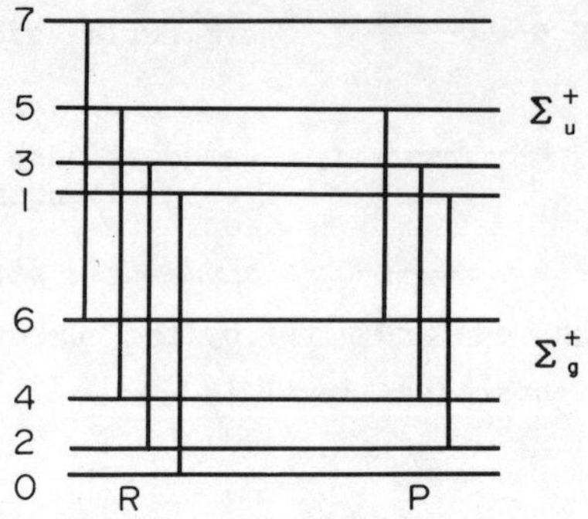


*Fig. 12.15. Rotational structure of the $00^01$-$10^00$ transitions*

branch transitions. Sobolev and Sokovikov (1967) have reported that the following lines have been observed

$00^01$-$02^00$ band $P_{4,6,8}$---$P_{56}$; $R_{4,6,8}$---$R_{54}$

$00^01$-$10^00$ band $P_{4,6,8}$---$P_{60}$; $R_{4,6,8}$---$R_{52}$

Laser transitions have also been observed in $CO_2$ on the following transitions in the 11-18 μm range:

$14^00$-$05^10$; $14^00$-$13^10$; $21^10$-$12^20$; $03^11$-$02^21$; $24^00$-$23^10$; $10^00$-$01^10$; $02^00$-$01^10$.

Let us now consider the important excitation and de-excitation mechanisms for the upper and lower laser levels.

12.4.1 Upper Laser-level Excitation Mechanisms

At the moment there are at least three processes which are considered to be important in exciting the upper laser level. These are as follows:

12.4.1.1 Direct electron-impact excitation. In Fig. 3.24, the cross sections for electron impact excitation of $CO_2$ were shown as presented by Lowke, Phelps and Irwin (1973). Figure 12.16 summarizes again the cross sections for vibrational excitation. The cross sections for excitation of the upper level (001) is considerably greater than that for excitation of the lower laser levels (020 + 100) over most of the electron energy ranges of interest.

12.4.1.2 Electron impact pumping of nitrogen and resonant energy transfer. Even though $N_2$ is a homonuclear molecule with vibrational states that are not optically connected to ground and which do not readily convert vibrational to translational energy, there is still a large cross section for electron impact excitation of the vibrational levels v = 1 to 8. (This occurs through the formation of a compound state.) The cross section for electron impact excitation was found by Schulz (1964) for each vibrational level (as was previously shown in Fig. 3.23). The sum of the cross sections for excitation to v = 1 to 8 of $N_2$ is

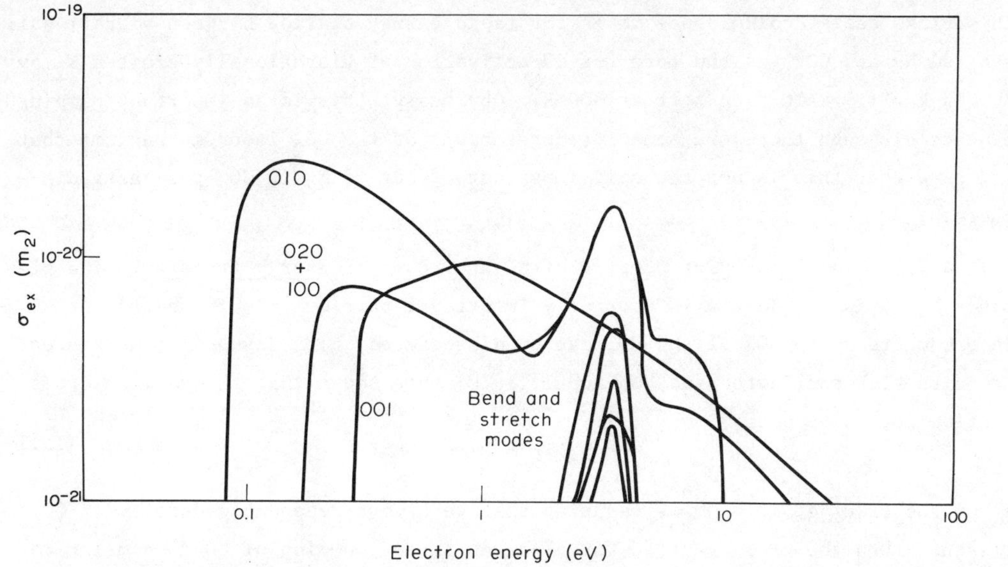


*Fig. 12.16. Cross sections for vibrational excitation in $CO_2$ (from Lowke, Phelps and Irwin, 1973)*

shown in Fig. 12.17. It is estimated that 10-30% of the $N_2$ in a discharge can be in a vibrationally excited state. The interest now lies in the fact that there is a very close energy coincidence between $N_2$(v = 1 to 6) and $CO_2$ ($00^o1$ to $00^o6$).

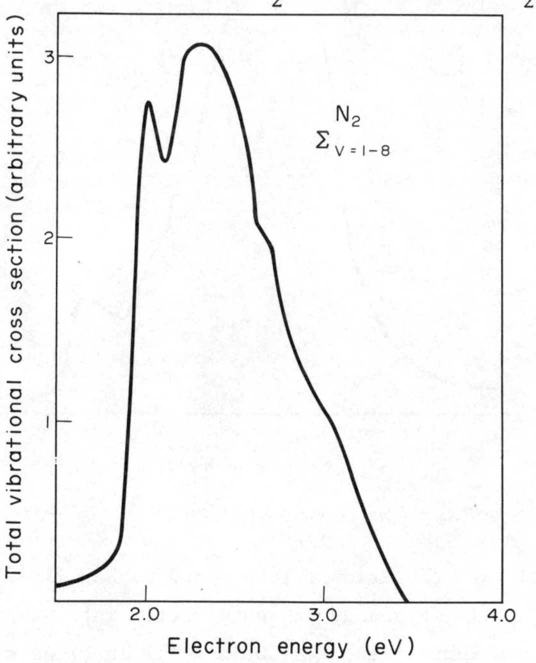


*Fig. 12.17. Cross section for excitation of $N_2$ vibrational levels (from Schulz, 1964)*

This close energy coincidence makes for rapid energy sharing between vibrationally excited $N_2$ and $CO_2$ and the rate for de-activation of vibrationally excited $N_2$ by $CO_2$ is $k = 1.9 \times 10^4$/sec-torr at 300 K. Obviously, this is an important pumping process although there are some interpretations of the $CO_2$ laser mechanisms that indicate that this is not the only important effect of $N_2$ in $CO_2$ gas laser discharges.

12.4.1.3 Electron impact pumping of CO and resonant energy transfer. The presence of CO in $CO_2$ lasers is extremely important, particularly in sealed off $CO_2$ lasers where up to 80% of the $CO_2$ can be dissociated. The dissociation rate can be quite high and Barton and Von Engel (1970) have shown that the maximum cross section for the process

$$e + CO_2 \rightarrow CO + O + e \tag{12.16}$$

is of the order of $10^{-21}$ $m^2$. Assuming that we have a reasonable density of CO present, then the cross section for electron impact pumping of CO is similar to that for $N_2$ except that it has a lower threshold and a higher peak. The cross section obtained by Schultz (1964) is shown in Fig. 12.18. A sizeable fraction of

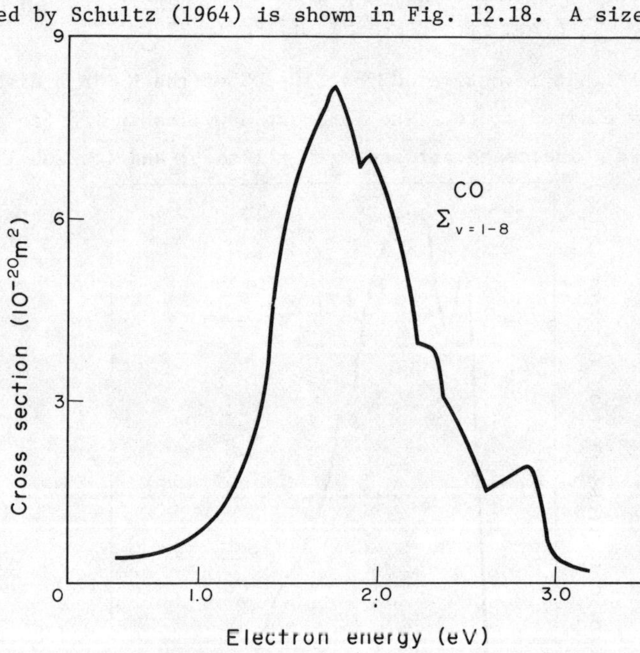


*Fig. 12.18. Cross section for excitation of CO vibrational levels (from Schulz, 1964)*

the CO will be excited into vibrational levels and Nighan (1969) has shown that up to 15% of the electron kinetic energy can go directly into pumping the vibrational levels of CO. The vibrationally excited CO is still in close energy agreement with the vibrational levels of $CO_2$, although the energy difference is greater than for $N_2$-$CO_2$ collisions. This fact and the allowed radiative decay of vibrationally excited CO reduces its effectiveness in transferring energy to $CO_2$, however, the large densities of vibrationally excited CO that are possible in the discharge ensure that this pumping process plays an important role.

### 12.4.2 Laser Level Relaxation Mechanisms

The laser levels are de-populated by radiative and collisional processes, the collisional processes usually being much more dominant. Under lasing conditions, the radiative process dominates and the change in $00^o1$ population between lasing and non-lasing conditions is coupled tightly to the population of the nitrogen excited atoms and can cause marked changes in the electron distribution function.

Let us consider the laser level relaxation methods individually.

12.4.2.1 Radiative relaxation. The radiative decay coefficients for the $00^o1$ level are in the range of A = .2 to 2 $sec^{-1}$. The radiative decay to ground $00^o1$-$00^o0$ is trapped, resulting in an effective A ~ 10 $sec^{-1}$. Statz, Tang and Koster (1966) present the transition probabilities (A coefficients) in $sec^{-1}$ shown in Table 12.1. We include here some of the decay coefficients for lower laser levels as well. The trapped radiative decay coefficients assume a temperature of 400 K. a tube radius of 1 cm and a pressure of 1 torr. In all cases, the radiative de-excitation is orders of magnitude smaller than the collisional relaxation except for the upper state under conditions of stimulated emission.

*Table 12.1. Einstein A Coefficients for Transitions in $CO_2$ (from Statz, Tang and Koster, 1966)*

| Transition | | Untrapped A | Trapped A |
|---|---|---|---|
| $00^o1$-$00^o0$ | R | $1.9 \times 10^2$ | 8.9 |
| | P | $2.0 \times 10^2$ | 10.2 |
| $00^o1$-$02^o0$ | R | 0.34 | |
| | P | 0.36 | |
| $00^o1$-$10^o0$ | R | 0.23 | |
| | P | 0.24 | |
| $10^o0$-$01^10$ | R | 0.24 | |
| | P | 0.28 | |
| | Q | 0.53 | |
| $01^10$-$00^o0$ | R | 0.55 | |
| | P | 0.52 | |
| | Q | 1.07 | |

12.4.2.2 Collisional relaxation. The collisional relaxation of the upper and lower laser levels is the dominant kinetic process in the $CO_2$ laser and so has attracted considerable attention. The papers by Yardley and Moore (1967) and by Moore, Wood, Hu and Yardley (1967) are good references. A good review article has also been published by Taylor and Bitterman (1969).

Some of the important processes that relax the upper laser level are:

$$CO_2(00^o1) + N_2(v = 0) \rightarrow CO_2(00^o0) + N_2(v = 1) \ , \tag{12.17}$$

$$CO_2(00^o1) + CO_2(00^o0) \rightarrow CO_2(00^o0) + CO_2(11^10) \ , \tag{12.18}$$

$$CO_2(00^o1) + H_2O(000) \rightarrow CO_2(01^10) + H_2O(010) \ , \tag{12.19}$$

$$CO_2(00^o1) + H_e \rightarrow CO_2(04^o0) + H_e \ , \tag{12.20}$$

$$CO_2(00^o1) + CO(v = 0) \rightarrow CO_2(00^o0) + CO(v = 1) \ . \tag{12.21}$$

In the collisional relaxation of the lower laser levels, we consider first the relaxation to the $CO_2$ $(01^10)$ state and then the relaxation from $CO_2$ $(01^10)$ to ground.

We indicated in the notation that the lower laser levels are in Fermi resonance so the conversion between these levels is extremely fast.

$$CO_2[10^o0,02^o0]_I + CO_2 \rightleftarrows CO_2[10^o0,02^o0]_{II} + CO_2 \ . \qquad k > 10^6 \ sec^{-1} \ torr^{-1} \tag{12.22}$$

The subsequent vibrational relaxation down to the $01^10$ level is also very fast:

$$CO_2[10^o0,02^o0]_I + CO_2 \rightarrow CO_2(02^20) + CO_2 \ , \qquad k = 6 \times 10^5 \ sec^{-1} \ torr^{-1} \tag{12.23}$$

$$CO_2(02^20) + CO_2 \rightarrow CO_2(01^10) + CO_2(01^10) \ . \qquad k = 1.3 - 3.6 \times 10^6 \ sec^{-1} \ torr^{-1} \tag{12.24}$$

The limiting step is the conversion of vibrational into translational energy in relaxing the $01^10$ level to ground. Some of the important processes are

$$CO_2(01^10) + CO_2 \rightarrow CO_2(00^o0) + CO_2(00^o0) \ , \tag{12.25}$$

$$CO_2(01^10) + H_2O \rightarrow CO_2 + H_2O \ , \tag{12.26}$$

$$CO_2(01^10) + H_e \rightarrow CO_2 + H_e \ , \tag{12.27}$$

$$CO_2(01^10) + H_2 \rightarrow CO_2 + H_2 \ . \tag{12.28}$$

From Yardley and Moore (1967), Moore, Wood, Hu and Yardley (1967), Taylor and Bitterman (1969) and Tyte (1970), the collisional relaxation rates are presented in Table 12.2.

*Table 12.2. Collisional Relaxation Rates of $CO_2$ in $sec^{-1}$ $torr^{-1}$ at 300 K*

| Relaxing gas | $00^01$ Upper level | $01^10$ Lower level |
|---|---|---|
| $CO_2$ | 350-365 | 175-210 |
| $N_2$ | 106-110 | <200 [a] |
| $O_2$ | 110 | ~40 |
| $H_2$ | $(3.8-4.1) \times 10^3$ | $(7 \pm 3) \times 10^4$ |
| $D_2$ | 271 | $(1.5-2.3) \times 10^4$ |
| CO | 993 | $4 \times 10^3$ |
| $H_2O$ | $(2.4-3.3) \times 10^4$ | $(3.5-4.9) \times 10^5$ |
| He | <85 | $(3.4-5.3) \times 10^3$ |
| Ne | 57.5 | 110 |
| Ar | 53 | <44 |
| Xe | ~30 | <44 |

[a] It is generally felt that $N_2$ is 1/10 to 1/20 less effective than $CO_2$ in relaxing the lower level (see Taylor and Bitterman, 1969).

One factor that must be clearly kept in mind since it has a profound influence on $CO_2$ laser characteristics is that all of the collisional relaxation rates are very sensitive functions of gas temperature as shown, for example, in Fig. 12.19 for the process

$$CO_2\ (01^10) + CO_2 \rightarrow CO_2 + CO_2 \ .$$

The overall effect is that $CO_2$ laser gain and available power drops sharply with increasing gas temperature.

Referring back to Table 12.2, we can now examine the role of some additives in $CO_2$ discharges. Obviously $N_2$ is not there for its relaxation properties. Its primary importance is due to its collisional energy transfer to the upper $CO_2$ level and, possibly, its effect on the electron distribution function. The presence of $D_2$ would appear to be desirable in terms of collisional relaxation. $H_2$ and $H_2O$ are beneficial in sealed off $CO_2$ lasers since they help to reverse the dissociation of $CO_2$ into CO. It does seem to be generally agreed that the real advantage in adding He is in cooling by de-activation of the $(01^10)$ level, the great improvement in thermal conductivity and in the improvement in discharge characteristics

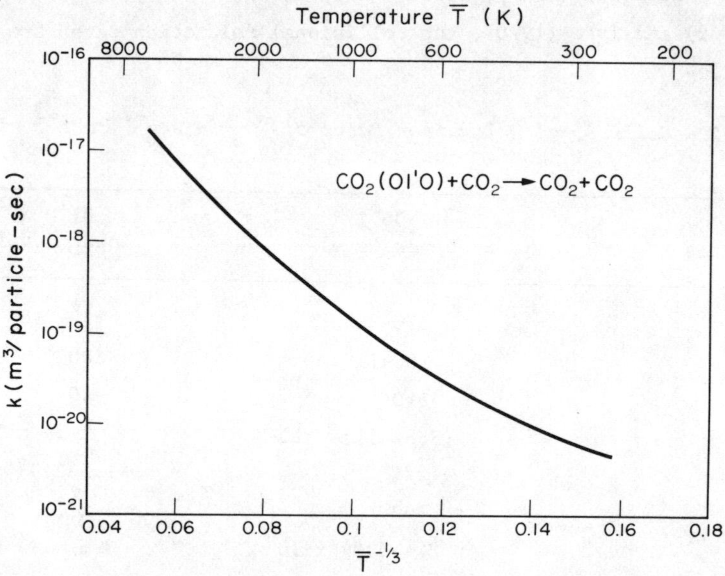


*Fig. 12.19. Rate constant for relaxation of $CO_2$ $(01^10)$ by $CO_2$ as a function of gas temperature (from Taylor and Bitterman, 1969)*

(especially in the high pressure TEA lasers). Since $CO_2$ laser action is very sensitive to gas temperature (power decreasing with higher temperature) this improved thermal conductivity is advantageous. Ne, Ar and Xe play no significant collisional role, although Xe, because of its low ionization energy, can play a useful role as a gas additive.

### 12.4.3 Electron Energy Distributions and Fractional Power Transfer

As was shown in Section 4.7.3, low energy inelastic collisions in molecular gases tend to make the electron distribution function markedly non-Maxwellian. It then becomes necessary to determine the shape of the distribution function reasonably accurately if we are to be able to compute electron collisional excitation and ionization rates. Figure 4.19 showed the electron energy distribution as calculated by Nighan (1970) for $CO_2$. Similar calculations for $CO_2$ laser mixtures have been performed, and Fig. 12.20 shows the distribution functions derived by Lowke, Phelps and Irwin (1973) for a $CO_2$ : $N_2$ : He mixture of 1 : 2 : 3. Of course, once the cross sections and distribution functions are known, then the fractional power transfer can be calculated as a function of E/N. Figure 12.21 shows the results of such a calculation by Lowke, Phelps and Irwin for the same conditions as used for Fig. 12.20. Obviously, if the laser can operate at an E/N corresponding to the peak in the energy transfer to the upper laser levels, then quite high efficiencies can be expected. Actually, it may be more desirable to work where

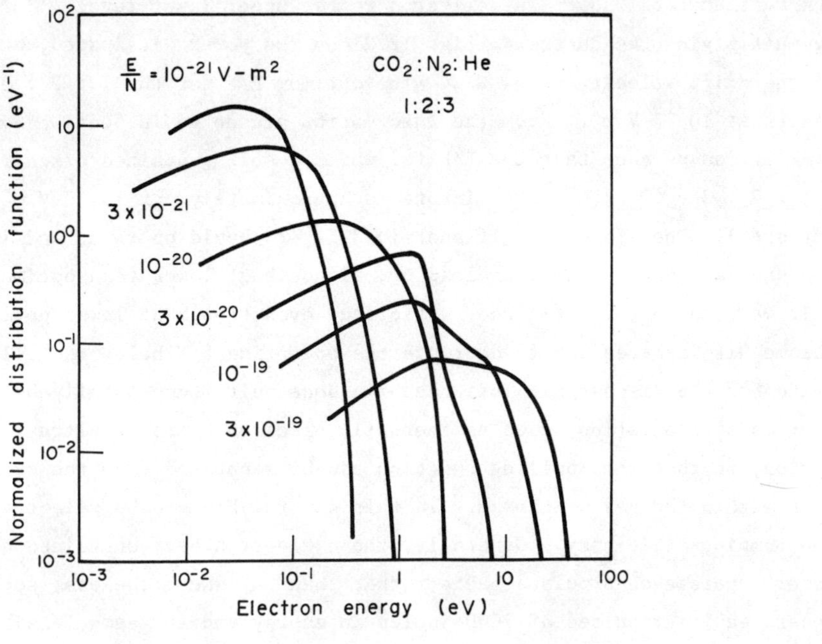


*Fig. 12.20. Derived distribution functions of electrons for $CO_2 : N_2 : He = 1 : 2 : 3$ for various values of E/N (from Lowke, Phelps and Irwin, 1973)*

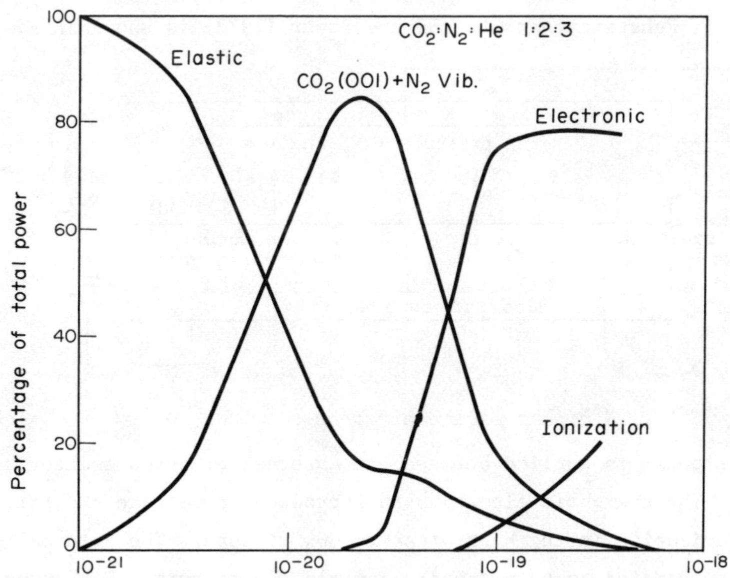


*Fig. 12.21. Percentage of power lost to various collisional processes as a function of E/N (from Lowke, Phelps and Irwin, 1973)*

the maximum total power is delivered to the upper laser level. This is obtained by multiplying the curves in Fig. 12.21 by the power dissipated, or by the product of the drift velocity times E/N. The optimum E/N for the 1 : 2 : 3 mixture then occurs at $10^{-19}$ V $m^2$. From the information presented in Section 8.2 and in Fig. 8.6, it can be seen that the E/N for which a self-sustained discharge operates in a 1 : 2 : 3 = $CO_2$ : $N_2$ : He mixture is approximately 5 x $10^{-20}$ V $m^2$ (at moderate currents). Therefore, a self-sustained laser should operate, and indeed does operate, at quite good power loading, although at lower than optimum efficiency.

If we wish to work at higher efficiency even though at lower powers per unit volume, it is necessary to decrease the operating E/N below the self-sustaining value for the discharge. This has been done quite successfully by using auxiliary sources of ionization, such as energetic electron beams or ultraviolet photoionization, so that the ionizing function can be separated from the excitation function within the gas discharge. In this way the E/N can be selected to optimize the pumping efficiency. Generally, these electron beam sustained and similar lasers operate on a pulsed basis rather than CW. Nevertheless, such devices have generated laser pulses of 2000 joules in energy and it seems feasible to anticipate pulsed $CO_2$ lasers delivering tens to hundreds of kilojoules on a pulsed basis.

A possible configuration for an electron beam controlled $CO_2$ laser (similar to that reported by Fenstermacher, Leland and Boyer (1972) is shown in Fig. 12.22.

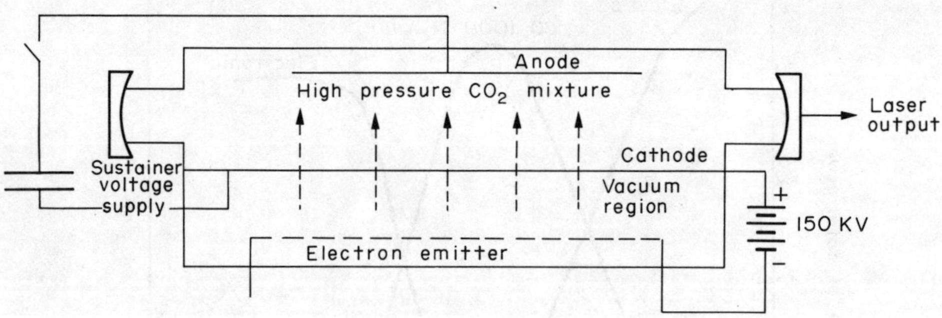


*Fig. 12.22. A high pressure electron-beam-controlled* $CO_2$ *laser*

A high voltage pulse is applied between the external electron emitter and the thin foil cathode. The energetic electrons thus produced penetrate the foil cathode and produce ionization in the high pressure $CO_2$ mixture. The external sustainer voltage is then applied to pump the pre-ionized gas mixture. The proper E/N can therefore be chosen to optimize the pumping efficiency and to obtain maximum power output. As we mentioned before, pulse energies of 2000 joules have been reported using such a device.

## 12.4.4 Laser Kinetics Model

The previously discussed information on excitation and relaxation rates can be combined into a simple kinetics model that can be used to give quite accurate predictions of laser level populations, gain and power output. A basic feature of the kinetics model to be described is that each vibrational mode of the molecules can be considered to be in local Boltzmann equilibrium at its own vibrational temperature, and the vibrational modes interact with each other by collisions and by photon emission and absorption. This model was described by Moore, Wood, Hu and Yardley (1967) and can be characterized by Fig. 12.23 as presented by DeMaria (1973).

In this model, the $CO_2$ discharge containing $N_2$ can be described in terms of three vibrational temperatures: $T_1$, the symmetric stretch and bending mode temperature (lower laser level), $T_2$ the asymmetric stretch temperature (upper laser level) and $T_v$, the nitrogen vibrational temperature. The important processes as shown are:

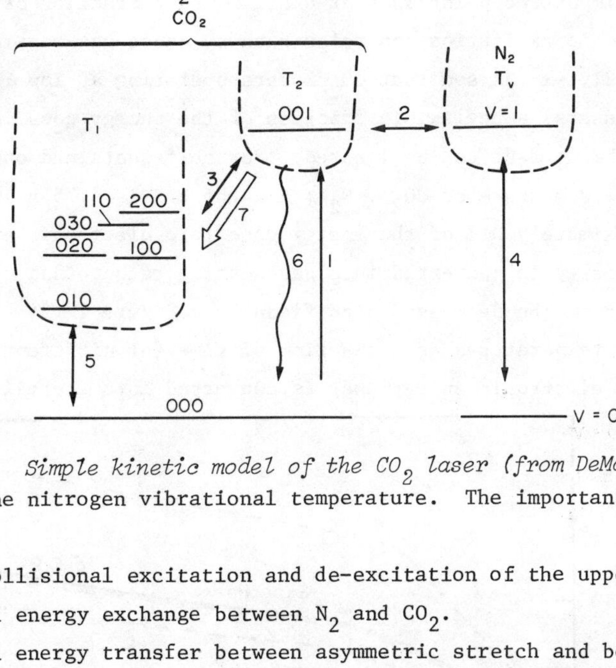


*Fig. 12.23. Simple kinetic model of the $CO_2$ laser (from DeMaria, 1973)*

1. Electron collisional excitation and de-excitation of the upper laser level.
2. Vibrational energy exchange between $N_2$ and $CO_2$.
3. Collisional energy transfer between asymmetric stretch and bending modes.
4. Electron collisional excitation and de-excitation of nitrogen.
5. Collisional excitation and de-excitation of the bending modes.
6. Spontaneous emission from the upper laser level.
7. Laser emission (and absorption) at 10.6 μm.

Using this simple model, a relatively good feel for the dependence of laser performance conditions can be obtained.

Denes, Babcock, George, Kline, Liberman, Liebermann, Partlow, Taylor and Weaver (1975) have used such a model to analyze a pulsed, high pressure $CO_2$ laser including the effects of uniform gas flow (to cool the gas), the effects of $H_2O$ additive, (0.4%), the temperature dependence of the collisional relaxation mechanism and the electric field dependence of the electrical excitation. Because of the great sensitivity of the collisional rates on gas temperature, an important feature of the analysis is the solution of the nonstationary equation of thermal conductivity to determine the time dependence of the neutral gas temperature. Energy defect heating from collisions in which vibrational energy is exchanged (V-V), heating from collisions in which vibrational energy is converted to translational energy by relaxing to the ground state (V-T) and heating by the electrical discharge have all been taken into account. In a 1 : 2 : 3 mixture of $CO_2$ : $N_2$ : He, the V-V collisions contribute about 10% of the total heating while V-T heating can contribute between 40-80% depending whether or not laser action is occurring.

An additional important point is that a significant fraction of the energy that goes into electronic excitation can relax back to cause gas heating. Referring back to Fig. 12.21, we can see that $CO_2$ lasers operating at low E/N (electron beam sustained lasers) a negligible fraction of the energy goes into electronic excitation, so this effect can be ignored. For self-sustained operation at 600 torr of a 1 : 2 : 3 mixture of $CO_2$ : $N_2$ : He, the E/N is $4.75 \times 10^{-20}$ V $m^2$. At this value, approximately 31% of the energy goes into electronic excitation and if all of this energy is converted into gas heating rather than photon flux, the gain of the laser can be decreased significantly. Figure 12.24 shows theoretical and measured gas temperatures as a function of time for different assumed fractions, ξ, of the electronic energy that is converted into thermal energy. Note

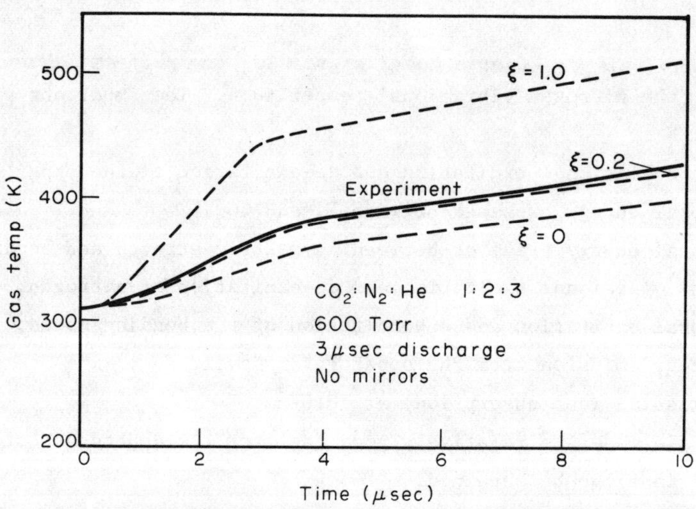


*Fig. 12.24. Gas temperature as a function of time for various values of electronic gas heating (from Denes et al., 1975)*

that the discharge pulse starts at 300 nsec and ends at 3 μsec. The measured and calculated gain is shown in Fig. 12.25. A very good fit with experiment is found when it is assumed that 20% of the electronic energy is converted into gas heating.

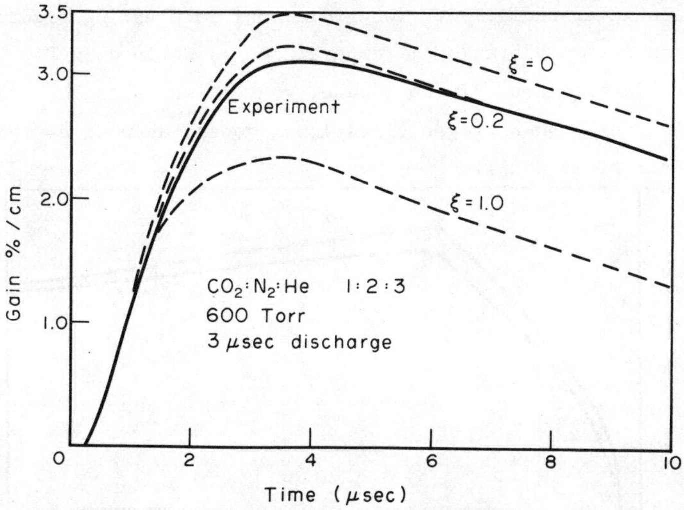


*Fig. 12.25. Small signal gain as a function of time for various values of electronic gas heating (from Denes et al., 1975)*

When the laser equations are solved assuming that mirrors are on the system and laser action can occur, some interesting changes in the predicted behavior occur. Figure 12.26 shows the change in the time dependence of the gas temperature. In Fig. 12.26 laser action begins at 1 μsec and decays quickly after 3 μsec. When laser action occurs, the $CO_2$ (001) mode is radiatively coupled to the $CO_2$ (100)

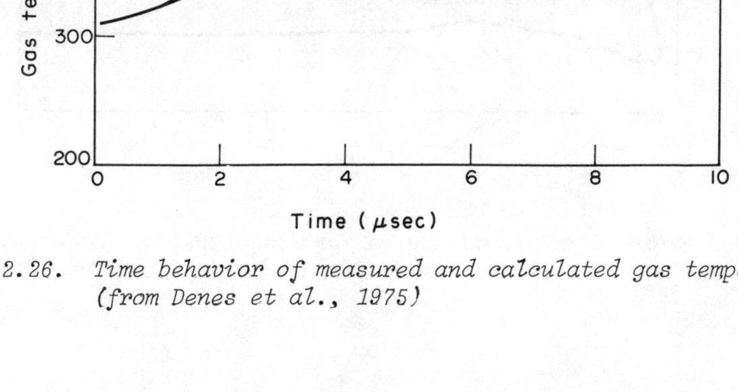


*Fig. 12.26. Time behavior of measured and calculated gas temperatures (from Denes et al., 1975)*

mode. The increased heating comes from the increased V-T relaxation of the $CO_2$ bending mode. Eventually, of course, the gas temperature for the laser case would be less than for the non-laser case since some of the energy that would heat the gas is lost as laser radiation. Two additional interesting figures present the time variation in the vibrational temperatures as defined in Fig. 12.23 for the laser and non-laser cases. In this case, you can see that laser action significantly "cools" both the nitrogen vibrational modes, as well as the $CO_2$ upper levels due to their tight coupling.

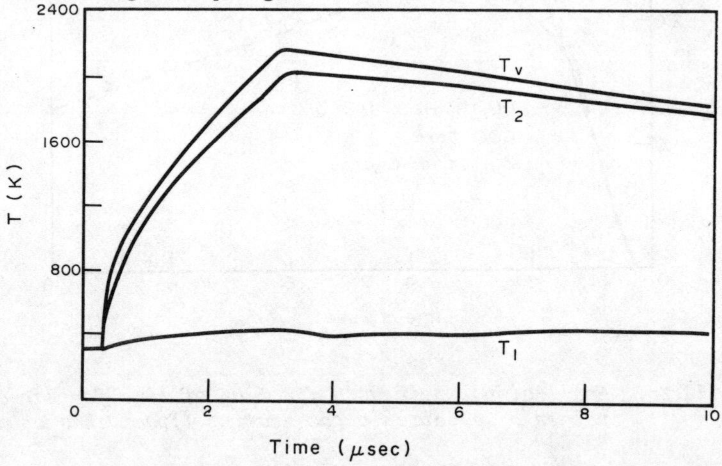


*Fig. 12.27. Vibrational temperatures, non-laser (from Denes et al., 1975)*

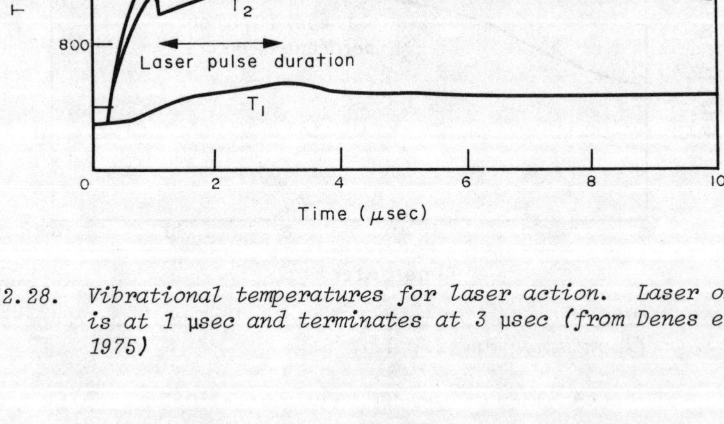


*Fig. 12.28. Vibrational temperatures for laser action. Laser onset is at 1 μsec and terminates at 3 μsec (from Denes et al., 1975)*

A factor of considerable importance is the overall laser efficiency. Figure 12.29 shows the calculated stored energy efficiency for the $CO_2$ laser system under discussion. The efficiency is the percentage of the input energy that is stored in the $N_2$ vibrational levels plus the $CO_2$ upper laser level and is, therefore, extractable as laser energy. As can be seen, efficiencies of 25% are possible and in the absence of lasing, that energy remains available for a considerable period of time. When laser action occurs, some of this energy is extracted and so the stored energy decreases as expected. The efficiency figures shown are approximately what you would expect since, from Fig. 12.21, 50-80% of the electrical energy can be put into the upper laser level plus the $N_2$ reservoir. The quantum efficiency of the 10.6 μm transition is 41%, therefore 20-35% of the total electrical energy should be available for conversion to optical energy. Typical experimental values are in the 10-20% range.

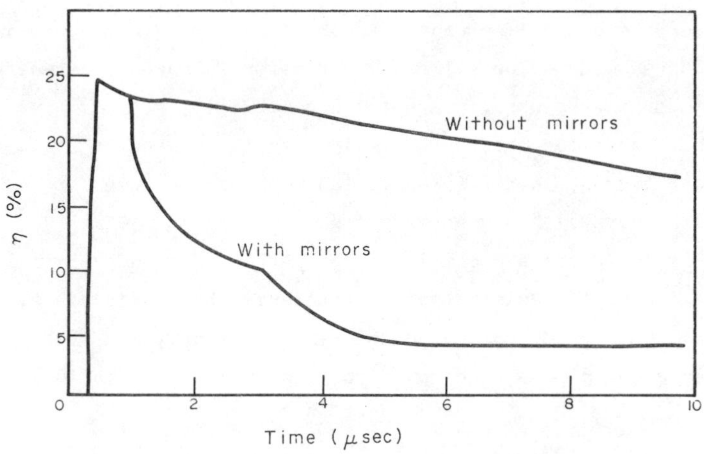


*Fig. 12.29. Stored energy efficiency (from Denes et al., 1975)*

## 12.5 EXCIMER LASERS

The search for high efficiency, high power lasers has resulted in extensive study of the "excimer" laser systems over the last few years. The distinguishing feature of the molecules used in these lasers is that the excited molecule exists in a stable configuration but the ground state molecule does not. The basic molecular structure is shown in Fig. 12.30. In such a system it should be possible to readily achieve a population inversion between the stable excited molecular species and the unstable ground state molecular species. Some of the implications of such a system are as follows:

(a) Since the transition is from a bound molecular state to a free molecular state (bound-free), laser action should be tunable over a wide range.

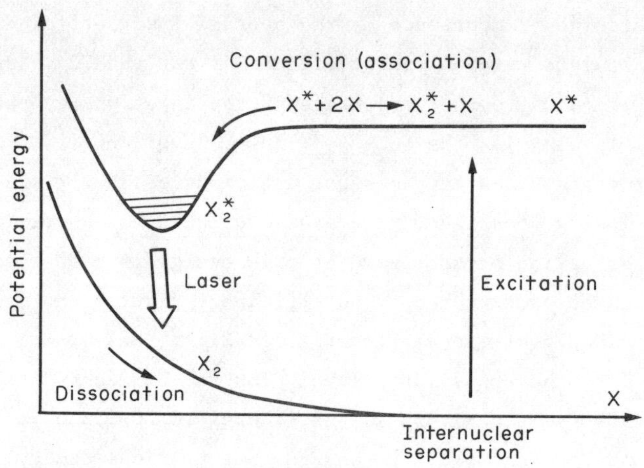


*Fig. 12.30. Basic molecular structure for excimers*

(b) Because of the broad line width (tunability) the gain on the laser transition should be low, allowing a substantial amount of energy to be stored in the upper state (high power).

(c) Since the transition occurs from a low excited molecular state to a state near ground, the efficiency (laser energy out over excitation energy) should be very high.

Figure 12.30 also indicates a possible energy pathway for an excimer laser. Since the ground state molecules exist almost entirely in the atomic state, excitation could occur from the ground state atom to a low excited atomic state. The excited molecule is then created by a three body conversion (or association) process.

$$X^* + 2X \rightarrow X_2^* + X \quad . \tag{12.29}$$

Obviously, the higher the pressure, the more rapid the conversion process and the higher the pumping rate for the upper laser level. Once the excimer is formed in sufficient numbers, stimulated emission will occur and the excited molecule will radiatively relax to a ground state molecule. The ground state molecule is unstable (created on the repulsive portion of the potential curve) and so it rapidly dissociates into the ground state atomic system. Because of the importance of the association and dissociation processes, excimer lasers are also referred to as association or dissociation lasers.

There are a considerable number of systems that have bound-free transitions including the rare gas molecules, metals such as $Hg_2$ and $Cd_2$, alkali-rare gas compound molecules and rare gas-halogen molecules. Extensive investigations have been made of most of these systems and laser action has been observed in high pressure rare gases (Xe, Kr, Ar) as well as rare gas mixtures (Ar/Xe). Laser action has also been observed in compound molecules, such as XeO, and a considerable amount of work has been done on the rare gas-halogen lasers (XeBr, XeCl, XeF, KrF, etc.). Many of these systems are still not fully understood, but general characteristics of excimer lasers are becoming clear and will be discussed in the following sections.

### 12.5.1 Rare-gas Excimers

From the work of Mulliken (1970), it was known that the heavier rare gases, particularly xenon, had a molecular structure similar to that shown in Fig. 12.30 and were suitable candidates for laser action. Basov, Danilychev and Popov (1971) used 700 keV electrons at current densities of up to 300 A/cm$^2$ to excite liquid xenon, and observed stimulated emission at 1760 Å. Subsequently stimulated emission or laser action at 1730 Å was observed in electron beam excited high pressure ($10^3$ - $2 \times 10^4$ torr) xenon gas by Koehler, Ferderber, Redhead and Ebert (1972, 1974), Hoff, Swingle and Rhodes (1973a) and Gerardo and Johnson (1973a). Hoff, Swingle and Rhodes (1973b) also observed laser emission in krypton at 1457 Å and in xenon/argon mixtures (on the xenon line). In fact, superfluorescent emission from xenon has been observed as well (Gerardo and Johnson, 1973b).

Figure 12.31 from Gerardo and Johnson (1973a) gives an indication of the type of efficiency that might be expected from electron-beam excited xenon lasers. This curve is based on the energy deposited by a 1.5 MeV electron beam in xenon as the gas pressure is varied and the solid line assumes that 6.2% of the input energy is radiated by $Xe_2^*$. Roughly 75% of the beam energy is deposited in the gas as ionization, excitation and gas heating. From Fig. 12.31 it can be seen that at pressures greater than $10^4$ torr, the efficiency begins to decrease, possibly due to

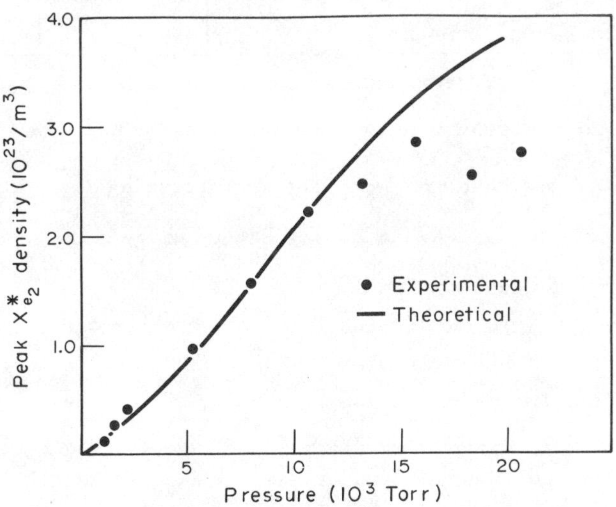


*Fig. 12.31. Measured peak density of excited molecules that radiate in the VUV as a function of gas pressure (from Gerardo and Johnson, 1973)*

losses caused by collisional quenching, photo-ionization, and excimer-excimer collisions, as will be discussed later.

From experimental observations it is possible to establish a reasonable model for the excitation and de-excitation processes in the xenon laser. Since laser action occurs in the afterglow as well as during the electron beam excitation pulse, we must include the important role played by recombination as well as by direct excitation in forming the excimers. Figure 12.32 shows an energy-level diagram including most of the basic states of importance. George and Rhodes (1973) and Rhodes (1974) have indicated many of the kinetic processes that can be important.

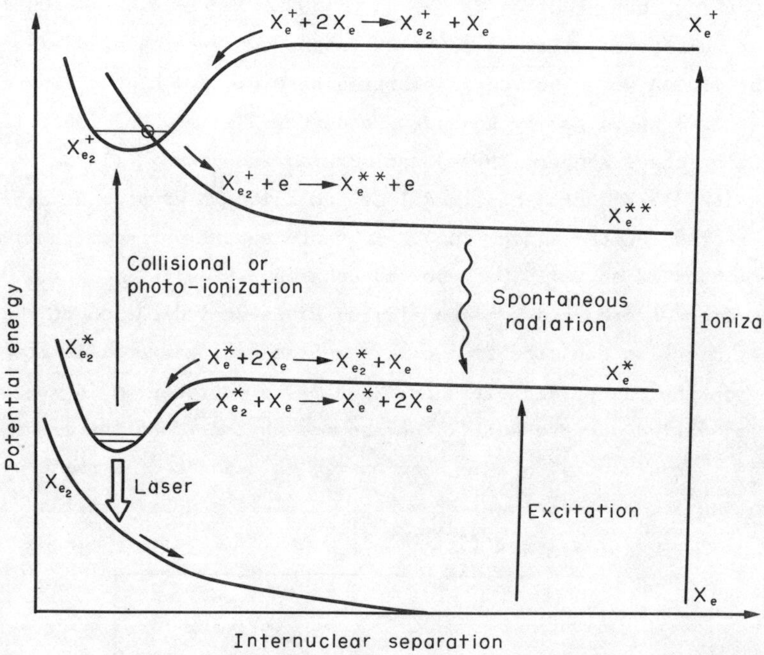


*Fig. 12.32. Energy level structure for the Xe system and important kinetic processes for the $Xe_2{}^*$ laser*

There are two processes that are important for producing $Xe^*$. The direct excitation process,

$$e + Xe \to Xe^* + e \tag{12.30}$$

and the recombination process consisting of

$$\text{ionization} \quad e + Xe \to Xe^+ + e + e \quad , \tag{12.31}$$

$$\text{ion conversion} \quad Xe^+ + 2Xe \to Xe_2{}^+ + Xe \ , \tag{12.32}$$

$$\text{dissociative recombination} \quad e + Xe_2{}^+ \to Xe + Xe^{**} \quad , \tag{12.33}$$

$$\text{and radiative decay} \quad Xe^{**} \to Xe^* + h\nu \tag{12.34}$$

$$\text{(or collisional decay} \quad Xe^{**} \to Xe_2{}^{**} \to Xe^* .) \tag{12.35}$$

Most of these processes are well known, as have been discussed earlier, but remember that dissociative recombination, in particular, is quite sensitive to temperature.

Probably the most important kinetic process to consider is the three body conversion from the excited atom to the excited molecule

$$Xe^{*} + 2Xe \rightarrow Xe_2^{**} + Xe \quad . \tag{12.36}$$

For the laser, this process determines the pumping rate for the upper laser level and hence the available power. Since this is a three body process, it scales as the pressure squared, so high pressure, as was seen before, is desirable.

Many of the appropriate rate coefficients for three body conversion in the rare gases are known and are summarized in Table 12.3 from Ku, Verdeyen, Cherrington, and Eden (1973), Boucique and Mortier (1970), and Gerardo and Johnson (1976).

*Table 12.3. Three-body Conversion Rates for Excited Rare Gas Dimers,* $X^* + 2X \rightarrow X_2^* + X$

| Excited atomic species $X^*$ | Rate | |
|---|---|---|
| | $m^6/sec$ | $sec^{-1}$(p in torr) |
| He($2^3S$) | $0.25 \times 10^{-45}$ | $0.31\ p^2$ |
| Ne(1s5) | $0.5 \times 10^{-45}$ | $0.63\ p^2$ |
| Ar(1s5) | $17 \times 10^{-45}$ | $21\ p^2$ |
| Ar(1s3) | $11.4 \times 10^{-45}$ | $14\ p^2$ |
| Kr(1s5) | $38 \times 10^{-45}$ | $48\ p^2$ |
| Kr(1s3) | $53.6 \times 10^{-45}$ | $67\ p^2$ |
| Xe | $(17\text{-}78) \times 10^{-45}$ | $(21\text{-}97)\ p^2$ |

There is quite a variation in the measured rates for $Xe^* + 2Xe \rightarrow Xe_2^* + Xe$, which may be attributed, in part, to the different experimental approaches taken to measure the conversion rate. The difference may also be due to the fact that there are two states of $Xe_2^*$, the $^1\Sigma$ and the $^3\Sigma$, that must be taken into account. For instance, Werner, George, Hoff and Rhodes (1977) attribute the following three-body conversion rates:

$$Xe^* + 2Xe \rightarrow Xe_2^* (^1\Sigma) + Xe, \quad k = 17 \times 10^{-45}\ m^6/sec \tag{12.37}$$

and

$$Xe^* + 2Xe \rightarrow Xe_2^* (^3\Sigma) + Xe, \quad k = 30 \times 10^{-45}\ m^6/sec \quad . \tag{12.38}$$

In some cases, xenon lasers have been operated with large quantities of other rare gases added. Three body association collisions of the form

$$Xe^* + Xe + M \rightarrow Xe_2^* + M \tag{12.39}$$

are then possible, where M is a ground state rare gas atom. Rice and Johnson (1975), Johnson and Rice (1976) and Gerardo and Johnson (1976) have measured the conversion coefficients for different rare gases and their results are shown in Table 12.4.

*Table 12.4. Three-body Conversion Rates for Excited Xenon Atoms with a Ground-state Xenon Atom plus a Rare Gas Atom, M.* $Xe^* + Xe + M \rightarrow Xe_2^* + M$

| Rare gas atom M | Rate ($m^6$/sec) |
|---|---|
| He | $14\text{-}17 \times 10^{-45}$ |
| Ne | $18 \times 10^{-45}$ |
| Ar | $23\text{-}28 \times 10^{-45}$ |
| Kr | $47 \times 10^{-45}$ |

At higher temperatures, of course, the inverse process, $Xe_2^* + Xe \rightarrow Xe^* + 2Xe$, is possible. The balance of the rates depends on the binding energy of the excited molecule and can be inferred from detailed balance arguments if the equilibrium constant for the molecule is known (Hedges, Drummond and Gallagher (1972), Appendix A).

Once the excited molecule is formed, stimulated emission and optical gain is possible. In this case, the radiative lifetimes of the singlet and triplet state are quite different. Keto, Gleason, Bonifield and Walters (1976) have given the following radiative lifetimes:

$$Xe_2^*(^1\Sigma) \rightarrow 2Xe + h\nu \ , \ \tau = 6.2 \times 10^{-9} \text{ sec} \quad , \qquad (12.40)$$

and

$$Xe_2^*(^3\Sigma) \rightarrow 2Xe + h\nu \ , \ \tau = 1 \times 10^{-7} \text{ sec} \quad . \qquad (14.41)$$

They have also shown that there is a strong neutral collisional coupling between these states.

$$Xe_2^*(^3\Sigma) + Xe \underset{k_u}{\overset{k_d}{\rightleftarrows}} Xe_2^*(^1\Sigma) + Xe \qquad (12.42)$$

where $k_d = 1.25 \pm 0.68 \times 10^{-19}$ $m^3$/sec
and $k_u = 6.6 \pm 1.8 \times 10^{-21}$ $m^3$/sec .
Similarly, electron collisions can convert the triplet to the singlet,

$$Xe_2^*(^3\Sigma) + e \rightarrow Xe_2^*(^1\Sigma) + e \quad , \qquad (12.43)$$

with a rate coefficient of

$$k = (1.9 \pm 0.5) \times 10^{-12} \ m^3/\text{sec} \ .$$

The large difference in the radiative lifetimes and hence optical gains on the two molecular states makes it important that the two states be treated separately in any comprehensive analysis of the xenon laser and that the collisional coupling terms be taken into account. The importance of the difference in optical gains becomes especially evident when optical absorption due to photoionization is considered.

Since laser action occurs at 1730 Å, the photons have sufficient energy to photo-ionize the molecules.

$$Xe_2^* + h\nu \rightarrow Xe_2^+ + e \qquad (12.44)$$

with a cross section of

$$\sigma_{ph} = 1.9 \times 10^{-22}\ m^2 \quad . \qquad (12.45)$$

This turns out to be a serious optical loss since the stimulated cross section of the $^3\Sigma$ state is $\sigma_\nu \simeq 1 \times 10^{-22}\ m^2$ and the stimulated emission cross section of the $^1\Sigma$ state is $\sigma_\nu \simeq 1.5 \times 10^{-21}\ m^2$. (See Werner, George, Hoff and Rhodes, 1977.) It is only the $^1\Sigma$ state that has a net optical gain. This turns out to be a primary limiting factor of the xenon laser at high pressures where triplet production is favored (see equation 12.42) and the lower electron densities are less effective in maintaining a balance in singlet and triplet densities according to their statistical weights.

Three additional loss processes that play an important role in limiting attainable power are electron collisional ionization, quenching by atomic xenon and excited state-excited state collisions. Electron collisional ionization

$$Xe^* + e \rightarrow Xe^+ + e + e \qquad (12.46)$$

requires only fairly low electron energies, so could be very important in the active discharge and possibly in the afterglow unless the electron energy decreases rapidly with time. The rates for these processes are not well known, but can be reasonably well estimated (see Section 3.3.1). Johnson and Gerardo (1973) have evaluated the neutral quenching reaction and the excited state-excited state collision rate. Collisional quenching by atomic xenon,

$$Xe_2^* + Xe \rightarrow \text{products} \quad , \qquad (12.47)$$

was found to have a cross section of $\sigma = 3.3 \times 10^{-22}\ m^2$. The excimer-excimer collisional destruction process,

$$Xe_2^* + Xe_2^* \rightarrow Xe_2^+ + e + 2Xe \quad , \qquad (12.48)$$

was found to have a rate coefficient of

$$k = (3.5 \pm 1.4) \times 10^{-16}\ m^3\ sec^{-1} \quad .$$

Zamir, Werner, Lapatovich and George (1975) also studied the excimer-excimer collision process and deduced a value of

$$k = 8 \times 10^{-17}\ m^3\ sec^{-1} \quad .$$

This discrepancy may be due to the fact that Johnson and Gerardo's (1973) analysis did not consider the triplet and singlet separately.

The destruction of excited molecules by excimer-excimer collisions is a very important limitation on the xenon laser and if the density of $Xe_2^*$ is limited to around $10^{23}$ $m^{-3}$ (see Fig. 12.31) then, as Rhodes (1974) has pointed out, this would limit the energy density to around 250 joules/litre for this laser and would restrict the use of this laser as a high energy source.

As one last point, George and Rhodes (1973) have modeled the excimer density in a high pressure xenon discharge as shown in Fig. 12.33. Their curves predict

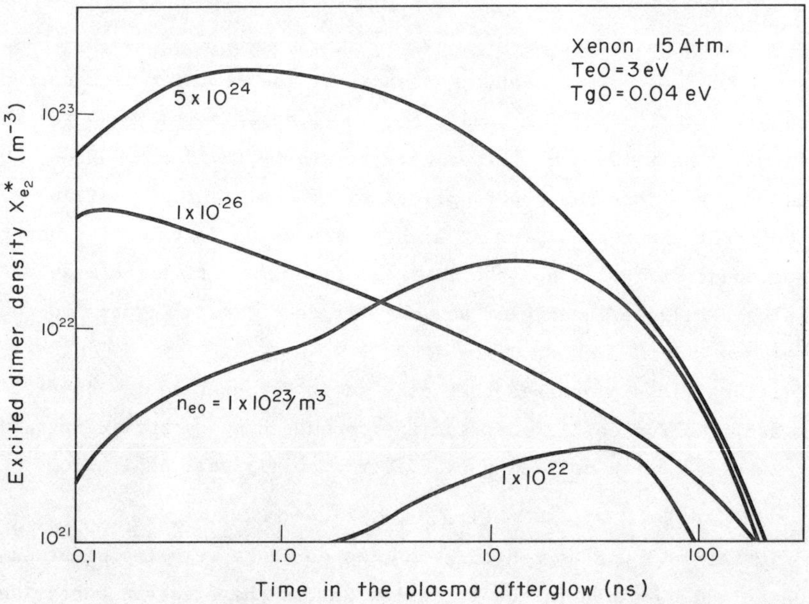


*Fig. 12.33. Time evolution of excited dimer density for various values of initial electron density (from George and Rhodes, 1973)*

continuum emission which is in fairly good agreement with experiment. Werner, George, Hoff and Rhodes (1977) have performed a more comprehensive modeling of the xenon laser that also gives quite good agreement with experiment. A considerable amount of caution is necessary, however, in modeling systems where all of the rate processes are not well known. Good agreement with experiment is encouraging, but it must not be assumed that a good fit with the data means that the model is correct. It is not uncommon to have a wide variation in possible rate coefficients that will give the same fit to experimental data. Such modeling is probably at best an indicator that you are on the right track.

### 12.5.2 Rare Gas - Monohalide Excimers

Laser action has been observed on the $\Sigma - \Sigma$ transitions in quite a number of rare gas-monohalide systems. These are:

XeBr at 2818 Å (Searles and Hart, 1975)
XeCl at 3080 Å
KrF at 2490 Å (Ewing and Brau, 1975)

XeF at 3511, 3531 Å (Ault, Bradford and Bhaumik, 1975; Brau and Ewing, 1975)
ArF at 1933 Å (Hoffman, Hays and Tisone, 1976)
KrCl at 2220 Å (Murray and Powell, 1976; Eden and Searles, 1976a)
ArCl at 1750 Å (Waynant, 1977).

The molecular structure is similar to that for the $Xe_2$ laser system as shown in Fig. 12.34 for the KrF system. A point to be noticed here is that the excited molecule has a substantial binding energy, so it should be quite stable.

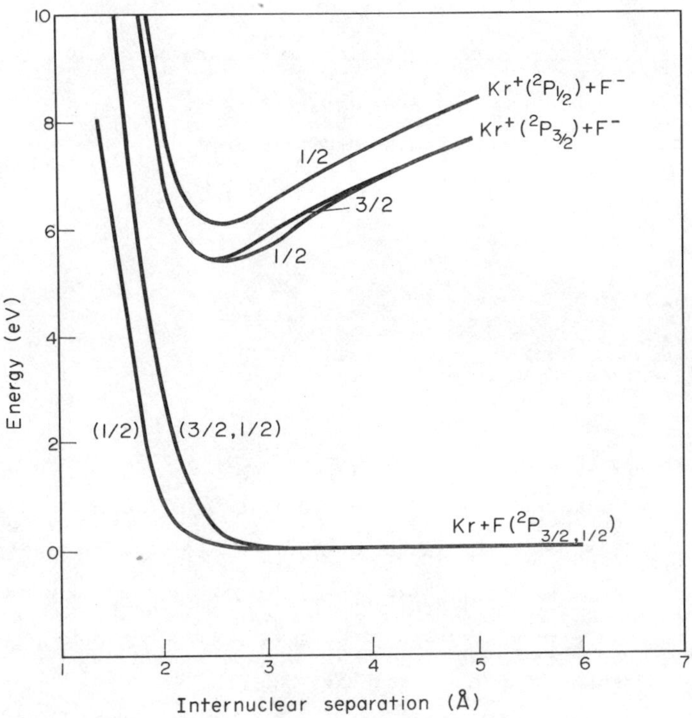


*Fig. 12.34. Potential energy curves for low lying electronic states of KrF (from Dunning and Hay, 1976)*

Rather extensive investigations of the rare gas-halide lasers have been conducted over the last few years, with most of the attention focused on XeF and KrF systems where overall efficiencies approaching 10% and power loadings of around 12 joules/litre have been achieved. Although not all aspects of these systems are completely understood, the major processes seem to be fairly well defined. The paper by Rokni, Mangano, Jacob and Hsia (1978) is an excellent review of the major processes. The paper by Nighan (1978) gives a good feel for the relative role of the various production and loss processes under typical discharge conditions and Greene and Brau (1978) provide an interesting example of modeling of KrF and ArF lasers.

The major kinetic processes in the KrF laser, for example, can be broken down into two broad classifications: the pure electron beam pumped system where negative ion-positive ion recombination is the dominant pumping mechanism and the discharge pumped system where metastable reactions with fluorine bearing compounds (the harpoon reaction) are the dominant pumping mechanisms.

In electron beam pumped KrF lasers, the typical gas mixture would consist of 90% Ar, < 10% Kr and ≃ 0.5% $F_2$ at about 1 to 3 atmospheres. The energetic electrons interact with the majority gas, argon, to produce positive atomic ions.

$$e(\text{fast}) + Ar \rightarrow Ar^{+} \quad . \tag{12.49}$$

The cold secondary electrons attach efficiently to form negative ions

$$e(\text{slow}) + F_2 \rightarrow F^{-} \quad . \tag{12.50}$$

Ion-ion recombination occurs through the atomic states

$$Ar^{+} + F^{-} \rightarrow ArF^{*} \quad (k = 10^{-12}\ m^{3}/sec) \tag{12.51}$$

or through the molecular states

$$Ar^{+} + 2Ar \rightarrow Ar_2^{+} \quad , \tag{12.52}$$

$$Ar_2^{+} + F^{-} \rightarrow ArF^{*} \quad (k = 10^{-12}\ m^{3}/sec) \quad . \tag{12.53}$$

The excited $ArF^{*}$ molecule can then react to form $KrF^{*}$,

$$ArF^{*} + Kr \rightarrow KrF^{*} \quad (k = 1.6 \times 10^{-15}\ m^{3}/sec) \quad , \tag{12.54}$$

or $KrF^{*}$ can be formed through the Kr system

$$Ar_2^{+} + Kr \rightarrow Kr^{+} \quad (k = 7.5 \times 10^{-16}\ m^{3}/sec) \quad , \tag{12.55}$$

$$Kr^{+} + F^{-} \rightarrow KrF^{*} \quad (k = 1 \times 10^{-12}\ m^{3}/sec) \quad , \tag{12.56}$$

$$Kr_2^{+} + F^{-} \rightarrow KrF^{*} \quad (k = 1 \times 10^{-12}\ m^{3}/sec) \quad .$$

In discharge pumped KrF lasers, the metastable species of Ar and Kr are produced with a much greater efficiency than are the ions

$$e + Ar \rightarrow Ar^{*} \quad , \tag{12.57}$$

$$e + Kr \rightarrow Kr^{*} \quad . \tag{12.58}$$

$ArF^{*}$ is then produced through the harpoon reaction

$$Ar^{*} + F_2 \rightarrow ArF^{*} + F \quad (k = 7.5 \times 10^{-16}\ m^{3}/sec) \tag{12.59}$$

and $ArF^{*}$ reacts with Kr to give $KrF^{*}$,

$$ArF^{*} + Kr \rightarrow KrF^{*} \quad . \tag{12.60}$$

Similarly, the Kr metastables undergo a harpoon reaction

$$Kr^{*} + F_2 \rightarrow KrF^{*} \quad (k = 7.2 \times 10^{-16}\ m^{3}/sec) \quad . \tag{12.61}$$

As before, negative ion formation by cool electrons is efficient

$$e + F_2 \rightarrow F^{-} \quad , \tag{12.62}$$

and $KrF^{*}$ can be formed through the ion channel, although the metastable channel dominates in discharge pumped lasers.

The important limiting processes for the KrF laser involve collisional quenching and photo absorption. Consider the quenching reactions first. $KrF^*$ is collisionally quenched by all of the major constituents of the gas discharge. Collisional quenching by $F_2$,

$$KrF^* + F_2 \rightarrow \text{products} \quad (k = 7.8 \times 10^{-16}\ m^3/sec) \quad , \tag{12.63}$$

is particularly severe, often being comparable to spontaneous emission as a loss process. This requires that the fluorine donor concentration be low in order to minimize the quenching effect which, in turn, limits the number of excited molecules that can be formed in the first place. $F_2$ also quenches $ArF^*$,

$$ArF^* + F_2 \rightarrow \text{products} \quad (k = 1.9 \times 10^{-15}\ m^3/sec) \quad . \tag{12.64}$$

Other significant collisional loss processes involve de-excitation by the rare gases

$$KrF^* + 2Ar \rightarrow \text{products} \quad (k = 7 \times 10^{-44}\ m^6/sec) \quad , \tag{12.65}$$

and three body trimer formation

$$KrF^* + Kr + M \rightarrow Kr_2F^* + M \quad (k = 6.5 \times 10^{-43}\ m^6/sec) \quad , \tag{12.66}$$

where the third body, M, can be argon or krypton.

The stimulated emission cross section for $KrF^*$ is quite large

$$\sigma_\nu = 2.5 \times 10^{-20}\ m^2 \quad , \tag{12.67}$$

however quite a number of discharge constituents absorb at the laser wavelength. $F_2$ photo dissociation, $F^-$ photo detachment and $Kr_2F^*$ photo dissociation are the primary effects. The photo absorption cross sections for KrF are

$$F^- \quad \sigma = 5.6 \times 10^{-22}\ m^2 \quad ,$$

$$Kr_2F^* \quad \sigma = 1.6 \times 10^{-22}\ m^2 \quad ,$$

and

$$F_2 \quad \sigma = 2 \times 10^{-24}\ m^2 \quad .$$

Other absorbing species that do not have such a large net effect are

$$Kr_2^+ \quad \sigma = 1.6 \times 10^{-22}\ m^2 \quad ,$$

$$Ar_2^+ \quad \sigma = 1.3 \times 10^{-21}\ m^2 \quad ,$$

and highly excited argon or krypton

$$Ar^{**},\ Kr^{**} \quad \sigma = 6 \times 10^{-22}\ m^2 \quad .$$

One method of overcoming the photo absorption loss is to use a fluorine donor, such as $NF_3$, that does not absorb at the KrF wavelength. This approach has been used in KrF discharge lasers, however higher fluorescence and laser efficiencies are still obtained using $F_2$ since $Kr^+$ can charge transfer to $NF_3$ and this reduces the pumping rate to $KrF^*$. $NF_3$ does turn out to be useful, however, in XeF lasers.

One unfortunate aspect of the rare gas-monohalide lasers is that discharges containing electro-negative gases (halides) tend to constrict and become unstable.

Daugherty, Mangano and Jacob (1976) considered the stability of such discharges and determined that the discharge would be stable if the attachment rate exceeded twice the ionization rate in the medium. Since the dominant ionization mechanism is by electron collision with the rare gas metastables, this establishes another limitation on such lasers since a large metastable density will lead to a large two step ionization rate and instability.

The whole picture of the kinetics of XeF or KrF lasers is becoming progressively clearer, however some of the potentially important rate coefficients are unknown. Still, the desirable spectral range and efficiencies make the rare gas-monohalides a very interesting class of lasers and further exploration will surely clear up any unknowns.

### 12.5.3 Mercury-Halide Lasers

Although not strictly excimer lasers in the same sense as $Xe_2$ or KrF, investigations of the mercury halide lasers grew out of the rare gas-monohalide work so we will discuss them here.

Laser action has been observed in:

HgCl at 5576 Å, 5583 Å (Parks, 1977a; Eden, 1977)
HgBr at 5018 Å (Parks, 1977b)
and HgI at 4430 Å (Burnham, 1978).

The experiments of Parks (1977a, 1977b) and Eden (1977) used electron-beam pumping to excite high pressure mixtures of Ar (85%) - Xe (11%) - Hg (2-4%) - $CCl_4$ (0.2 - 1.1%) or Ar (86.4%) - Xe (10.8%) - Hg (2.0%) - HBr (0.8%). Burnham (1978) used a UV preionized transverse electric discharge in atmospheric pressure helium with about 10% $N_2$ and roughly 1 to 10 torr of $HgCl_2$, $HgBr_2$ and $HgI_2$ to obtain laser action in all three halides. Tang, Hunter, Oldenettel, Howton, Huestis, Eckstrom, Perry and McCusker (1978) also achieved laser action in HgCl with an electron beam controlled discharge in 1500 torr Ar, 10 torr Hg and 8 torr $Cl_2$. Eden (1978) was also able to photolytically pump the HgCl laser by using $Xe_2^*$ emission to irradiate a mixture of $HgCl_2$ and helium.

The understanding of the mercury-halide lasers is still in its infancy at this time, so no attempt will be made to enumerate the proposed excitation and de-excitation processes. The reader should instead watch the current literature to see how these studies develop.

## 12.6 REFERENCES

Ali, A. W., and Kepple, Paul C. (1972), "$H_2$ Lyman and Werner Bands Laser Theory," Appl. Optics 11, 2591-2596.

Ali, A. W., and Kolb, A. C. (1968), "Hydrogen Molecular Vacuum Ultraviolet Laser Theory," Appl. Phys. Letters, 13, 259-261.

Ali, A. W., Kolb, A. C., and Anderson, A. D. (1967), "Theory of the Pulsed Molecular Nitrogen Laser," Appl. Optics 6, 2115-2119.

Ault, E. R., Bradford, R. S., Jr., and Bhaumik, M. L. (1975), "High Power Xenon Fluoride Laser," Appl. Phys. Letters 27, 413-415.

Barton, M. J., von Engel, A. (1970) "Electric Dissociation of $CO_2$," Phys. Letters 32A, 173-174.

Basov, N. G., Danilychev, V. A., and Popov, Y. M. (1971), "Stimulated Emission in the Vacuum Ultraviolet Region," Sov. J. Quant. Electr., 1, 18-22.

Boucique, R., and Mortier, P. (1970), "On the Production and the Decay of Delayed Molecular Ultraviolet Radiation in Rare Gas Townsend Discharges," J. Phys. D, 3, 1905-1911.

Brau, C. A., and Ewing, J. J. (1975), "354 nm Laser Action on XeF," Appl. Phys. Letters 27, 435-437.

Burnham, R. (1978), "Discharge Pumped Mercuric Halide Dissociation Lasers" Appl. Phys. Letters, 33, 156-159.

Cartwright, D. C., Trajmar, S., Chutjian, A., and Williams, W. (1977), "Electron Impact Excitation of the Electronic States of $N_2$, II Integral Cross Sections at Incident Energies from 10 to 50 eV," Phys. Rev. A16, 1041-1051.

Daugherty, J. D., Mangano, J. A., and Jacob, J. H. (1976), "Attachment Dominated Electron-beam-ionized Discharges," Appl. Phys. Letters 28, 581-583.

DeMaria, Anthony J. (1973), "Review of CW High-power $CO_2$ Lasers," Proc. IEEE 61, 731-748.

Denes, L. J., Babcock, R. V., George, T. V., Kline, L. E., Liberman, I., Liebermann, R. W., Partlow, W. D., Taylor, L. H., and Weaver, L. A. (1975), "Ultraviolet Initiated $CO_2$ Laser Research," Technical Report AFWL-TR-74-106, Air Force Weapons Laboratory, Air Force Systems Command, Kirtland Air Force Base.

Dreyfus, R. W., and Hodgson, R. T. (1972), "Electron-beam excitation of the nitrogen Laser," Appl. Phys. Letters 20, 195-197.

Dunning, T. H., Jr. and Hay, P. J. (1976), "Electronic States of KrF," Appl. Phys. Letters 28, 649-651.

Eden, J. Gary (1977), "Green HgCl ($B^2\Sigma^+ \rightarrow X^2\Sigma^+$) Laser" Appl. Phys. Letters, 31, 448-450.

Eden, J. Gary (1978), "VUV-pumped HgCl Laser" Appl. Phys. Letters, 33, 495-497.

Eden, J. G., and Searles, S. K. (1976a), "Observation of Stimulated Emission in KrCl," Appl. Phys. Letters 29, 350-352.

Ewing, J. J., and Brau, C. A. (1975), "Laser Action on the ${}^2\Sigma^+_{1/2} - {}^2\Sigma^+_{1/2}$ Bands of KrF and XeCl," Appl. Phys. Letters 27, 350-352.

Fenstermacher, C. A., Leland, W. T., and Boyer, K. (1972), "Electron-Beam-Controlled Electrical Discharge as a Method of Pumping Large Volumes of $CO_2$ Laser

Media at High Pressure," Appl. Phys. Letters 20, 56-60.

George, E. V., and Rhodes, C. K. (1973), "Kinetic Model of Ultraviolet Inversions in High-Pressure Rare-Gas Plasmas," Appl. Phys. Letters 23, 139-141.

Gerardo, J. B., and Johnson, A. W. (1973a), "High-pressure Xenon Laser at 1730 Å," IEEE J. Quantum Electronics 9, 748-755.

Gerardo, J. B., and Johnson, A. W. (1973b), "1730 Å Radiation Dominated by Stimulated Emission from High Pressure Xenon," J. Appl. Phys. 44, 4120-4124.

Gerardo, J. B., and Johnson, A. W. (1976), "Formation and Decay of Xenon Dimers in Photo Excited Xenon and Xenon-noble Gas Mixtures," Bull. Am. Phys. Soc. 21, 170.

Gerry, E. T. (1965), "Pulsed Molecular Nitrogen Laser Theory," Appl. Phys. Letters 7, 6-8.

Greene, A. E., and Brau, C. A. (1978), "Theoretical Studies of UV-Preionized Transverse Discharge KrF and ArF Lasers" IEEE J. Quantum Electronics, 14, 951-957.

Hedges, R. E. M., Drummond, D. L. and Gallagher, A. (1972), "Extreme-wing Line Broadening and Cs-inert-gas Potentials," Phys. Rev. A 6, 1519-1544.

Herzberg, G., Molecular Spectra and Molecular Structure in three volumes:

I. Spectra of Diatomic Molecules

II. Infrared and Raman Spectra of Polyatomic Molecules

III. Electronic Spectra and Electronic Structure of Polyatomic Molecules

D. Van Nostrand Co., Princeton.

Hodgson, R. T. (1970), "Vacuum-Ultraviolet Laser Action Observed in the Lyman Bands of Molecular Hydrogen," Phys. Rev. Letters 25, 494-497.

Hodgson, R. T., and Dreyfus, R. W. (1970), "Vacuum-UV Laser Action Observed in $H_2$ Werner Bands: 1161-1240 Å," Phys. Rev. Letters 28, 536-539.

Hoff, P. W., Swingle, J. C., and Rhodes, C. K. (1973a), "Demonstration of Temporal Coherence, Spatial Coherence, and Threshold Effects in the Molecular Xenon Laser," Opt. Commun. 8, 128-131.

Hoff, P. W., Swingle, J. C., and Rhodes, C. K. (1973b), "Observations of Stimulated Emission from High Pressure Krypton and Argon/Xenon Mixtures," Appl. Phys. Letters, 23, 245-246.

Hoffman, J. M., Hays, A. K., and Tisone, G. C. (1976), "High-power UV Noble-gas-halide Lasers," Appl. Phys. Letters, 28, 538-539.

Johnson, A. W., and Gerardo, J. B. (1973), "De-excitation Rates for Excited Xenon Molecules," J. Chem. Phys. 59, 1738-1741.

Johnson, A. W., and Rice, J. K. (1976), "Formation of Xenon Dimers in Electron-beam-excited Xenon and Xenon-Noble Gas Mixtures," Bull. Am. Phys. Soc. 21, 169-170.

Kasuya, Takahiro, Lide, David, Jr. (1967), "Measurements on the Molecular Nitrogen Pulsed Laser," Appl. Optics 6, 69-80.

Keto, J. W., Gleason, R. E., Jr., Bonifield, T. D., and Walters, G. K. (1976), "Collisional Mixing of the Lowest Bound Molecular States in Xenon and Argon" Chem. Phys. Letters, 42, 125-128.

Koehler, H. A., Ferderber, L. J., Redhead, D. L., and Ebert, P. J. (1972), "Stimulated VUV Emission in High-pressure Xenon Excited by High-current Relativistic Electron Beams," Appl. Phys. Letters 21, 198-200.

Koehler, H. A., Ferderber, L. J., Redhead, D. L., and Ebert, P. J. (1974), "Vacuum-ultraviolet emission from High-pressure Xenon and Argon Excited by High-current Relativistic Electron Beams," Phys. Rev. A 9, 768-778.

Ku, R. T., Verdeyen, J. T., Cherrington, B. E., and Eden, J. G. (1973), "Decay of Krypton $1s_2$ and $1s_3$ Excited Species in the Late Afterglow," Phys. Rev. A 8, 3123-3130.

Leonard, D. A. (1965), "Saturation of the Molecular Nitrogen Second Positive Laser Transition," Appl. Phys. Letters 7, 4-6.

Lowke, J. J., Phelps, A. V., and Irwin, B. W. (1973), "Predicted Electron Transport Coefficients and Operating Characteristics of $CO_2$-$N_2$-He Laser Mixtures," J. Appl. Phys. 44, 4664-4671.

Mathias, L. E. S., and Parker, J. T. (1963), "Stimulated Emission in the Band Spectrum of Nitrogen," Appl. Phys. Letters 3, 16-18.

Moore, C. Bradley, Wood, Robert E., Hu, Bei-Lok, Yardley, James T. (1967), "Vibrational Energy Transfer in $CO_2$ Lasers," J. Chem. Phys. 16, 4222-4231.

Mulliken, R. S. (1970), "Potential Curves of Diatomic Rare-gas Molecules and their Ions, with Particular Reference to $Xe_2$," J. Chem. Phys. 52, 5170-5180.

Murray, J. R., and Powell, H. T. (1976), "KrCl Laser Oscillation at 222 nm," Appl. Phys. Letters, 29, 252-253.

Newman, L. A., and DeTemple, T. A. (1976), "Electron Transport Parameters and Excitation Rates in $N_2$," J. Appl. Phys. 47, 1912-1915.

Nighan, W. L. (1969), "Effect of Molecular Dissociation and Vibrational Excitation on Electron Energy Transfer in $CO_2$ Laser Plasmas," Appl. Phys. Letters 15, 355-357.

Nighan, W. L. (1970), "Electron Energy Distributions and Collision Rates in Electrically Excited $N_2$, CO and $CO_2$," Phys. Rev. A 2, 1989-2000.

Nighan, W. L. (1978), "Plasma Processes in Electron-Beam Controlled Rare-Gas Halide Lasers," IEEE J. Quantum Electronics, 14, 714-726.

Parks, J. A. (1977a), "Laser Action on the $B^2\Sigma^+_{1/2} - X^2\Sigma^+_{1/2}$ Band of HgCl at 5576 Å," Appl. Phys. Letters, 31, 192-194.

Parks, J. A. (1977b), "Laser Action on the $B^2\Sigma^+_{1/2} \rightarrow X^2\Sigma^+_{1/2}$ Band of HgBr at 5018 Å," Appl. Phys. Letters, 31, 297-300.

Patterson, E. L., Gerardo, J. B., Johnson, A. Wayne (1972), "Intense Electron Beam Excitation of the 3371 Å $N_2$ Laser System" Appl. Phys. Letters 21, 293-295.

Rhodes, C. K. (1974), "Review of Ultraviolet Laser Physics," IEEE J. Quantum Electronics 10, 153-174.

Rice, James K. and Johnson, A. Wayne (1975), "Enhancement by Helium and Argon of the Formation Rate of the 1720 Å Radiating States of $Xe_2^*$ Excited by an E Beam" J. Chem. Phys., 63, 5235-5237.

Rokni, M., Mangano, J. A., Jacob, J. H., and Hsia, J. C. (1978), "Rare Gas Fluoride Lasers" IEEE J. Quantum Electronics, 14, 464-481.

Schulz, G. J. (1964), "Vibrational Excitation of $N_2$, CO and $H_2$ by Electron Impact," Phys. Rev. 135A, 988-994.

Searles, S. K. (1974), "Superfluorescent Laser Emission from Electron-beam-pumped Ar-$N_2$ Mixtures," Appl. Phys. Letters 25, 735-737.

Searles, S. K., Hart, G. A. (1975), "Stimulated Emission at 281.8 nm from XeBr," Appl. Phys. Letters, 27, 243-245.

Shipman, J. D., Jr. (1967), "Traveling Wave Excitation of High Power Gas Lasers," Appl. Phys. Letters 10, 3-4.

Sobolev, N. N., and Sokovikov, V. V. (1967), "$CO_2$ Lasers," Sov. Phys. Uspekhi 10, 153-170.

Statz, H., Tang, C. L., and Koster, G. F. (1966), "Transition Probabilities Between Laser States in Carbon Dioxide," J. Appl. Phys. 37, 4278-4284.

Tang, Kenneth Y., Hunter, R. O., Jr., Oldenettel, J., Howton, C., Huestis, D., Eckstrom, D., Perry, B., and McCusker, M. (1978), "Electron-beam-controlled Discharge $HgCl^*$ Laser," Appl. Phys. Letters, 32, 226-228.

Taylor, Raymond L., and Bitterman, Steven (1969), "Survey of Vibrational Relaxation Data for Processes Important in the $CO_2$-$N_2$ Laser System," Rev. Mod. Phys. 41, 26-47.

Tyte, D. C. (1970), "Carbon Dioxide Lasers," Advances in Quantum Electronics 1, 129-198.

Waynant, R. W. (1972), "Observations of Gain by Stimulated Emission in the Werner Band of Molecular Hydrogen," Phys. Rev. Letters 28, 533-535.

Waynant, Ronald W. (1977), "A Discharge-Pumped ArCl Superfluorescent laser at 175.0 nm" Appl. Phys. Letters, 30, 234-235.

Waynant, R. W., Ali, A. W., and Julienne, P. S. (1971), "Experimental Observations and Calculated Band Strengths for the $D_2$ Lyman Band Laser," J. Appl. Phys. 42, 3406-3408.

Waynant, R. W., Shipman, J. D., Jr., Elton, R. C., and Ali, A. W. (1970/1971) "Vacuum Ultraviolet Laser Emission from Molecular Hydrogen," Appl. Phys. Letters 17, 383-384; also "Laser Emission in the Vacuum Ultraviolet from Molecular Hydrogen," Proc. IEEE 59, 679-684.

Werner, C. W., George, E. V., Hoff, P. W. and Rhodes, C. K. (1977), "Radiative and

Kinetic Mechanisms in Bound-Free Excimer Lasers" IEEE J. Quantum Electronics, 13, 769-783.

Yardley, James T., and Moore, C. Bradley (1967), "Intramolecular Vibration-to-vibration Energy Transfer in Carbon Dioxide," J. Chem. Phys. 46, 4491-4495.

Zamir, E., Werner, C. W., Lapatovich, W. P., and George, E. V. (1975), "Temporal Evolution of the Electron Density in High Pressure Electron-Beam-Excited Xenon Plasmas," Appl. Phys. Letters, 27, 56-58.

# Appendix A

# EXPANSION OF THE BOLTZMAN EQUATION IN SPHERICAL HARMONICS

In equation (4.33) the electron distribution function was expressed in terms of a sum of spherical harmonics

$$f(\bar{v},\bar{r},t) = \sum_{k=0}^{\infty} P_k\ (\cos\theta_1)\ f_k(v,\bar{r},t) \tag{A.1}$$

where $\theta_1$ is the angle between the velocity vector $\bar{v}$ and the applied electric field $\bar{E}$, $P_k\ (\cos\theta_1)$ are the Legendre polynomials and $f_k$ depends only on the magnitude of the velocity. The Boltzmann equation,

$$\frac{\partial f}{\partial t} + \bar{v}\cdot\nabla_r f - \frac{e}{m}\bar{E}\cdot\nabla_v f = S \quad , \tag{A.2}$$

then becomes

$$\frac{d}{dt}\ \sum_k P_k f_k + \bar{v}\cdot\nabla_r \sum_k P_k f_k - \frac{e}{m}\bar{E}\cdot\nabla_v\ \sum P_k f_k = S \quad . \tag{A.3}$$

In order to solve for the values of $f_k$, we can make use of the orthogonality properties of Legendre polynomials.

First of all, let us simplify equation (A.3)

$$\nabla_v = \hat{v}\frac{\partial}{\partial v} + \frac{\hat{\theta}_1}{v}\frac{\partial}{\partial\theta_1} + \frac{\hat{\phi}_1}{v\sin\theta_1}\frac{\partial}{\partial\phi_1} \quad . \tag{A.4}$$

Therefore,

$$\nabla_v\ f_k\ P_k = \hat{v}\ P_k\frac{\partial f_k}{\partial v} + \frac{\hat{\theta}_1}{v}\ f_k\ \frac{\partial P_k\ (\cos\theta_1)}{\partial\cos\theta_1}\cdot\frac{\partial\cos\theta_1}{\partial\theta_1}$$

$$+ \frac{\hat{\phi}_1}{v\sin\theta_1}\ f_k\ \frac{\partial P_k\ (\cos\theta_1)}{\partial\cos\theta_1}\ \frac{\partial\cos\theta_1}{\partial\phi_1} \quad . \tag{A.5}$$

Now

$$\frac{\partial\cos\theta_1}{\partial\theta_1} = -\sin\theta_1 \text{ and } \frac{\partial\cos\theta_1}{\partial\phi_1} = 0 \quad . \tag{A.6}$$

Therefore,

$$\nabla_v f_k P_k = \hat{v} P_k \frac{\partial f_k}{\partial v} - \frac{\hat{\theta}_1}{v} f_k \sin\theta_1 \frac{\partial P_k(\cos\theta_1)}{\partial \cos\theta_1} \quad . \tag{A.7}$$

Also

$$\bar{E} \cdot \hat{\theta}_1 = - E \sin\theta_1 \quad ,$$
$$\bar{E} \cdot \hat{v} = E \cos\theta_1 \quad . \tag{A.8}$$

Therefore,

$$\bar{E} \cdot \nabla_v f_k P_k = E \left\{\cos\theta_1 P_k \frac{\partial f_k}{\partial v} + \frac{\sin^2\theta_1}{v} f_k \frac{\partial P_k(\cos\theta_1)}{\partial\cos\theta_1}\right\} \quad . \tag{A.9}$$

We can now express the Boltzmann equation as

$$\Sigma P_k \frac{\partial f_k}{\partial t} + \Sigma P_k \bar{v} \cdot \nabla_r f_k - \frac{e E}{m} \left[ \Sigma x P_k \frac{\partial f_k}{\partial v} + \Sigma (1 - x^2) \frac{f_k}{v} \frac{\partial P_k}{\partial x} \right] = S \tag{A.10}$$

where

$$x = \cos\theta_1 \quad .$$

We may now use the recursion relations for Legendre polynomials

$$(2k + 1)\, x P_k = (k + 1) P_{k+1} + k P_{k-1} \tag{A.11}$$

and

$$(x^2 - 1) \frac{\partial P_k}{\partial x} = (k + 1)(P_{k+1} - x P_k) \quad , \tag{A.12}$$

to solve for

$$(1 - x^2) \frac{\partial P_k}{\partial x} = \frac{k(k + 1)}{2k + 1} [P_{k-1} - P_{k+1}] \tag{A.13}$$

and

$$x P_k = \frac{(k + 1) P_{k+1} + k P_{k-1}}{2k + 1} \quad . \tag{A.14}$$

Keeping in mind that $\bar{E} \,||\, \hat{Z}$ so that

$$P_k \bar{v} \cdot \nabla_r f_k = P_k v_z \frac{\partial f_k}{\partial z} = P_k v \cos\theta_1 \frac{\partial f_k}{\partial z} = P_k x v \frac{\partial f_k}{\partial z} \quad , \tag{A.15}$$

we may now write

$$\sum_k \left\{ P_k \frac{\partial f_k}{\partial t} + v \frac{\partial f_k}{\partial z} \left[ \frac{k + 1}{2k + 1} P_{k+1} + \frac{k}{2k + 1} P_{k-1} \right] \right. \tag{A.16}$$
$$\left. - \frac{e E}{m} \left[ \left( \frac{\partial f_k}{\partial v} - \frac{k f_k}{v} \right) \frac{k + 1}{2k + 1} P_{k+1} + \left( \frac{\partial f_k}{\partial v} + \frac{(k + 1) f_k}{v} \right) \frac{k}{2k + 1} P_{k-1} \right] \right\} = S \quad .$$

We may now use the orthogonality conditions and multiply the previous equation by $P_k'$ and integrate $\int P_k' [\quad] \, d\Omega_1$ where $d\Omega_1 = \sin\theta_1 \, d\theta_1 \, d\phi_1$ .

The normalization condition

$$\int P_k^2 \, d\Omega_1 = \frac{4\pi}{2k+1} \quad \text{comes from} \quad \int_{-1}^{+1} P_k^2 (x) \, dx = \frac{2}{2k+1} \; .$$

We then obtain

$$\frac{\partial f_k}{\partial t} + v \left[ \frac{k}{2k-1} \frac{\partial f_{k-1}}{\partial Z} + \frac{k+1}{2k+3} \frac{\partial f_{k+1}}{\partial Z} \right]$$

$$- \frac{e E_z}{m} \left\{ \frac{k}{2k-1} \left[ \frac{\partial f_{k-1}}{\partial v} - \frac{(k-1)}{v} f_{k-1} \right] + \frac{k+1}{2k+3} \left[ \frac{\partial f_{k+1}}{\partial v} + \frac{(k+2) f_{k+1}}{v} \right] \right\} = S_k \tag{A.17}$$

where

$$S_k = \frac{2k+1}{4\pi} \int P_k \, S \, d\Omega_1 \; . \tag{A.18}$$

Using the condition that

$$\frac{1}{v^{k+2}} \frac{\partial}{\partial v} v^{k+2} f_{k+1} = \frac{\partial f_{k+1}}{\partial v} + \frac{(k+2) f_{k+1}}{v} \tag{A.19}$$

and

$$v^{k-1} \frac{\partial}{\partial v} \frac{f_{k-1}}{v^{k-1}} = \frac{\partial f_{k-1}}{\partial v} - \frac{(k-1) f_{k-1}}{v} \tag{A.20}$$

we obtain

$$\frac{\partial f_k}{\partial t} + v \left[ \frac{k}{2k-1} \frac{\partial f_{k-1}}{\partial z} + \frac{k+1}{2k+3} \frac{\partial f_{k+1}}{\partial z} \right]$$

$$- \frac{e E_z}{m} \left[ \frac{k}{2k-1} v^{k-1} \frac{\partial}{\partial v} \frac{f_{k-1}}{v^{k-1}} + \frac{k+1}{2k+3} \frac{1}{v^{k+2}} \frac{\partial}{\partial v} v^{k+2} f_{k+1} \right]$$

$$= \frac{2k+1}{4\pi} \int P_k \, S \, d\Omega_1 \tag{A.21}$$

which was previously given as equation (4.35).

# INDEX

## *OTHER TITLES IN THE SERIES IN NATURAL PHILOSOPHY*

Vol. 1. Davydov–Quantum Mechanics
Vol. 2. Fokker–Time and Space, Weight and Inertia
Vol. 3. Kaplan–Interstellar Gas Dynamics
Vol. 4. Abrikosov, Gor'kov and Dzyaloshinskii–Quantum Field Theoretical Methods in Statistical Physics
Vol. 5. Okun'–Weak Interaction of Elementary Particles
Vol. 6. Shklovskii–Physics of the Solar Corona
Vol. 7. Akhiezer *et al.*–Collective Oscillations in a Plasma
Vol. 8. Kirzhnits–Field Theoretical Methods in Many-body Systems
Vol. 9. Klimontovich–The Statistical Theory of Nonequilibrium Processes in a Plasma
Vol. 10. Kurth–Introduction to Stellar Statistics
Vol. 11. Chalmers–Atmospheric Electricity (2nd Edition)
Vol. 12. Renner–Current Algebras and their Applications
Vol. 13. Fain and Khanin–Quantum Electronics, Volume 1–Basic Theory
Vol. 14. Fain and Khanin–Quantum Electronics, Volume 2–Maser Amplifiers and Oscillators
Vol. 15. March–Liquid Metals
Vol. 16. Hori–Spectral Properties of Disordered Chains and Lattices
Vol. 17. Saint James, Thomas and Sarma–Type II Superconductivity
Vol. 18. Margenau and Kestner–Theory of Intermolecular Forces (2nd Edition)
Vol. 19. Jancel–Foundations of Classical and Quantum Statistical Mechanics
Vol. 20. Takahashi–An Introduction to Field Quantization
Vol. 21. Yvon–Correlations and Entropy in Classical Statistical Mechanics
Vol. 22. Penrose–Foundations of Statistical Mechanics
Vol. 23. Visconti–Quantum Field Theory, Volume 1
Vol. 24. Furth–Fundamental Principles of Theoretical Physics
Vol. 25. Zheleznyakov–Radioemission of the Sun and Planets
Vol. 26. Grindlay–An Introduction to the Phenomenological Theory of Ferroelectricity
Vol. 27. Unger–Introduction to Quantum Electronics
Vol. 28. Koga–Introduction to Kinetic Theory: Stochastic Processes in Gaseous Systems
Vol. 29. Galasiewicz–Superconductivity and Quantum Fluids
Vol. 30. Constantinescu and Magyari–Problems in Quantum Mechanics
Vol. 31. Kotkin and Serbo–Collection of Problems in Classical Mechanics
Vol. 32. Panchev–Random Functions and Turbulence
Vol. 33. Talpe–Theory of Experiments in Paramagnetic Resonance
Vol. 34. Ter Haar–Elements of Hamiltonian Mechanics (2nd Edition)
Vol. 35. Clarke and Grainger–Polarized Light and Optical Measurement
Vol. 36. Haug–Theoretical Solid State Physics, Volume 1
Vol. 37. Jordan and Beer–The Expanding Earth
Vol. 38. Todorov–Analytical Properties of Feynman Diagrams in Quantum Field Theory
Vol. 39. Sitenko–Lectures in Scattering Theory
Vol. 40. Sobel'man–Introduction to the Theory of Atomic Spectra
Vol. 41. Armstrong and Nicholls–Emission, Absorption and Transfer of Radiation in Heated Atmospheres
Vol. 42. Brush–Kinetic Theory, Volume 3
Vol. 43. Bogolyubov–A Method for Studying Model Hamiltonians
Vol. 44. Tsytovich–An Introduction to the Theory of Plasma Turbulence
Vol. 45. Pathria–Statistical Mechanics
Vol. 46. Haug–Theoretical Solid State Physics, Volume 2
Vol. 47. Nieto–The Titius-Bode Law of Planetary Distances: Its History and Theory
Vol. 48. Wagner–Introduction to the Theory of Magnetism
Vol. 49. Irvine–Nuclear Structure Theory
Vol. 50. Strohmeier–Variable Stars
Vol. 51. Batten–Binary and Multiple Systems of Stars
Vol. 52. Rousseau and Mathieu–Problems in Optics
Vol. 53. Bowler–Nuclear Physics
Vol. 54. Pomraning–The Equations of Radiation Hydrodynamics
Vol. 55. Belinfante–A Survey of Hidden Variables Theories
Vol. 56. Scheibe–The Logical Analysis of Quantum Mechanics
Vol. 57. Robinson–Macroscopic Electromagnetism
Vol. 58. Gombas and Kisdi–Wave Mechanics and its Applications
Vol. 59. Kaplan and Tsytovich–Plasma Astrophysics
Vol. 60. Kovacs and Zsoldos–Dislocations and Plastic Deformation
Vol. 61. Auvray and Fourrier–Problems in Electronics